**Frontiers of
Consciousness** WITHDRAWN

A gathering at All Souls College of some of the participants: John Davis (Warden of All Souls), David Milner, Lawrence Weiskrantz, Adam Zeman, Chris Frith, Cecilia Heyes, Joseph LeDoux.

Frontiers of Consciousness
Chichele Lectures

Edited by

Lawrence Weiskrantz and
Martin Davies
University of Oxford

OXFORD
UNIVERSITY PRESS

Great Clarendon Street, Oxford OX2 6DP

Oxford University Press is a department of the University of Oxford.
It furthers the University's objective of excellence in research, scholarship,
and education by publishing worldwide in

Oxford New York

Auckland Cape Town Dar es Salaam Hong Kong Karachi
Kuala Lumpur Madrid Melbourne Mexico City Nairobi
New Delhi Shanghai Taipei Toronto

With offices in

Argentina Austria Brazil Chile Czech Republic France Greece
Guatemala Hungary Italy Japan Poland Portugal Singapore
South Korea Switzerland Thailand Turkey Ukraine Vietnam

Oxford is a registered trade mark of Oxford University Press
in the UK and in certain other countries

Published in the United States
by Oxford University Press Inc., New York

© Oxford University Press 2008

The moral rights of the author have been asserted
Database right Oxford University Press (maker)

First published 2008

All rights reserved. No part of this publication may be reproduced,
stored in a retrieval system, or transmitted, in any form or by any means,
without the prior permission in writing of Oxford University Press,
or as expressly permitted by law, or under terms agreed with the appropriate
reprographics rights organization. Enquiries concerning reproduction
outside the scope of the above should be sent to the Rights Department,
Oxford University Press, at the address above

You must not circulate this book in any other binding or cover
and you must impose this same condition on any acquirer

A catalogue record for this title is available from the British Library

Data available

Library of Congress Cataloguing in Publication Data

Frontiers of consciousness / edited by Lawrence Weiskrantz and Martin Davies.
  p. ; cm. — (The Chichele lectures; 2008)
  Includes bibliographical references and index.
  1. Consciousness—Physiological aspects. 2. Neuropsychology.
  [DNLM: 1. Consciousness—physiology. 2. Neuropsychology—methods.
  3. Philosophy. 4. Unconscious (Psychology) WL 705 F925 2008]
  I. Weiskrantz, Lawrence. II. Davies, Martin, 1950-
    QP411.F77 2008 612.8—dc22   2008017221

Typeset by Cepha Imaging Private Ltd., Bangalore, India
Printed in Great Britain
on acid-free paper by
Biddles Ltd., King's Lynn, UK

ISBN 978–0–19–923315–1

10 9 8 7 6 5 4 3 2 1

To the memory of
Susan Hurley
16 September 1954–16 August 2007

ALL SOULS COLLEGE AND THE McDONNELL CENTRE
FOR COGNITIVE NEUROSCIENCE

CHICHELE LECTURES 2006
FRONTIERS OF CONSCIOUSNESS

The symposium consists of six seminars and six lectures. The lectures will be of broad general interest, the seminars more specialised.

Seminars on Thursdays at 5 p.m. in the Old Library, All Souls College

27 April	Professor David Milner (Wolfson Research Institute, Durham)	The Dorsal Stream: its role in automatic visual processing for action
4 May	Professor Joseph LeDoux (Center for Neural Science, New York)	Why is it hard to be happy? The fearful brain is the problem and the solution
11 May	Professor Adam Zeman (Peninsula Medical School, Exeter)	Sidestepping the problem of Consciousness: building bridges between neurology, psychology and psychiatry
18 May	Professor Chris Frith (Institute of Neurology, London)	What has Brain imaging revealed about the neural basis of Consciousness?
25 May	Professor Cecilia Heyes (University College London)	Mentalising, Metacognition and Animal Consciousness
1 June	Professor Martin Davies (Corpus Christi College, Oxford)	Experience and Belief: the Role of Consciousness in the Aetiology of Delusions

Lectures on Fridays at 5 p.m. in the University Museum of Natural History

12 May	Professor Martin Davies	Explaining Consciousness: theories and challenges
19 May	Professor Joseph LeDoux	The Conscious and the Unconscious Brain: fear as a test case
26 May	Professor David Milner	Sight Unseen: an unconscious visual processing system in the human brain
2 June	Professor Chris Frith	What is Consciousness for? Sharing experiences
9 June	Professor Cecilia Heyes	Beast Machines? The question of Animal Consciousness
16 June	Professor Adam Zeman	Does Consciousness spring from the Brain?

ALL ARE WELCOME

Foreword

All Souls was delighted to welcome six 'consciousness' Visiting Fellows in the summer of 2006. Apart from the company of philosophers, medics, ethnologists, biologists, and the chance to hear them speak about their expertise, it was a valuable experiment for the College.

The College offers 27 terms-worth of Visiting Fellowships each year. Normally, they apply in response to advertisement, and a committee chooses among the 200 or so applicants. In 2005 we decided to be proactive and to dedicate six Fellowships to specialists in one broadly defined topic. After discussion in the University, particularly with Colin Blakemore, we agreed that getting a group of scholars and researchers on consciousness would be of significant value to disciplines in the University. With the help of Susan Hurley, Larry Weiskrantz, Edmund Rolls, and Susan Iversen, we drew up a list of people who—given sufficient notice—might be able to accept our invitation. They would be asked to give one lecture and one seminar during Trinity Term 2006. In the event, nearly all our prospected visitors were able to come. The success of the experiment is partly measured by the papers in this volume. We also noted that the Fellows set up their own regular informal discussion group: a sign perhaps that the term was useful to the visitors.

So far as All Souls is concerned, we enjoyed the company of men and women from disciplines which are not continuously represented in College: many of our minds were broadened certainly in the lectures, and also through conversation at lunch and dinner. The College (not lacking in mathematical and theoretical physicists) had not had such a concentration of scientists in residence since the seventeenth century, and it was good to be reminded of that history, and to experience the civilized adventurousness of our guests. Finally, the idea of earmarking a number of Visiting Fellowships to assemble a cluster of scholars and researchers working on a topic of contemporary importance may commend itself in the future: it was undoubtedly a good way to focus our expenditure of resources.

The Oxford McDonnell Centre for Cognitive Neuroscience was privileged to join All Souls in helping to define the programme for the Visiting Fellows and to support their stay in Oxford. Each of the Visiting Fellows invited to All Souls also became a member of the Oxford McDonnell Centre. The public lectures given by our Visiting Fellows were outstandingly well attended, particularly given the warm summer weather that we enjoyed. The quality of the lectures was superb and the subsequent interaction between the audience and the speakers represented debate at the highest level. We are very grateful to the Visiting Fellows for this contribution to the McDonnell Centre's activities. We hope that, in exchange, these experiences have promoted deeper analysis and understanding that is reflected in the contents of this volume.

John Davis, Warden of All Souls College, Oxford
Andrew Parker, Director, Oxford McDonnell Centre for Cognitive Neuroscience
April 2007

Preface

In Trinity Term (April–June) 2006, a series of Chichele lectures and seminars was held in All Souls College, Oxford, under the heading 'Frontiers of Consciousness'. The Chichele lectures are named for Henry Chichele (1362?–1443) Archbishop of Canterbury who, with King Henry VI, founded the College. The 2006 Chichele Lectures were organized, in collaboration with the Oxford McDonnell Centre for Cognitive Neuroscience, by a committee comprising John Davis (Warden of All Souls College), Andrew Parker (Director of the Oxford McDonnell Centre), Susan Hurley, Susan Iversen, Edmund Rolls, and Larry Weiskrantz. The six lecturers—five scientists ranging over psychology, neuroscience and neurology, and a philosopher—were invited to spend the term in Oxford as Visiting Fellows of the College.

The study of consciousness has become a topic of great interest in recent years, and many books and reviews have appeared. The present volume has some claim to uniqueness, stemming from the way in which the contributions emerged. Each lecturer delivered a public lecture and also a seminar for a more specialized audience. The seminar talks included more technical details and each talk was followed by a response from an invited discussant and then open discussion. The lecturers, along with Larry Weiskrantz, Susan Hurley and Nick Shea, also met each week to talk about central topics in recent work on consciousness. Around the end of term, there were occasional discussions with the philosopher David Rosenthal, whose 'higher-order thought' theory of consciousness figures in several of the chapters in this volume.

The resulting contributions from the six lecturers each encompass material ranging from large-scale questions about the study of consciousness to some of the details of the lecturer's own research. The discussants were invited to develop their contributions (which follow each lecturer's chapter) beyond their specific comments on the seminar talks, to bring in other relevant material including, of course, aspects of their own research. In this way, the range of the volume was extended beyond that covered by the lecturers themselves, although that is already quite broad. It encompasses the apparent explanatory gap between science and consciousness, our conscious experience of emotions such as fear and of willed actions by ourselves and others, subjective differences between two ways in which visual information guides behaviour, scientific investigation of consciousness in non-human animals, challenges that the mind–brain relation presents for clinical practice as well as for theories of consciousness; and it involves the methods of philosophy, experimental psychology, functional imaging of the brain, neuropsychology, neuroscience, and clinical neurology. We believe that the result is a volume that is distinctive in its accessibility, coverage, and depth.

***

Susan Hurley died on 16 August 2007 at the age of 52. As a Fellow of All Souls College and a member of the Board of the Oxford McDonnell Centre for Cognitive Neuroscience, she played a pivotal role in planning the 2006 Chichele Lectures, she chaired the seminars, and she organized and led weekly discussions with the lecturers and others. She was a powerful, independent, and influential thinker who made lasting contributions to the interdisciplinary study of consciousness. In her book *Consciousness in Action* (Harvard University Press, 1998) and in subsequent papers, she developed distinctive accounts of the location of the conscious self, embodied and embedded in the natural order, and of the relationship between perception and action. This volume is dedicated to her memory.

LW and MKD

Contents

Contributors *xiii*

In Memoriam: Susan Hurley *xv*
Andy Clark

1. Consciousness and explanation *1*
   Martin Davies

2. Explanatory gaps and dualist intuitions *55*
   David Papineau

3. Emotional colouration of consciousness: how feelings come about *69*
   Joseph LeDoux

4. Emotion, higher-order syntactic thoughts, and consciousness *131*
   Edmund T. Rolls

5. Conscious and unconscious visual processing in the human brain *169*
   A. David Milner

6. Vision, action, and awareness *215*
   Manos Tsakiris and Patrick Haggard

7. The social functions of consciousness *225*
   Chris D. Frith

8. Are we studying consciousness yet? *245*
   Hakwan C. Lau

9. Beast machines? Questions of animal consciousness *259*
   Cecilia Heyes

10. Why a rat is not a beast machine *275*
    Anthony Dickinson

11. Does consciousness spring from the brain? Dilemmas of awareness in practice and in theory *289*
    Adam Zeman

12. On the ubiquity of conscious–unconscious dissociations in neuropsychology *323*
    Lawrence Weiskrantz

Index *335*

Chapter 1

Consciousness and explanation

Martin Davies

1.1 Two questions about consciousness: 'what?' and 'why?'

Many aspects of our mental lives are conscious—an ache in tired muscles; the sight, smell, and taste of a glass of wine; feelings of happiness, love, anxiety or fear; trying to work out how best to test a hypothesis or structure an argument. It seems beyond dispute that at least some sensations, perceptions, emotional episodes, and bouts of thinking are conscious. But equally, there is much in our mental lives that is not conscious. It is a central idea in cognitive science that there can be unconscious information processing. It is also plausible that there can be unconscious thought and unconscious emotions; there are cases of 'perception without awareness'; and perhaps even bodily sensations can sometimes be unconscious.[1] What, then, is the difference between conscious and unconscious mental states? Is there, for example, something distinctive about the neural underpinnings of conscious mental states? An answer to this 'what?' question could be called (in some sense) an explanation of consciousness.

We might, however, expect rather more from an explanation of consciousness than just a principle or criterion that sorts conscious mental states from unconscious ones. Suppose that we were told about a neural condition, NC, that was met by conscious mental states but not by unconscious ones. Suppose that this was not just an accidental correlation. Suppose that the difference between meeting this neural condition and not meeting it really was the difference that

---

[1] Claims about unconscious thoughts and emotions are common in, but not restricted to, the psychoanalytic tradition. In ordinary life, it sometimes seems that we arrive at a solution to a problem by processes of thinking that do not themselves surface in consciousness, although their product does. For the conception of emotion systems as unconscious processing systems whose products are sometimes, but not always, available to consciousness, see LeDoux (1996, this volume). The term 'perception without awareness' is applied to a wide range of phenomena (Merikle *et al.* 2001) including blindsight (Weiskrantz 1986, 1997). For the proposal that unconscious mental states may include even sensations, such as pains, see Rosenthal (1991, 2005); for a recent discussion, see Burge (2007, pp. 414–419; see also 1997, p. 432).

makes the difference. There would remain the question *why* mental states that meet condition NC are conscious. Even if condition NC were to make the difference, it would not be *a priori* that it makes the difference. It would seem perfectly conceivable that condition NC might have been met in the absence of consciousness. So—we would ask—why, in reality, in the world as it actually is, is this condition sufficient for consciousness? The problem with this 'why?' question is that, once it is allowed as legitimate, it is apt to seem unanswerable.

The intractability of the 'why?' question is related to our conception of consciousness, a conception that is grounded in the fact that we ourselves are subjects of conscious mental states. The situation would be quite different if our conception of consciousness were a third-person conception, exhausted by structure and function—if it were a physical-functional conception. Our conception of a neurotransmitter, for example, *is* a physical-functional conception. There is a 'what?' question about neurotransmission: What chemicals make the difference? But, once we know the structure and function of GABA or dopamine, its role in relaying, amplifying, and modulating electrical signals between neurons, there is no further question why it is a neurotransmitter. That is just what being a neurotransmitter means. Similarly, if our conception of consciousness were a physical-functional conception then lessons about the nature of condition NC and about its role in the overall neural economy, about its constitution and connectivity, could persuade us that neural condition NC was consciousness—or, at least, that NC played the consciousness role in humans—because it had the right structure and function.

As things are, however, our conception of consciousness does not seem to be exhausted by structure and function and the 'why?' question remains. A neuroscientific answer to the 'what?' question would be of great interest but it would not render consciousness intelligible in neuroscientific terms. Consciousness would remain a brute fact. Between neural (or, more generally, physical) conditions and consciousness there is an *explanatory gap* (Levine 1983).

1.1.1 Positions in the philosophy of consciousness

Sometimes, the explanatory gap is presented as licensing a conclusion about the nature of reality itself, and not just about our conceptions of reality. It is argued that the existence of an explanatory gap between the physical sciences and consciousness supports the conclusion that consciousness is metaphysically or ontologically distinct from the world that the physical sciences describe. It would be no wonder that consciousness could not be explained in terms of the physical sciences if consciousness were something quite different from the physical world. The conclusion that consciousness falls outside the physical order is sometimes dramatized as the claim that there could, in principle, be a creature physically just like one of us yet lacking consciousness—a zombie (Chalmers 1996; Kirk 2006), or even a complete physical duplicate of our world from which consciousness was totally absent—a zombie world. In line with this claim, David Chalmers proposes that 'a theory of consciousness requires the addition of *something* fundamental to our ontology, as everything in physical theory is compatible with the absence of consciousness' (1995, p. 210).

While Chalmers argues that consciousness is not wholly physical, it is more common (at least within academic philosophy) to assume or argue that some version of physicalism is true, so that consciousness must be part of the physical world (Papineau 2002).[2] Contemporary physicalists reject the duality of material and mental substances that Descartes proposed and also reject the duality of material and mental properties or attributes. According to physicalism, conscious mental states, processes, and events are identical to physical (specifically, neural) states, processes, and events. Furthermore, the phenomenal properties of conscious mental states (what being in those states is like for the subject) are the very same properties as physical properties of neural states or—if the claim of identity between phenomenal and physical properties seems too bold—the phenomenal properties are strongly determined by physical properties. The idea of strong determination in play here is that the phenomenal properties are necessitated by the physical properties. The phenomenal properties do not and could not vary independently of the physical properties; they *supervene* on the physical properties.[3]

Physicalist approaches to the philosophy of consciousness come in two varieties. Chalmers (1996) calls the two kinds of approach *type-A materialism* and *type-B materialism*. Some physicalists (type-A materialists) deny that there is an explanatory gap and maintain, instead, that consciousness can be fully and satisfyingly explained in physical terms. This option is, of course, mandatory for physicalists who agree with anti-physicalists like Chalmers that there is a good argument from the existence of an explanatory gap to the conclusion that consciousness falls outside the physical order.

Other physicalists (type-B materialists) allow that there is an explanatory gap but deny that there is a good argument from the gap to the anti-physicalist conclusion. In his development of the notion of an explanatory gap, Joseph Levine (1993) distinguishes two senses in which it might be said that the physical

---

[2] See Zeman (this volume, section 11.2.3) for some data about public understanding of the mind. In a survey of undergraduate students, '64% disputed the statement that "the mind is fundamentally physical"' (p. 294).

[3] We shall return (section 1.7.2) to the distinction between the strict version of physicalism (phenomenal properties are identical to physical properties) and the relaxed version (phenomenal properties are determined by, or supervene on, physical properties).

sciences *leave out* consciousness, the epistemological sense and the metaphysical sense. The claim that the physical sciences leave out consciousness in the epistemological sense is the claim that there is an explanatory gap. The claim that the physical sciences leave out consciousness in the metaphysical sense is the claim that consciousness falls outside the physical order. Levine says that the distinction between epistemological leaving out and metaphysical leaving out 'opens a space for the physicalist hypothesis' (1993, p. 126). Type-B materialist approaches typically involve two claims. First, the explanatory gap results from our distinctively subjective *conception* of consciousness. Second, there can be both scientific conceptions and subjective conceptions of the same physical reality (just as, in the familiar case of Hesperus and Phosphorus, there can be two concepts of a single object, the planet Venus). Type-B materialists maintain that there can be a duality of conceptions without a duality of properties.

1.1.2 Outline

This chapter begins with the subjective conception of consciousness that gives rise to the explanatory gap and the intractability of the 'why?' question. Next, there is a discussion of the approach to the study of consciousness that was adopted by Brian Farrell (1950), an approach that frankly rejects the subjective conception in favour of a broadly behaviourist one.[4] Farrell's approach serves as a model for subsequent type-A materialists.

The second half of the chapter is organized around Frank Jackson's *knowledge argument*—an argument for the anti-physicalist claim that phenomenal properties of conscious mental states are not physical properties. The knowledge argument 'is one of the most discussed arguments against physicalism' (Nida-Rümelin 2002) and the philosophical literature of the last 25 years contains many physicalist responses to the argument. Perhaps the most striking response is Jackson's own, for he now rejects the knowledge argument, adopting a type-A materialist approach and denying that there is an explanatory gap.

The type-A materialism that Jackson shares with Farrell denies that there is anything answering to our conception of consciousness to the extent that the conception goes beyond structure and function. In that respect, type-A materialism 'appears to deny the manifest' (Chalmers 2002, p. 251), and is probably the minority approach amongst philosophers who defend physicalism

---

[4] Brian Farrell was Wilde Reader in Mental Philosophy in the University of Oxford from 1947 to 1979. He died in August 2005, at the age of 93. The institutional and historical setting of the lecture on which this chapter is based (Oxford in the spring of 2006) invited extended reflection on Farrell's paper.

(although it is the approach adopted by such influential figures as Daniel Dennett and David Lewis). The more popular approach is type-B materialism, accepting that there is an explanatory gap but denying that this leads to the anti-physicalist conclusion (Chalmers 1999, p. 476): 'It simultaneously promises to take consciousness seriously (avoiding the deflationary excesses of type-A materialists) and to save materialism (avoiding the ontological excesses of the property dualist).' By considering the knowledge argument and responses to it, we shall be in a position to assess the costs and benefits of some of the most important positions in contemporary philosophy of consciousness.

1.2 The subjective conception of consciousness

We have distinguished two questions about the explanation of consciousness, the 'what?' question and the 'why?' question. The question *what* makes the difference between conscious and unconscious mental states seems to be a tractable question and many scientists and philosophers expect an answer in broadly neuroscientific terms—an answer that specifies the *neural correlates* of consciousness (Chalmers 2000; Block 2005; Lau, this volume). The question *why* this neuroscientific difference makes the difference between conscious and unconscious mental states is more problematic. As Thomas Nagel put the point over 30 years ago, in his paper, 'What is it like to be a bat?' (1974/1997, p. 524):

> If mental processes are indeed physical processes, then there is something that it is like, intrinsically, to undergo certain physical processes. What it is for such a thing to be the case remains a mystery.

Ned Block expressed a similar view in terms of qualia, the subjective, phenomenal, or 'what it is like' properties of conscious mental states (1978, p. 293):

> *No* physical mechanism seems very intuitively plausible as a seat of qualia, least of all a brain. ... Since we know that *we are brain-headed systems*, and that *we* have qualia, we know that brain-headed systems can have qualia. [But] we have no theory of qualia which explains how this is *possible*.

1.2.1 Nagel's distinction: subjective and objective conceptions

Nagel's announcement of mystery was not based on gratuitous pessimism about the progress of science but on an argument. The starting point was the thought that we cannot conceive what it is like to be a bat. The conclusion was that, although we can (of course) conceive what it is like to be a human, we cannot explain, understand, or account for (our) conscious mental states in terms of the physical operation of (our) brains. We should take a moment to review the steps that led Nagel from the alien character of bat consciousness to the mystery of human consciousness.

The initial thought about bats can be extended to a distinction between two types of conception. We cannot conceive what it is like to be a bat and likewise a bat or a Martian, however intelligent, could not conceive what it is like to be a human—for example, what it is like to undergo the conscious mental states that you are undergoing now. These limitations reflect the fact that conceptions of conscious mental states as such are *subjective*; they are available from some, but not all, points of view. Roughly, the conscious mental states that we can *conceive* are limited to relatively modest imaginative extensions from the conscious mental states that we ourselves *undergo*. We cannot conceive what it is like to be a bat although we can conceive what it is like to be human. We cannot conceive what it is like for a bat to experience the world through echolocation although we can conceive what it is like for a human being to experience the red of a rose or a ripe tomato.

While grasping what a conscious mental state is like involves deployment of subjective conceptions, the physical sciences aim at objectivity in the sense that the conceptions deployed in grasping theories in physics, chemistry, biology, or neuroscience are accessible from many different points of view. The physical theories that we can grasp are limited, not by our sensory experience, but by our intellectual powers; and the conceptions that are required are, in principle, no less available to sufficiently intelligent bats and Martians than to humans.

1.2.2 Knowing what it is like

Nagel said (1974/1997, p. 521; emphasis added), 'I want to *know what it is like* for a bat to be a bat', and he went on to point out that the expression 'knowing what it is like' has two different, though related, uses (*ibid.* p. 526, n. 8; see also Nida-Rümelin 2002, section 3.3). In one use, knowing what a particular type of experience is like is *having a subjective conception* of that type of experience. There is a partial analogy between knowing what a type of experience is like and the 'knowing which' that is required for thought about particular objects (Evans 1982).[5] Knowing what a type of experience is like is similar to knowing

---

[5] Gareth Evans's (1982) theorizing about object-directed thoughts was guided by *Russell's Principle*, which says that in order to think about a particular object a thinker must *know which* object it is that is in question. Evans interpreted the principle as requiring discriminating knowledge, that is, the capacity to discriminate the object of thought from all other things. Initially, this may sound so demanding as to make object-directed thought an extraordinary achievement. But Evans's examples of ways of meeting the 'knowing which' requirement make it seem more tractable.

which object is in question in being a kind of discriminatory knowledge. In the case of thought about particular objects, there are many ways of meeting the 'knowing which' requirement: for example, presently perceiving the object, being able to recognize it, or knowing discriminating facts about it. In the case of thought about types of experience, there may also be many ways of meeting the 'knowing which' requirement. Having a subjective conception of a type of experience is meeting the 'knowing which' requirement in virtue (roughly) of being the subject of an experience of the type in question. (We shall refine this shortly.)

In a second use, knowing what it is like is *having propositional knowledge* about a type of experience, conceived subjectively. It is not easy to provide a philosophical account of having a conception or concept, but a subject who has a conception of something has a cognitive capacity to think about that thing. A subject who has a conception of a type of experience can deploy that conception in propositional thinking and may achieve propositional knowledge about that type of experience. He might know that he himself is having an experience of that type, or that he has previously had such an experience; and he may know something of the circumstances in which other people have experiences of that type. In the latter case, the subject knows what it is like for people to be in those circumstances.

It is plausible that a subjective conception of a type of experience can be deployed in thought even when the subject is not having an experience of the type in question. If that is right, then it must be possible for a subject to meet the 'knowing which' requirement in respect of a type of experience without concurrently being the subject of an experience of that type. On some accounts of having a subjective conception, it might be that remembering being the subject of an experience of the type in question would be sufficient to meet the 'knowing which' requirement. (Perhaps having a veridical apparent memory would suffice.) Alternatively, it might be proposed that meeting the 'knowing which' requirement involves being able to imagine being the subject of an experience of the type in question or being able to recognize other token experiences of which one is the subject as being of the same type again. (We shall return to these abilities to remember, imagine, and recognize in section 1.10.2.)

Michael Tye (2000) suggests that there are two different ways in which a subject can meet the requirements for having a subjective conception of a type of experience. In the case of a relatively coarse-grained experience type, such as the experience of red, a subject might meet the 'knowing which' requirement on the basis of long-standing abilities to remember, imagine, and recognize experiences of that type. In the case of a very fine-grained experience type, such as the experience of a specific shade of red, the limitations of human memory may prevent a subject from reliably discriminating later experiences of that precise type from others. Nevertheless, it seems that a subject who is actually having an experience of that shade of red (and whose attention is not occupied elsewhere) has a subjective conception of that fine-grained experience type and knows what it is like to experience that specific shade of red. In such a case, the subject meets the 'knowing which' requirement in virtue of being the subject of an experience of the fine-grained type in question even if possession of the subjective conception lasts no longer than the experience itself.

1.2.3 Nagel's conclusion: physical theories and the explanation of consciousness

With these two uses of 'knowing what it is like' in mind, we can distinguish two claims that are immensely plausible in the light of Nagel's distinction between subjective and objective conceptions. The first claim is that subjective conceptions cannot be constructed from (are not woven out of) the objective conceptions that are deployed in grasping theories in the physical sciences. A subject might be able to deploy all the objective conceptions needed to grasp physical theories about colour vision without having any subjective conception of the experience of red. The second claim that is plausible in the light of Nagel's distinction is that there is no *a priori* entailment from physical truths to truths about conscious mental states conceived subjectively.

The second claim is not an immediate consequence of the first (Stoljar 2005; Byrne 2006) because *a priori* entailment of subjective truths by physical truths does not require that subjective conceptions should be constructible from physical conceptions. The second claim says that a subject who was able to deploy objective conceptions of physical states and who also *possessed the subjective conception of a particular type of experience* would not, just in virtue of having those conceptions, be in a position to know that a person in such-and-such a physical state in such-and-such a physical world would have an experience of that particular type.

If these claims are correct then physical theories, to the extent that they achieve the objectivity to which science aspires, will not say anything about conscious mental states conceived subjectively. We know what it is like to undergo various conscious mental states, but the conceptions that constitute or figure in that knowledge have no place in our grasp of objective physical theory. Nor will the content of our distinctively subjective propositional knowledge about conscious experience be entailed *a priori* by physical theory.

Once we grant the contrast between subjective conceptions and the objective conceptions that are deployed in grasping physical theories, the conclusion of Nagel's argument is compelling. We cannot explain conscious mental states *as such*—that is, conceived subjectively—in terms of the physical operation of brains conceived objectively.

In a similar spirit to the Nagelian argument, Colin McGinn says (2004, p. 12):

> any solution to the mind-body problem has to exhibit consciousness as *conservatively emergent* on brain processes: that is, we must be able to explain how consciousness emerges from the brain in such a way that the emergence is not *radical* or *brute*.

And (*ibid.*, p. 15):

> What the theory has to do is specify some property of the brain from which it follows *a priori* that there is an associated consciousness .... *A priori* entailments are what would do the trick.

But *a priori* or conceptual entailments will not be available precisely because of the 'vastly different concepts' (p. 19) that figure, on the one hand, in the physical sciences of the brain and, on the other hand, in our knowledge of what it is like to undergo conscious mental states.

1.2.4 Subjective conceptions and physicalism

According to Nagel's argument, the explanatory gap is a consequence of the distinction between subjective conceptions and the objective conceptions that are deployed in grasping physical theories. On the face of it, this duality of conceptions is consistent with the truth of physicalism and, indeed, at the end of his paper, Nagel says (1974/1997, p. 524): 'It would be a mistake to conclude that physicalism must be false.'

If physicalism is true and conscious mental states fall within the physical order then they are part of the subject matter of objective physical theory. Similarly, if thinking about things, or conceiving of things, falls within the physical order then the activity of deploying conceptions—even deploying subjective conceptions—is part of the subject matter of objective physical theory. Thus, when we grasp physical theories by deploying objective conceptions, we may think about a physical event or process that is, in fact, the deployment of a subjective conception. But this does not require us to be in a position, nor does it put us into a position, to deploy that subjective conception ourselves. Even on a physicalist view of what there is in the world, grasping physical theories is one thing and deploying subjective conceptions is another. (In sections 1.11 and 1.12, we shall consider arguments that this duality of conceptions is not, in fact, consistent with physicalism.)

Tye argues that the explanatory gap presents no threat to physicalism because, really, there is no gap (1999/2000, p. 23): 'it is a cognitive illusion'. By claiming that there is no gap, Tye does not mean that there really are *a priori* entailments from physical truths to truths about conscious mental states conceived subjectively. He agrees with Nagel that the distinction between subjective and objective conceptions guarantees that there are no such entailments. But he argues that it is a mistake to describe the absence of such entailments as a *gap* (*ibid.*, p. 34):

> [T]he character of phenomenal [subjective] concepts and the way they differ from third-person [objective] concepts conceptually guarantees that the question [*why* it is that to be in physical state P is thereby to have a feeling with this phenomenal character] has no answer. But if it is a conceptual truth that the question can't be answered, then there can't be an explanation of the relevant sort, *whatever* the future brings. Since an explanatory *gap* exists only if there is something unexplained that needs explaining, and something needs explaining only if it can be explained (whether or not it lies within the power of *human beings* to explain it), there is again no gap.

There are at least two important points to take from this bracing passage. First, if there are distinctively subjective conceptions of types of experience then there will be truths about conscious experience that are not entailed *a priori* by physical truths. So, a philosopher who maintains that all truths about conscious experience *are* entailed *a priori* by physical truths (a type-A materialist) must deny that there are distinctively subjective conceptions of the kind that Nagel envisages. Second, the absence of *a priori* entailment from physical truths to truths about conscious experience (subjectively conceived) is conceptually guaranteed (Sturgeon 1994). So, it is not an absence that will be overcome by progress in the physical sciences.

I shall not, myself, put these important points in Tye's way. Instead of saying, with Tye, that there is no explanatory gap, I shall say that there is an explanatory gap if there is no *a priori* entailment from physical truths to truths about conscious mental states conceived subjectively. The difference from Tye is terminological. I am prepared to allow that an explanatory gap exists even though what is unexplained is something which, as a matter of conceptual truth, cannot be explained.

1.3 Farrell on behaviour and experience: Martians and robots

In discussions of Nagel's (1974) paper, it is often noted that the 'what it is like' terminology and, indeed, the example of the bat, occurred in a paper by Brian Farrell, 'Experience', published in the journal *Mind* in 1950. I shall come in a moment to the use that Farrell made of the bat example. Before that, I need to describe the problem that Farrell was addressing—a problem which, he said, troubled physiologists and psychologists, even if not 'puzzle-wise professional philosophers' (1950, p. 174).

The problem is that scientific accounts of 'what happens when we think, recognize things, remember, and see things' *leave something out*, namely, the experiences, sensations, and feelings that the subject is having (*ibid.*, p. 171).[6] The experimental psychologist, for example, gathers data about a subject's 'responses and discriminations', dealing with 'behaviour' but not with 'experience'. Thus (p. 173): 'while psychology purports to be the scientific study of experience,... the science, in effect, does not include experience within its purview'. The problem that troubled the physiologists and psychologists was, in short, that the sciences of the mind leave out consciousness.

Farrell argues that there is really *no such problem* as the physiologists and psychologists take themselves to face. He asks us to consider the sentence (1950, p. 175):

> If we merely consider all the differential responses and readinesses, and such like, that X exhibits towards the stimulus of a red shape, we are leaving out the experience he has when he looks at it.

He argues that this is quite unlike ordinary remarks, such as:

> If you merely consider what Y says and does, you leave out what he really feels behind that inscrutable face of his.

The difference between the two cases is said to be this (p. 176): 'What we leave out [in the second sentence] is something that Y can tell us about [whereas] what is left out [in the first sentence] is something that X cannot in principle tell us about'. But why is it that X cannot tell us about what seems to be left out by a description of responses and readinesses, namely, his experience? Farrell answers (*ibid.*):

> He has already given us a lengthy verbal report, but we say that this is not enough. We want to include something over and above this, *viz.*, X's experience. It is useless to ask X to give us further reports and to make further discriminations if possible, because these reports and discriminations are mere behaviour and leave out what we want.

A critic of Farrell's argument might object at this point. For, even granting that X's report itself would be a piece of behaviour, it does not yet follow that what X would tell us *about* would be mere behaviour. On the contrary, it seems that X might tell us about the phenomenal properties of the experience that he had when presented with a red shape. So we need to be provided with a reason why X's apparent description of an experience should not be taken at face value.

---

[6] Farrell does not distinguish between epistemological and metaphysical 'leaving out' claims.

A major theme in Farrell's argument is the apparent contrast between 'behaviour' and 'experience', in terms of which the problem is raised. Farrell points out that, in ordinary unproblematic cases where behaviour is contrasted with experience, the term 'behaviour' is restricted to *overt* behaviour. But in the case of the putatively problematic contrast—where the sciences of the mind are supposed to leave out experience—the notion of behaviour is stretched to include 'the covert verbal and other responses of the person, his response readinesses, all his relevant bodily states, and all the possible discriminations he can make' (p. 177). Farrell insists that, once the notion of behaviour is extended in this way, we cannot simply assume that it continues to *contrast* with experience rather than *subsuming* experience. This theme is developed in discussion of two classic philosophical examples, Martians and robots.

1.3.1 Wondering what it is like: Martians, opium smokers, and bats

In the example of 'the man from Mars' (1950, p. 183), Farrell asks us to imagine that physiologists and psychologists have found out all they could find out about a Martian's sensory capacities and yet they still wonder what it would be like to be a Martian. He says that the remark, 'I wonder what it would be like to be a Martian', seems to be sensible because it superficially resembles other remarks, such as 'I wonder what it would be like to be an opium smoker' and 'I wonder what it would be like to be, and hear like, a bat' (*ibid.*).

If, in an ordinary *unproblematic* context, I wonder what it would be like to be an opium smoker, then I may suppose or imagine that I take up smoking opium and that I thereby come to learn how the addiction develops, for example. What I would learn in the hypothetical circumstances of being an opium smoker might, Farrell says, outrun what could be learned by the 'clumsy' scientific methods available at a given time. But it would not be different in principle from what could be learned from third-person observation. Thus (pp. 172–3):

> Quite often [a psychologist] places himself in the role of subject. ... What is important to note is that by playing the role of observer-subject, he does not add anything to the discoveries of psychological science that he could not in principle obtain from the observation of X [another subject] alone.

According to Farrell, what I would learn about the experience of the opium smoker from the point of view of the observer-subject would not fall under the term 'behaviour' in the sense restricted to overt behaviour, but it would fall under the term in its extended sense that includes covert responses, response readinesses, discriminations, and so on.

In a similar way, I could unproblematically wonder what it would be like to be a bat. I could suppose that a witch turns me into a bat and that, from the privileged position of observer-subject, I learn something about the being's discriminations and response readinesses. But, on Farrell's view, if I were to spend a day or so as a bat then what I would learn would not outrun developed bat physiology and psychology. And he is quite explicit that it would require no distinctively subjective concepts or conceptions (p. 173):

> [N]o new concepts are required to deal with what [the psychologist's] own subject-observation reveals which are not also required by what was, or can be, revealed by his [third-person] observation of [another subject].

In this unproblematic kind of wondering what it is like to be an opium smoker or a bat, what I would learn about would be covert responses and internal discriminations, behaviour in the extended and inclusive sense of that term. This would also be the case if I unproblematically wondered what it would be like to be a Martian (p. 185): 'the "experience" of the Martian would … be assimilable under "behaviour"'.

The example of the Martian began, however, with a kind of wondering that was supposed to be quite different from this unproblematic wondering about behaviour in the inclusive sense of the term. It was supposed to be a *problematic* wondering about something that would inevitably be left out by the sciences of the Martian mind—a wondering about experience as contrasted, not only with overt behaviour, but even with behaviour in the extended and inclusive sense of the term. Farrell's point is that, while unproblematic wondering is 'sensible', this putatively problematic wondering is 'pointless' (p. 185). We have no right to assume that this contrast—between experience and behaviour in the inclusive sense—is legitimate.

A critic of Farrell's argument might concede this point but also insist on another. We cannot simply assume that behaviour in the inclusive sense *contrasts* with experience; but equally we cannot simply assume that it *subsumes* experience. Until we have a positive argument for subsumption, the relationship between behaviour and experience should remain an open question. We shall come to Farrell's positive arguments shortly (section 1.4); but, before that, we review the second of the two classic philosophical examples, the robots.

1.3.2 Robots—and the criteria for having a sensation

The question under discussion in the example of the robot is whether we need to retain the contrast between behaviour and experience in order to say (1950, p. 189): 'If a robot were to behave just like a person, it would still not have any sensations, or feelings.'

Farrell's answer to the question comes in two stages. First, in ordinary talk about robots, the unproblematic contrast between experience and overt behaviour is adequate for the purpose. A robot, in the ordinary sense of the term, duplicates the overt behaviour of a human being but not the covert responses, bodily states, internal discriminations, and so on. So, second, if the example of the robot is to present a problem for Farrell's view then we must be thinking of a robot that duplicates, not only our overt behaviour, but all our covert responses and internal discriminations as well. But then, Farrell says, he has already argued that we cannot presume upon a contrast between experience and behaviour in this extended and inclusive sense.

In order to avoid the 'muddle' that results, according to Farrell, from this 'unobserved departure from the ordinary usage of "robot"' (p. 190), we could set aside that term for the time being. Then there are two ways that we might describe a mechanical system that duplicates the overt and covert, external and internal, behaviour of a person. On the one hand, we might allow, in line with what Farrell regards as our 'usual criterion' for having a sensation, that the mechanical system has sensations. On the other hand, we might adopt a more demanding criterion for having a sensation and deny that the mechanical system has sensations on the grounds that it is not a living thing.

Does either way of describing the mechanical system present a problem for Farrell's view about experience? If, on the one hand, we allow that a system that produces the right external and internal behaviour has experience then clearly the example provides no reason to retain a contrast between experience and behaviour in the inclusive sense. If, on the other hand, we insist that, while mechanical systems produce behaviour, only a living thing has experience then, of course, we do retain a kind of contrast between experience and behaviour in the inclusive sense. This more demanding criterion allows us to deny experience to inanimate robots. But the contrast between mechanical systems and living things has no relevance to questions about the mental lives of human beings, Martians, or bats. Farrell thus concludes that the example of the robot does not present a problem for the *behaviourist psychology of organisms*.

Bringing his discussion of robots even closer to contemporary philosophy of consciousness, Farrell invites us to consider a series of imaginary examples of robots that duplicate our external and internal behaviour and are increasingly like living things. He suggests that, as we progress along this series, it will be increasingly natural to allow that the robots have experience—sensations and feelings: (p. 191):

> General agreement [to allow the attribution of experience] would perhaps be obtained when we reach a machine that exhibited the robot-like analogue of reproduction, development and death.

of physics) and that instantiations of non-physical properties have no causal consequences in the physical order (dualist interactionism is false). As a consequence, he accepted that phenomenal properties of experience, if they are not physical properties, are epiphenomenal. Against such a view, the causal argument for physicalism makes no headway (because premise 1 is not accepted). From the mid-1990s, Jackson came to argue (1998b, 2005a; Braddon-Mitchell and Jackson 1996) that, since the denial of physicalism involves epiphenomenalism about qualia, the knowledge argument is undermined by its own conclusion.

Jackson says that his reason for changing his mind about the knowledge argument is that (1998b/2004, p. 418): 'Our knowledge of the sensory side of psychology has a causal source.' When Mary emerges from her black-and-white room and sees something red, she undergoes a change—from not knowing what it is like to see red to knowing what it is like to see red. This is, or involves, a physical change. The physical change is caused by something and, by the completeness of physics, it has a full physical cause. If the phenomenal properties of Mary's experience of seeing a red rose or a ripe tomato are non-physical, and so epiphenomenal, then those properties of Mary's experience can play no part in the causation of Mary's coming to know what it is like to see red. Thus (Braddon-Mitchell and Jackson 1996, p. 134): 'Mary's discovery ... of something important and new about what things are like is in no sense due to the properties, the qualia, whose alleged instantiation constituted the inadequacy of her previous picture of the world.'

As Jackson came to see the situation, the conclusion of the knowledge argument has the consequence that phenomenal properties are epiphenomenal, and this undermines the intuition that Mary gains new knowledge on her release as a result of experiencing for herself what it is like to see red.[16] This was enough to persuade Jackson that 'there must be a reply' to the knowledge argument (Braddon-Mitchell and Jackson 1996, p. 143), 'it *must* go wrong' (Jackson 2005a, p. 316).

1.8.2 Jackson's version of physicalism

Physicalism as Jackson conceives it is not the strict version. It is not committed to the claim that phenomenal properties are identical to properties that figure

---

[16] This objection against the knowledge argument was raised by Michael Watkins (1989): 'if Jackson's [1982] epiphenomenalism is correct, then we cannot even know about our own qualitative experiences' (p. 158); 'Jackson's epiphenomenalism provides us with no avenues by which we might justifiably believe that there are qualia. If epiphenomenalism is correct, then Mary, the heroine of Jackson's knowledge argument against physicalism, gains no new knowledge when she leaves her black and white room' (p. 160).

in the present or future science of physics, nor even to the claim that phenomenal properties are identical to properties that figure in the physical sciences conceived more broadly, to include physics, chemistry, biology, and neuroscience. The reason is that physicalism is 'a theory of everything in space-time' (2006, p. 231) and 'the patterns that economics, architecture, politics and very arguably psychology, pick out and theorize in terms of, include many that do not figure in the physical sciences' (*ibid.*, p. 234). The properties in terms of which Jackson's physicalism is defined are not just physical properties 'in the core sense' (properties that figure in the physical sciences) but also include physical properties 'in an extended sense' (p. 233).

Jackson's version of physicalism is not the relaxed version either. The reason is this (2006, p. 243): 'A live position for dual attribute theorists is that psychological properties, while being quite distinct from physical properties, are necessitated by them.' So, supervenience physicalism characterized without some additional requirement is not properly distinguished from 'a necessitarian dual attribute view' (*ibid.*). According to Jackson, if supervenience physicalism is not to be 'a dual attribute theory in sheep's clothing' (p. 227) then the determination of supervening properties by core physical properties must be necessary *and a priori*.

Thus, Jackson proposes that physical properties in the extended sense are properties whose distribution is determined *a priori* by the distribution of physical properties in the core sense. Two simple examples may help to make this idea clearer. First, while the property of being silver and the property of being copper both figure in the science of chemistry, it is not clear that chemistry or any other physical science has a use for the disjunctive property of being either silver or copper. So the disjunctive property might not be a physical property in the core sense. But it is a physical property in the extended sense because whether something instantiates the disjunctive property is determined *a priori* by whether it is silver and whether it is copper. Second, while jewellers talk about sterling silver it is not clear that there is a science of things that are made up of 92.5% silver and 7.5% copper. So the property of being sterling silver might not be a physical property in the core sense. But it is a physical property in the extended sense because the distribution of sterling silver is determined *a priori* by the distributions of silver and of copper.

1.8.3 The case for *a priori* physicalism

Jackson's version of physicalism is *a priori physicalism* and, in fact, the notion of the *a priori* enters twice over. First, *a priori* physicalism requires that all properties should be physical properties, defined as properties that are *determined a priori* by properties that figure in the physical sciences (section 1.8.2).

Second, *a priori* physicalism requires that all the facts, particularly the psychological facts, should be *entailed a priori* by the physical facts (section 1.6.1). The two requirements are not obviously equivalent. If there were subjective conceptions of physical properties then the requirement of *a priori* entailment would not be met (there would be an explanatory gap) although the requirement of *a priori* determination of properties could still be met. In both cases, however, the *a priori* element in the account is promoted as distinguishing physicalism from 'a dual attribute theory in sheep's clothing' (2006, p. 227) or from 'some covert form of dual attribute theory of mind' (1995/2004, p. 414; see above, section 1.6.2).[17]

Jackson places his *a priori* physicalism in the tradition of Australian materialism—the materialism of J.J.C. Smart (1959) and David Armstrong (1968)—according to which 'spooky properties are rejected along with spooky substances' (2006, p. 227). He is opposed to all dual attribute theories, including even the 'necessitarian' dual attribute theory that says that phenomenal properties are distinct from, but strongly determined (necessitated) by, physical properties. The problem with dual attribute theories, Jackson says, is that 'spooky properties… would be epiphenomenal and so both idle and beyond our ken' (*ibid.*). As we saw (section 1.8.1), it was because of this problem that Jackson rejected the knowledge argument and its conclusion that the phenomenal properties of experience are 'spooky', non-physical, properties.

It may be, however, that non-physical properties need not be epiphenomenal. As Terence Horgan puts the point (1984/2004, p. 308, n. 6): 'Indeed, even if qualia are nonphysical they may not be epiphenomenal. As long as they are supervenient upon physical properties, I think it can plausibly be argued that they inherit the causal efficacy of the properties upon which they supervene.' (In section 1.7.1, we noted that Papineau (2002, p. 32) makes a similar proposal.) The possibility of non-physical, but causally potent, properties raises two potential worries about Jackson's position. First, it allows a response to Jackson's specific reason for rejecting the knowledge argument. Second, it raises the question whether there is any good objection to dual attribute theories of the necessitarian variety.

---

[17] Jackson says that the thesis about the *a priori* determination of physical properties is *a priori* physicalism 'understood as a doctrine in metaphysics, understood *de re* if you like' (2006, p. 229). This thesis is already sufficient to distinguish *a priori* physicalism from a necessitarian dual attribute theory. The thesis about *a priori* entailment is *a priori* physicalism understood *de dicto*. Although *a priori* physicalism understood *de dicto* seems to go beyond *a priori* physicalism understood *de re*, Jackson (2005b, p. 260) advances an argument 'that takes us from the *de re* thesis to the *de dicto* thesis'.

Jackson could respond to these worries by maintaining that his *a priori* physicalism is a better, because more austere, theory than any dual attribute view. He characterizes *a priori* physicalism as 'bare' physicalism in a passage that manifests something of W. V. O. Quine's (1953; see Jackson 2005b, p. 257) 'taste for desert landscapes' (2003/2004, pp. 425–6):

> The *bare physicalism hypothesis* ... that the world is exactly as is required to make the physical account of it true in each and every detail but nothing more is true of this world in the sense that nothing that fails to follow *a priori* from the physical account is true of it ... is not *ad hoc* and has all the explanatory power and simplicity we can reasonably demand.

1.9 Physicalism and representationalism

Jackson now rejects the conclusion of the knowledge argument. As mentioned earlier, he thinks that 'we have no choice but to embrace some version or other of physicalism' (2004, p. xvi). The specific version of physicalism that he accepts is type-A materialism—also known as *a priori* physicalism, conceptually deflationist physicalism, or bare physicalism. Consequently, he rejects the epistemological first premise of the knowledge argument. This, he now says, is where the argument goes wrong (2004, p. xvii–xviii): '[Mary] learns nothing about what her and our world is like that is not available to her in principle while in the black and white room.'

This is what Farrell would say and Dennett does say.[18] It is what Jackson needs to say; but it is not easy to defend. It certainly contrasts sharply with what he said when he first put forward the knowledge argument (1986/2004, p. 52): '[I]t is very hard to believe that [Mary's] lack of knowledge could be remedied merely by her explicitly following through enough logical consequences of her vast physical knowledge.' In defence of his new position, Jackson needs to make it plausible that, when Mary 'knows what it is like to see red', what she really knows is something that is entailed *a priori* by the totality of physical facts about the world, facts that she already knew in her room.

1.9.1 Representationalism

I mentioned earlier (section 1.3.1) that Armstrong interprets Farrell's claim about experience being featureless as an anticipation of the representationalist view that the phenomenal properties of an experience are determined by its representational properties—that is, by how it represents the world as being. Similarly, Stoljar and Nagasawa (2004, p. 25, n. 11) see Farrell's idea of featureless

---

[18] See Dennett (1991, pp. 398–406, reprinted in Ludlow *et al.* 2004, pp. 59–68; 2005, Chapter 5, 'What RoboMary knows').

experience as related to the doctrine of the transparency or diaphanousness of experience, a doctrine that several contemporary philosophers take to stand in a close relationship to representationalism.[19] In any case, it is to representationalism that Jackson turns for his account of what Mary really knows when she knows what it is like to see red (2003/2004, p. 430): 'we have to understand the qualities of experience in terms of intensional [representational] properties'.

The starting point for representationalism is that 'experience is *essentially* representational ... it is impossible to have a perceptual experience without thereby being in a state that represents that things are thus and so in the world' (Jackson 2007, p. 57). As a claim about perceptual experience, such as Mary's experience of a red rose or a ripe tomato, this is highly plausible. It is of the nature of perception to represent how things are in the world. The claim that *all* experiences, including bodily sensations, are representational is less compelling but even strong opponents of representationalism may be prepared to grant it. Thus, for example, Ned Block says, 'I think that sensations—almost always—perhaps even always—have representational content' (2003, p. 165).

Going beyond this starting point, representationalism is usually formulated as a supervenience thesis. As between two conscious mental states, such as two perceptual experiences, there can be no difference in phenomenal properties without a difference in representational properties. The phenomenal character of a conscious mental state is determined by its representational content (Byrne 2001). Representationalism is an unclear thesis to the extent that the notion of representation itself is not well specified. It is also a controversial thesis. But the attraction of representationalism for a type-A materialist is that it promises physical-functional conceptions of types of experience.

In philosophy of mind over the last quarter-century or so, the topics of consciousness and representation have mainly been considered somewhat separately. As a result, even those who think that consciousness defies scientific explanation are apt to be confident that representation can be analysed in 'naturalistic', physical-functional, terms. It is against this background that Jackson says (2003/2004, p. 432):

> The project of finding an analysis of representation is not an easy one—to put it mildly. But ... the answers that have been, or are likely to be, canvassed are all answers

---

[19] It is difficult to spell out a compelling argument from the transparency or diaphanousness of experience to representationalism (Stoljar 2004; see also Burge 2003, pp. 405–7). Tye (2000, p. 45) says: 'I believe that experience is transparent. I also believe that its transparency is a very powerful motivation for the representationalist view. I concede, however, that the appeal to transparency has not been well understood.' Jackson (2007, p. 57) says: 'I conclude that the famous diaphanousness or transparency of experience is not *per se* the basis of an argument for representationalism.'

that would allow the fact of representation to follow *a priori* from the physical account of what our world is like.

1.9.2 Representationalism and phenomenology

Let us agree, for the sake of the argument, that the representational facts are entailed *a priori* by the physical facts, just as the water facts are entailed *a priori* by the H_2O facts. This is not yet sufficient for a physicalist account of what Mary really knows when she 'knows what it is like to see red'. For what Mary knows entails that *there is something that it is like* to see red; seeing red is a conscious mental state. But, even according to representationalism, it is *not* the case that the representational content of an experience 'suffices to make any state that has it conscious' (Seager and Bourget 2007, p. 263). Or, as Alex Byrne puts it (2001, p. 234): 'Intentionalism [representationalism] isn't much of a theory of consciousness.'

Representationalism says that the phenomenal character of a conscious mental state is determined by its representational content. But it does not say that the representational properties of a mental state determine that it is a *conscious* mental state. There can be representation without consciousness. So something needs to be added to representationalism if it is to provide an account of what Mary knows about the experience of seeing red. Jackson himself says that the nature of an experience, including the fact that it is a conscious mental state, is determined by 'the [representational] content of [the] experience *plus* the fact that the experience represents the content as obtaining *in the way distinctive of perceptual representation*' (2007, p. 58; also see 2005a, p. 323).

He goes on to list five features that are putatively distinctive of perceptual representation. The content of perceptual representation is rich, and inextricably rich; the representation is immediate; there is a causal element in its content; and perceptual experience plays a distinctive functional role in respect of belief. If a state has representational content with these five features then, Jackson says, 'we get the phenomenology for free' (2003/2004, p. 438). What is most important about these five features is that they can, let us suppose, be explicated in physical-functional terms.

The story of Mary generates the intuition that, on her release, Mary learns something new about the experiences of people who looked at red roses and ripe tomatoes while she was in her room. As required by type-A materialism, Jackson rejects this intuition. He says that what Mary really knows is that the people were in physical states with a particular representational property (roughly, representing something as being red) and meeting five further conditions. Since both representation and the further conditions can be explicated in physical-functional terms, this knowledge is not new but was, in principle, already available to Mary while she was in her room.

1.10 The epistemological intuition and the ability hypothesis

Type-A materialism is counter-intuitive. Accepting it commits Jackson to rejecting the epistemological intuition that Mary learns something new on her release. He needs to develop a critical agenda supporting that rejection. Like Farrell and Dennett, Jackson needs to undermine the idea that there is more to know about human experience than is entailed *a priori* by the totality of physical facts.

1.10.1 Representationalism and the epistemological intuition

Jackson stresses that representationalism highlights the distinction between a *representational* property and an *instantiated* property (2003, 2005a, 2007). Representationalism thus provides a reason to say that 'there is no such property' (2003/2004, p. 430) as the 'redness' of the experience of seeing a rose. Redness is not a property that experiences of roses *instantiate*; it is the property that experiences *represent* roses as instantiating.

Experiences do, of course, instantiate the property of representing things as being red. But this representational property of experiences is not a new property that was unknown to Mary before her release. It is a physical-functional property that Mary knew about (or could have known about) in her black-and-white room. What is new after her release is that Mary now has an experience instantiating this property (and meeting five further conditions). We must take care not to mistake a new instantiation of a representational, and therefore physical, property of experiences for the instantiation of a new, and therefore non-physical, property of experiences. A new instantiation of a property is not the instantiation of a new property.

This is an important point, but it is not clear that it undermines the intuition that supports the epistemological first premise of the knowledge argument (Alter 2007). Jackson says (2007, p. 61): 'The challenge from the knowledge argument is the intuition that the "red" of seeing red is a new sort of property.' But, on the face of it, the intuition that Jackson needs to undermine— the intuition that drives the knowledge argument—is not an explicitly metaphysical intuition that, on her release, Mary learns about a new property of experiences. It is the epistemological intuition that Mary gains new knowledge— that she comes to know a fact that is not entailed *a priori* by the totality of physical facts that she already knew in her room.

It is, of course, part of Jackson's overall position that new knowledge would have to be knowledge about new properties. That is what the second premise of the original knowledge argument says: if physicalism is true (no new properties) then the psychophysical condition is *a priori* (no new knowledge). But that connection between epistemology and metaphysics is not provided by representationalism about perceptual experiences. It depends on the assumption that type-B materialism can be rejected (section 1.6.3).

Representationalism may be developed in the service of physicalism about conscious mental states. It certainly plays a major role in contemporary philosophy of consciousness. But representationalism does not favour Jackson's conceptually deflationist physicalism (type-A materialism) over conceptually inflationist physicalism (type-B materialism). Tye (1995, 2000) develops a version of representationalism that is very similar to Jackson's.[20] Yet Tye maintains that, on her release, Mary gains new subjective conceptions of physical—specifically, representational—properties, deploys those conceptions in propositional thinking, and achieves new propositional knowledge.

1.10.2 The ability hypothesis

We have just seen that representationalism does not provide any independent motivation for rejecting the epistemological intuition that Mary learns something new on her release. But Jackson's physicalist account of conscious experience goes beyond representationalism.

According to representationalism, the properties of Mary's experience when she sees a red rose for the first time are physical properties. Specifically, they are properties of having such-and-such representational content and meeting further physical-functional conditions. They are not new properties but properties that Mary was already in a position to know about in her black-and-white room. What is *new* is that Mary now has an experience that *instantiates* those properties. Jackson describes Mary's situation as follows (2003/2004, p. 439):

> [S]he is in a new kind of representational state, different from those she was in before. And what is it to know what it is like to be in that kind of state? Presumably, it is to be able to recognize, remember, and imagine the state. ... We have ended up agreeing with Laurence Nemirow and David Lewis on what happens to Mary on her release.

---

[20] According to Tye's PANIC theory, phenomenal properties are determined by (indeed, are identical with) properties of having Intentional Content that meets three conditions: it is Poised (poised to have a direct impact on beliefs and desires), Abstract (does not involve particular objects), and Nonconceptual (in order for a state to have this kind of content it is not necessary for the subject of the state to be able to conceptualise the content). Thus, in Tye's account, P+A+N plays the role that the five features play in Jackson's account.

Here, Jackson goes beyond the basic claim of his response to the knowledge argument, namely, that Mary does not gain new propositional knowledge on her release. He concedes something to the epistemological intuition. He says that Mary does gain something new and he allows that this might be described as new 'knowing what it is like'. But following the *ability hypothesis response* to the knowledge argument proposed by Lewis (1988) and Nemirow (1980, 2007), he says that what Mary gains is not new propositional knowledge but only new abilities.

Papineau comments on the ability hypothesis (2002, pp. 59–60):

> Some philosophers are happy to accept that Mary acquires new powers of imaginative re-creation and introspective classification, yet deny that it is appropriate to view this as a matter of her acquiring any new phenomenal concepts. These are the sophisticated deflationists of the 'ability hypothesis'.

Jackson proposes to join these 'sophisticated' conceptually deflationist physicalists and, as Papineau says, their position—Mary gains new 'know how' or new abilities—certainly seems preferable to 'outright denial' that Mary comes to know anything new at all. Subtle type-A materialism seems preferable to the more straightforward type-A materialism of Farrell and Dennett. But, Papineau continues (*ibid.*, p. 61):

> Even so, the ability hypothesis does not really do justice to the change in Mary. If we look closely at Mary's new abilities, we will see that they are inseparable from her power to think certain new kinds of thoughts.

The abilities to which Jackson appeals in his description of Mary's new 'knowing what it is like' are the same abilities to which Tye (2000) appeals in describing a subject's possession of a subjective conception of a relatively coarse-grained type of experience, such as seeing red (section 1.2.2). The advocate of the ability hypothesis needs to say why having these abilities is not sufficient for possession of a subjective conception that can be deployed in propositional thinking about a type of experience.

In earlier writings, Jackson himself resists the ability hypothesis response to the knowledge argument. He argues that Mary's new abilities (to remember, imagine, recognize) are associated with a new cognitive capacity to engage in new propositional thinking and that Mary acquires 'factual knowledge about the experiences of others' (1986/2004, p. 55). For example, he says (*ibid.*, p. 54):

> Now it is certainly true that Mary will acquire abilities of various kinds after her release. She will, for instance, be able to imagine what seeing red is like, be able to remember what it is like ... But is it plausible that that is *all* she will acquire? ... On her release she sees a ripe tomato in normal conditions, and so has a sensation of red. Her first reaction is to say that she now knows more about the kind of experience others have when looking at ripe tomatoes.

Jackson now embraces the ability hypothesis but it seems to me that his earlier line of argument is still plausible.

Even if the ability hypothesis can be defended against these objections, its role is only to provide a more nuanced version of type-A materialism, reducing the counterintuitive impact of denying that Mary gains new propositional knowledge on her release. It does not provide any strong, independent reasons in favour of type-A materialism and against type-B materialism.[21] Thus, neither representationalism nor the ability hypothesis offers materials for the critical agenda that any type-A materialist needs to develop. They do nothing to undermine the idea that there is more to know about human experience than is entailed *a priori* by the totality of physical facts.

1.11 Physical properties in new guises?

If physicalism is true then phenomenal properties are physical properties. Specifically, if Jackson's *a priori* physicalism is true then phenomenal properties are physical properties either in the core sense (properties that figure in the physical sciences) or in the extended sense (properties whose distribution is determined *a priori* by the distribution of physical properties in the core sense). If Jackson's or Tye's representationalism is true then we can say more about phenomenal properties: they are representational properties. The leading idea of type-B materialism is that Nagelian subjective conceptions of types of experience are conceptions *of* physical properties. The explanatory gap is consistent with physicalism; there is habitable space between epistemological leaving out and metaphysical leaving out; there can be new knowledge that is not knowledge of new properties.

1.11.1 The 'old fact, new guise' response to the knowledge argument

Type-B materialism is conceptually inflationist physicalism. It involves a duality of objective and subjective conceptions without a metaphysical duality of physical and non-physical properties. The type-B materialist's response to the knowledge argument is to accept the epistemological first premise but deny that this leads to the metaphysical conclusion that physicalism is false. When Mary is released from her black-and-white room, her new knowledge involves a new subjective conception of a fact that she already knew under an old

---

[21] Yuri Cath (in press) argues that the ability hypothesis leads ultimately to the idea of new conceptions, and so turns out to be a version of the type-B materialist's 'old fact, new guise' response to the knowledge argument (see below, section 1.11).

objective conception. As Levine says (1993, p. 125): 'the case of Mary typifies the phenomenon of there being several distinguishable ways to gain epistemic access to the same fact'.

This is a common response to the knowledge argument (e.g. Horgan 1984), often known as the 'old fact, new guise' response (or the 'old fact, new mode' response, to suggest Gottlob Frege's (1892) notion of mode of presentation). Jackson rejects it, both early and late. When he first put forward the knowledge argument against physicalism, he already rejected the suggestion that the argument depends on 'the intensionality of knowledge' (1986/2004, p. 52).[22] He now accepts physicalism and is convinced that the knowledge argument '*must* go wrong'. But he still rejects the suggestion that the argument goes wrong in neglecting new conceptions, guises, or modes of presentation (Braddon-Mitchell and Jackson 2007, p. 137): 'This is the explanation of Mary's ignorance that is available to dual attribute theorists, not the explanation available to physicalists.'

There can certainly be multiple conceptions of the same physical property. But Jackson maintains that explaining Mary's new knowledge by appeal to new conceptions is incompatible with physicalism. In order to understand why, it is useful to recall the example of water and H_2O (section 1.6.2). On Jackson's view, it is a matter of conceptual analysis that water is the stuff that fills the water role. Knowledge that water is H_2O can only be arrived at *a posteriori*. But the fact that water is H_2O is entailed *a priori* by the fact that H_2O fills the water role. Now suppose that some physical stuff, S, fills two roles, R_1 and R_2, in the physical order. Then we can have two conceptions of S. We can think of S as the stuff that fills role R_1 or as the stuff that fills role R_2. It may very well be that examining these conceptions themselves will not tell us that they are two conceptions of the same physical stuff. It is likely that this knowledge can only be arrived at *a posteriori*. But if Mary, in her black-and-white room, knows the full story about the physical order, then she is already in a position to know that the stuff that fills role R_1 also fills role R_2. This kind of example of multiple conceptions is available to a physicalist, but it does not provide a model for Mary's gaining new knowledge on her release.

The situation would be different if S were to instantiate a non-physical property, N. Then we could have a third conception of S. We could think of S as the stuff that instantiates N. Even if Mary knew all there is to know about

---

[22] The intensionality of knowledge is illustrated by the fact that 'Nigel knows that Hesperus is a planet' may be true while 'Nigel knows that Phosphorus is a planet' is false even though Hesperus = Phosphorus.

the physical order, she might not be in a position to know that the stuff that fills role R_1 and role R_2 also instantiates N. In her black-and-white room, she might know nothing at all of property N. But, while this kind of example of multiple conceptions would provide a model for Mary's gaining new knowledge on her release, it is obviously not available to a physicalist (Jackson 2005b, p. 262).

1.11.2 Descriptive and non-descriptive conceptions

According to the conceptually inflationist physicalist (type-B materialist), Mary gains new knowledge on her release because she gains new subjective conceptions of physical properties. The problem for type-B materialism is that we have not been able to find a model for Mary's new conceptions that is consistent with physicalism.[23]

It is plausible that the source of this problem lies in the fact that we have considered only conceptions that pick out a kind of physical stuff *by description*. These are conceptions of the form 'the physical stuff that has property F', where the property in question might be a physical property or a non-physical property. Descriptive conceptions in which the descriptive property is physical or physical-functional (such as 'the physical stuff that *fills the water role*') are already available to Mary while she is in her room. So it may seem that new, distinctively subjective conceptions must be descriptive conceptions in which the descriptive property is non-physical (such as 'the physical stuff that has *property N*'). Thus, if all conceptions are descriptive then Nagel's duality of objective and subjective conceptions requires a metaphysical duality of physical and non-physical properties right from the outset. Since type-B materialism proposes a duality of conceptions without a duality of properties, it requires that subjective conceptions—including the conceptions that become available to Mary only on her release—are *not* descriptive conceptions.

A partial analogy for the distinction between objective conceptions and non-descriptive subjective conceptions is provided by the distinction between two kinds of conception of locations in space. One kind of conception specifies locations in terms of distances and directions from an objective point of origin. A location, L, might be specified as being 25 miles north-west from Carfax. Deploying that conception of L in thought, I might achieve propositional

---

[23] See Jackson (2005a, p. 318): '[T]he guises ... must all be consistent with physicalism if physicalism is true.... But then, it seems, Mary could know about their applicability when inside the room.'

knowledge (by looking at a map or reading a book, for example) that there is water at L—perhaps the book says that there is a pond with ducks. A different kind of conception of a location is made available to me when I am *at* that location. Without knowing how far I am from Carfax or in which direction, I might arrive at a location and decide to explore a little. What is going on here? I notice sheep grazing on the other side of a stone wall, some farm buildings further back, a tractor and, in the distance, trees. Then I see a pond with ducks. So, there is water here.

If I am, in fact, 25 miles north-west from Carfax then this is an example of a new instantiation of an old spatial property: I myself am at location L. In virtue of my new location, I gain new abilities: I can feed the ducks and, just by bending down, I can put my hand in the pond, I can see (and may later remember) things that I have never seen before. But I do not only gain new abilities. I also gain a new indexical or egocentric conception of a location that I already knew about under a different, map-based, 'distance and direction from origin' or allocentric conception. I have a new cognitive capacity: I can think of location L as 'here'. Deploying the new conception in propositional thinking, I achieve new propositional knowledge that (as I put it) 'there is water here'.

We should not rush from this partial analogy to the idea that subjective conceptions of types of experience are indexical conceptions, like the context-dependent conceptions of locations, times, and people expressed by 'here', 'now', and 'I'. In fact, there are reasons to reject the proposal that subjective conceptions are indexical conceptions (Tye 1999; Papineau 2007). One disanalogy is that at least some subjective conceptions can be deployed in thought by a subject who is not concurrently having an experience of the type in question. They seem to function like recognitional, rather than indexical, conceptions (Loar 1997). But, in the face of Jackson's objection to the 'old fact, new guise' response, even the partial analogy offers some encouragement to the conceptually inflationist physicalist.

1.12 Phenomenal concepts and physicalism

Non-descriptive subjective conceptions of phenomenal types of experience are often called *phenomenal concepts*. Papineau introduces the idea with three main points. First, he says (2002, p. 48): 'when we use phenomenal concepts, we think of mental properties, not as items in the material world, but in terms of *what they are like*'. Second, he stresses that 'as a materialist, I hold that even phenomenal concepts refer to material *properties*' (*ibid.*). Third, he insists that the advocate of phenomenal concepts must avoid the 'poisoned chalice' (p. 86) of considering phenomenal concepts as descriptive concepts. Phenomenal concepts refer 'directly, and not via some description' (p. 97).[24]

In an earlier and seminal paper, Brian Loar says (1997/2004, p. 219):

> On a natural view of ourselves, we introspectively discriminate our own experiences and thereby form conceptions of their qualities, both salient and subtle .... What we apparently discern are ways experiences differ and resemble each other with respect to *what it is like to have them*. Following common usage, I will call these experiential resemblances *phenomenal qualities*, and the conceptions we have of them, *phenomenal concepts*. Phenomenal concepts are formed 'from one's own case'.

Loar goes on to highlight the distinction between concepts and properties and to point to the possibility of accepting a duality of concepts or conceptions without a duality of properties (1997/2004, pp. 220–1):

> It is my view that we can have it both ways. We may take the phenomenological intuition at face value, accepting introspective concepts and their conceptual irreducibility, and at the same time take phenomenal qualities to be identical with physical-functional properties of the sort envisaged by contemporary brain science. As I see it, there is no persuasive philosophically articulated argument to the contrary.

1.12.1 A limitation on the promise of phenomenal concepts

Phenomenal concepts provide a model for subjective conceptions of types of experience—including the new conceptions that Mary gains on her release—and the model holds some promise of being consistent with physicalism. First, according to type-B materialism, subjective conceptions are conceptions *of* physical properties. Second, a phenomenal concept of a type of experience is a non-descriptive concept. It is a recognitional concept that a thinking subject possesses in virtue of having an experience of the type in question. Deploying a phenomenal concept in thought is not a matter of thinking of a physical property as the property that has such-and-such higher-order property (that is, such-and-such property of properties). So the type-B materialist need not face an objection along the lines that phenomenal concepts can only account for new knowledge if they involve *non-physical* higher-order properties (section 1.11.2). Nevertheless, the promise of phenomenal concepts is limited in an important way.

According to physicalism, conscious mental states are physical states and the phenomenal properties of conscious mental states are physical properties.

---

[24] In *Thinking About Consciousness* (2002), Papineau defends a 'quotational' or 'quotational-indexical' model of phenomenal concepts. More recently (2007), he acknowledges that this model faces some objections and he adopts a different view of phenomenal concepts as cases of, or at least as similar to, perceptual concepts—something like 'stored sensory templates' (2007, p. 114). This change leaves intact the three points in the main text.

In general, instantiating a physical property is not sufficient for gaining a conception of that property but, according to type-B materialism, the phenomenal properties of conscious mental states are special in this respect. Consider a subject who is, in general, able to form concepts and deploy them in thought. By being in a conscious mental state—having an experience of a particular type—such a subject can gain conceptions of certain physical properties of that state, namely, the phenomenal properties of that type of experience. These conceptions are direct, non-descriptive, subjective, phenomenal concepts and, intuitively, the subject gains a phenomenal concept of a physical property in such cases only because there is something that it is like to instantiate that property.

Now, recall Nagel's remark (1974/1997, p. 524):

> If mental processes are indeed physical processes, then there is something that it is like, intrinsically, to undergo certain physical processes. What it is for such a thing to be the case remains a mystery.

According to Nagel, if there is something that it is like to instantiate certain physical properties then we have no answer to the question *why* this is so.

We have just said that, intuitively, if we gain phenomenal concepts of certain physical properties by instantiating them then there must be something that it is like to instantiate those properties. Consequently, it seems, if we gain phenomenal concepts of certain physical properties by instantiating them then, ultimately, we have no answer to the question why that is so. Possessing phenomenal concepts of physical properties does not have a fully satisfying explanation in physical terms.[25]

1.12.2 An argument against type-B materialism?

This limitation on the promise of phenomenal concepts may offer the prospect of an argument against the type-B materialist's claim that phenomenal concepts are direct, non-descriptive concepts of physical properties.

A subject who has a phenomenal concept of a type of experience meets the requirement of *knowing which* type of experience is in question (section 1.2.2). The subject knows what that type of experience is like (in one use of 'knowing what it is like') in virtue of being, or having been, the subject of an experience of that type. But, a subject who knows which type of experience is in question need not think of that type of experience as the property with such-and-such physical-functional specification nor, indeed, as being a physical property at all.

---

[25] For discussion of what can reasonably be demanded of the phenomenal concept response to the knowledge argument, see Chalmers (2007), Levine (2007), Papineau (2007).

Furthermore, if Nagel is right then a subject who knows what a type of experience is like has no answer to the question why this is what it is like, or why there is anything at all that it is like, to instantiate a property whose nature is physical.

We might begin to wonder whether a subject can really possess a direct and non-descriptive concept *of a physical property*, and meet the requirement of knowing which physical property is in question, just in virtue of knowing what a particular type of experience is like. A subject who knows what a type of experience is like has a phenomenal concept. But we might wonder whether *thinking about a physical property* by deploying a phenomenal concept must be *indirect*, with the phenomenal concept embedded in a descriptive concept along the lines of: 'the physical property that it is *like this* to instantiate'.[26]

Developing, and then responding to, these inchoate concerns would require work on the metaphysics of properties—their individuation and their natures—and work on the 'knowing which' requirement that would inevitably lead into theories of reference in philosophy of language and thought. It is not obvious in advance what the outcome of this work would be. But suppose, for a moment, that it were to uncover a good argument for the claim that, if there are distinctively subjective phenomenal concepts of phenomenal properties, then these phenomenal properties are not identical with physical properties (they neither are, nor are determined *a priori* by, properties that figure in the physical sciences).

This would be an important argument. First, it would show that type-B materialism can be rejected—that a duality of objective and subjective conceptions requires a duality of physical and non-physical properties—and it would show this without simply relying on an assumption that all conceptions are descriptive (section 1.11.2). Second, by showing that type-B materialism can be rejected, the argument would provide the needed motivation for the second premise of the original knowledge argument (section 1.6.2) and—what comes to the same thing—it would license the transition from the limited conclusion of the simplified knowledge argument, that type-A materialism is false, to the more sweeping conclusion of the original knowledge argument, that physicalism is false (section 1.6.4).

Third, the argument would tie together the two requirements of Jackson's *a priori* physicalism (section 1.8.3). The first requirement is that all properties should be physical properties, defined as properties that are *determined*

---

[26] Chalmers (1999) and Horgan and Tienson (2001) argue against the claim that direct phenomenal concepts are concepts of physical-functional properties. Also recall McGinn's comment that 'if we know the essence of consciousness by means of acquaintance, then we can just see that consciousness is not reducible to neural or functional processes' (2004, p. 9).

a priori by properties that figure in the physical sciences. The second requirement is that all the facts should be *entailed a priori* by the physical facts. Suppose that the first requirement is met, so that physical properties are all the properties there are. It would follow from the envisaged argument that there are no distinctively subjective conceptions of any properties. But if all properties are physical properties and there are no subjective conceptions, then there is no impediment to *a priori* entailment of all the facts by the physical facts. So the second requirement would also be met (cf. Jackson 2005b, p. 260).

Fourth, the argument would figure as an item on the critical agenda that any type-A materialist needs to develop. Jackson needs to undermine the intuition that there is more to know about human experience than is entailed *a priori* by the totality of physical facts. While representationalism is consistent with physicalism, it does not reveal any error or confusion in the epistemological intuition that drives the knowledge argument (section 1.10.1). Nor does the ability hypothesis provide strong, independent reasons in favour of type-A materialism (section 1.10.2). By showing that type-B materialism can be rejected, the argument would reveal that, even if type-A materialism involves some cost to intuition, it is the only alternative to dualism.

1.12.3 Options for physicalism

Following Chalmers, we have divided physicalist approaches to the philosophy of consciousness into two varieties, type-A materialism (also known as conceptually deflationist physicalism) and type-B materialism (also known as conceptually inflationist physicalism). We have just considered, in a speculative way, a possible line of argument against type-B materialism. If there were to be a good argument of the envisaged kind then the options would seem to be severely limited. A physicalist, having rejected the dualist options of interactionism and epiphenomenalism, would seem bound to embrace the counterintuitive commitments of type-A materialism.

In fact, this is not quite right. At the beginning of this chapter, when I first contrasted dualism and physicalism (section 1.1.1), I said that, according to physicalism, phenomenal properties are either identical with physical properties or else strongly determined (necessitated) by physical properties. I also said that the causal argument for physicalism allows for both a strict identity version and a relaxed supervenience version of physicalism (section 1.7.2). In recent sections, however, I have adopted Jackson's terminology. His version of physicalism says that all properties are physical properties. He allows that physical properties include properties that do not themselves figure in the physical sciences but are determined *a priori* by properties that do figure there. He does *not* allow that properties that are determined or necessitated only *a posteriori* by physical properties are themselves physical. Some varieties of supervenience *physicalism* are now classified as varieties of *dualism* and, specifically, as necessitarian dual attribute theories.

This means that we need to reconsider the hypothetical situation if there were to be a good argument against the type-B materialist's claim that phenomenal concepts are direct, non-descriptive concepts of physical properties. If the dualist options of interactionism and epiphenomenalism were rejected there would still be *two* options available, not only type-A materialism, but also the necessitarian dual attribute view. Theorists who describe this view as a variety of supervenience physicalism, rather than dualism, will regard it as conceptually inflationist, rather than deflationist, physicalism. As a consequence, it will be grouped with type-B materialism. But it will, apparently, be left untouched by the line of argument against type-B materialism that we considered in section 1.12.2. According to the necessitarian dual attribute view, phenomenal concepts are not concepts of physical properties, but concepts of distinct phenomenal properties that supervene on physical properties.

The costs and benefits of the variety of supervenience physicalism also known as the necessitarian dual attribute view are not, of course, affected by a terminological decision between 'physicalism' and 'dualism'. In this chapter, we have seen only one argument against this option and that was an Ockhamist[27] argument in favour of the austerity of 'bare physicalism' (section 1.8.3). The benefits of austerity would have to be weighed against the costs to intuition of type-A materialism.

The less austere, but otherwise more intuitive, option is favoured by Edmund Rolls (this volume), who leaves it as an open question whether it is best described as 'physicalism'. According to his higher-order syntactic thought (HOST) theory of consciousness, conscious mental states are physical states of a system with a particular computational nature. The computational properties of the state necessitate phenomenal properties *a posteriori* (p. 154; some emphases added):

> [T]he present approach suggests that it *just is* a property of HOST computational processing with the representations grounded in the world that it feels like something. There is to some extent *an element of mystery* about why it feels like something, *why it is phenomenal* ... In terms of the physicalist debate, an important aspect of my proposal is that it is a *necessary* property of this type of (HOST) processing that it feels like something... and given this view, then it is *up to one to decide whether this view is consistent with one's particular view of physicalism or not.*

---

[27] The fourteenth-century philosopher, William of Ockham, is credited with a law of parsimony: 'Entities should not be multiplied beyond necessity.'

1.13 Conclusion

At the first choice point in the philosophy of consciousness, some philosophers deny that there is an explanatory gap and accept type-A materialism. We have seen that Jackson joins Farrell and Dennett in this group, rejecting the intuition that there is more to know about human experience than what is entailed *a priori* by a battery of physical fact and theory that can be grasped by Mary in her black-and-white room, or by a sufficiently intelligent Martian or bat. Philosophers who, at the first choice point, accept that there is an explanatory gap proceed to a second choice point. There, some opt for dualism, others for type-B materialism.

Jackson's knowledge argument and Chalmers's conceivability argument are arguments for dualism. If these arguments are correct then the phenomenal properties of experience are not physical properties. Philosophers who accept this conclusion—Jackson (at an earlier stage when he accepted the knowledge argument) and Chalmers (still)—face a further choice about the causal relationship between the phenomenal and the physical. One option is to accept dualist interactionism at the cost of rejecting the completeness of physics. Another is to accept epiphenomenalism at the cost of rendering phenomenal properties 'idle and beyond our ken', as Jackson (2006, p. 227) now puts it. These are not especially attractive views but Chalmers (2002) argues that these options, and others, should be taken seriously 'if we have independent reason to think that consciousness is irreducible' (2002, p. 263).[28]

Chalmers also commends a view, *Russellian* or *type-F monism* (Russell 1927), on which the most fundamental properties of the physical world are both protophysical and protophenomenal—the physical and the phenomenal are variations on a common theme (2002, p. 265–6): 'One could give the view in its most general form the name *panprotopsychism*, with either protophenomenal or phenomenal properties underlying all of physical reality.' This view is speculative and exotic, but Chalmers suggests that 'it may ultimately provide the best integration of the physical and the phenomenal within the natural world' (p. 267; see also Stoljar 2006).

According to type-B materialism, we can accept that consciousness is conceptually irreducible but reject dualism. This is an attractive option that is adopted by many contemporary philosophers of consciousness—probably the majority—including Block, Levine, Loar, Papineau and Tye. If type-B materialism can be defended then arguments for dualism are undermined and some

---

[28] Chalmers (2002) refers to dualist interactionism as *type-D dualism* and to epiphenomenalism as *type-E dualism*.

of the motivation for more exotic views, such as Russellian monism, is removed. Indeed, the knowledge argument against physicalism and in favour of dualism seems to rest on the assumption that type-B materialism is not a real option for the physicalist.

There are, however, arguments against type-B materialism—against the idea that we can have a duality of objective and subjective concepts, and an explanatory gap, without a duality of physical and non-physical properties. Some of these arguments seem to depend on the assumption that all concepts are descriptive and the dominant form of type-B materialism therefore appeals to direct, non-descriptive, subjective, phenomenal concepts of physical properties. But there are also arguments against this form of the view.

It may very well be that none of these arguments is, in the end, compelling and that Loar (1997) will turn out to be right in saying that we can 'have it both ways': irreducibly subjective phenomenal concepts are nevertheless concepts of physical properties of the kinds that figure in neuroscience. On the other hand, there may be a good argument against phenomenal concepts of physical properties and friends of type-B materialist may have to consider shifting to the necessitarian dual attribute view that phenomenal concepts are concepts of non-physical phenomenal properties that are determined or necessitated—but not *a priori*—by physical properties. Some philosophers may object to the departure from ontological austerity (Jackson 2003) and others may have concerns about a primitive relation of *a posteriori* necessitation between properties (strong necessities; see Chalmers 1996, 1999, 2002). But if, at the first choice point, there are good reasons to accept that there is an explanatory gap then, at the second choice point, the necessitarian dual attribute view should be taken at least as seriously as Russellian monism, dualist interactionism, or epiphenomenalism.

Acknowledgements

I am grateful to Tim Bayne, Ned Block, Tyler Burge, Alex Byrne, David Chalmers, Frank Jackson, David Papineau, Edmund Rolls, Nick Shea, Daniel Stoljar, and Larry Weiskrantz for comments and conversations.

References

Alter, T. (2007). Does representationalism undermine the knowledge argument? In Alter, T. and Walter, S. (eds) *Phenomenal Concepts and Phenomenal Knowledge: New Essays on Consciousness and Physicalism*, pp. 65–76. Oxford: Oxford University Press.

Armstrong, D.M. (1968). *A Materialist Theory of the Mind*. London: Routledge & Kegan Paul.

Armstrong, D.M. (1996). Qualia ain't in the head. Review of *Ten Problems of Consciousness: A Representational Theory of the Phenomenal Mind*, by Michael Tye. *Psyche* 2(31); http://psyche.cs.monash.edu.au/v2/psyche-2-31-armstrong.html.

Block, N. (1978). Troubles with functionalism. In Wade Savage, C. (ed.) *Perception and Cognition: Issues in the Foundations of Psychology, Minnesota Studies in the Philosophy of Science*, Volume 9, pp. 261–325. Minneapolis: University of Minnesota Press. Reprinted in Block, N. (ed.) *Readings in the Philosophy of Psychology*, Volume 1, pp. 268–306. Cambridge, MA: Harvard University Press, 1980.

Block, N. (2002). The harder problem of consciousness. *Journal of Philosophy* **99**, 391–425.

Block, N. (2003). Mental paint. In Hahn, M. and Ramberg, B. (eds) *Reflections and Replies: Essays on the Philosophy of Tyler Burge*, pp. 165–200. Cambridge, MA: MIT Press.

Block, N. (2005). Two neural correlates of consciousness. *Trends in Cognitive Sciences* **9**, 46–52.

Block, N. Flanagan, O., and Güzeldere, G. (eds) (1997). *The Nature of Consciousness: Philosophical Debates*. Cambridge, MA: MIT Press.

Braddon-Mitchell, D. and Jackson, F.C. (1996). *Philosophy of Mind and Cognition: An Introduction*. Oxford: Blackwell.

Braddon-Mitchell, D. and Jackson, F.C. (2007). *Philosophy of Mind and Cognition: An Introduction*, 2nd edn. Oxford: Blackwell.

Burge, T. (1997). Two kinds of consciousness. In Block, N., Flanagan, O., and Güzeldere, G. (eds) *The Nature of Consciousness: Philosophical Debates*, pp. 427–434. Cambridge, MA: MIT Press.

Burge, T. (2003). Qualia and intentional content: Reply to Block. In Hahn, M. and Ramberg, B. (eds) *Reflections and Replies: Essays on the Philosophy of Tyler Burge*, pp. 405–616. Cambridge, MA: MIT Press.

Burge, T. (2007). Reflections on two kinds of consciousness. In *Foundations of Mind: Essays by Tyler Burge*, Volume 2, pp. 392–419. Oxford: Oxford University Press.

Byrne, A. (2001). Intentionalism defended. *Philosophical Review* **110**, 199–240.

Byrne, A. (2006). Review of *There's Something About Mary*, by Peter Ludlow, Yujin Nagasawa, and Daniel Stoljar. *Notre Dame Philosophical Reviews* (20.1.2006) http://ndpr.nd.edu/review.cfm?id=5561.

Cath, Y. (in press). The ability hypothesis and the new knowledge-how. *Noûs*.

Chalmers, D.J. (1995). Facing up to the problem of consciousness. *Journal of Consciousness Studies*, **2**, 200–19. Reprinted as 'The hard problem of consciousness' (pp. 225–235) and 'Naturalistic dualism' (pp. 359–368) in Velmans, M. and Schneider, S. (eds) *The Blackwell Companion to Consciousness*. Oxford: Blackwell, 2007.

Chalmers, D.J. (1996). *The Conscious Mind: In Search of a Fundamental Theory*. Oxford: Oxford University Press.

Chalmers, D.J. (1999). Materialism and the metaphysics of modality. *Philosophy and Phenomenological Research* **59**, 473–496.

Chalmers, D.J. (2000). What is a neural correlate of consciousness? In Metzinger, T. (ed.) *Neural Correlates of Consciousness: Empirical and Conceptual Issues*, pp. 17–39. Cambridge, MA: MIT Press.

Chalmers, D.J. (2002). Consciousness and its place in nature. In Chalmers, D.J. (ed.) *Philosophy of Mind: Classical and Contemporary Readings*, pp. 247–272. Oxford: Oxford University Press.

Chalmers, D.J. (2007). Phenomenal concepts and the explanatory gap. In Alter, T. and Walter, S. (eds) *Phenomenal Concepts and Phenomenal Knowledge: New Essays on Consciousness and Physicalism*, pp. 167–194. Oxford: Oxford University Press.

Davies, M. and Humberstone, I.L. (1980). Two notions of necessity. *Philosophical Studies* **38**, 1–30.

Dennett, D.C. (1988). Quining qualia. In Marcel, A.J. and Bisiach, E. (eds) *Consciousness in Contemporary Science*, pp. 42–47. Oxford: Oxford University Press.

Dennett, D.C. (1991). *Consciousness Explained*. Boston: Little, Brown.

Dennett, D.C. (2005). *Sweet Dreams: Philosophical Obstacles to a Science of Consciousness*. Cambridge, MA: MIT Press.

Dennett, D.C. (2007). What RoboMary knows. In Alter, T. and Walter, S. (eds) *Phenomenal Concepts and Phenomenal Knowledge: New Essays on Consciousness and Physicalism*, pp. 15–31. Oxford: Oxford University Press.

Evans, G. (1982). *The Varieties of Reference*. Oxford: Oxford University Press.

Farrell, B.A. (1950). Experience. *Mind* **59**, 170–198.

Frege, G. (1892). On sense and reference. In Geach, P. and Black, M. (eds) *Translations from the Philosophical Writings of Gottlob Frege*, pp. 56–78. Oxford: Blackwell, 1970.

Horgan, T. (1984). Jackson on physical information and qualia. *Philosophical Quarterly* **34**, 147–152. Reprinted in Ludlow, P., Nagasawa, Y., and Stoljar, D. (eds) *There's Something About Mary: Essays on Phenomenal Consciousness and Frank Jackson's Knowledge Argument*, pp. 301–308. Cambridge, MA: MIT Press, 2004.

Horgan, T. and Tienson, J. (2001). Deconstructing new wave materialism. In Loewer, B. (ed.) *Physicalism and Its Discontents*, pp. 307–318. Cambridge: Cambridge University Press.

Jackson, F.C. (1982). Epiphenomenal qualia. *American Philosophical Quarterly* **32**, 127–36. Reprinted in Ludlow, P., Nagasawa, Y., and Stoljar, D. (eds) *There's Something About Mary: Essays on Phenomenal Consciousness and Frank Jackson's Knowledge Argument*, pp. 9–50. Cambridge, MA: MIT Press, 2004.

Jackson, F.C. (1986). What Mary didn't know. *Journal of Philosophy* **83**, 291–295. Reprinted in Ludlow, P., Nagasawa, Y., and Stoljar, D. (eds) *There's Something About Mary: Essays on Phenomenal Consciousness and Frank Jackson's Knowledge Argument*, pp. 51–56. Cambridge, MA: MIT Press, 2004.

Jackson, F.C. (1995). Postscript. In Moser, P.K. and Trout, J.D. (eds) *Contemporary Materialism*, pp. 184–189. London: Routledge. Reprinted in Ludlow, P., Nagasawa, Y., and Stoljar, D. (eds) *There's Something About Mary: Essays on Phenomenal Consciousness and Frank Jackson's Knowledge Argument*, pp. 409–415. Cambridge, MA: MIT Press, 2004.

Jackson, F.C. (1998a). *From Metaphysics to Ethics: A Defence of Conceptual Analysis*. Oxford: Oxford University Press.

Jackson, F.C. (1998b). Postscript on qualia. In Jackson, F.C. *Mind, Method, and Conditionals*, pp. 76–79. London: Routledge. Reprinted in Ludlow, P., Nagasawa, Y., and Stoljar, D. (eds) *There's Something About Mary: Essays on Phenomenal Consciousness and Frank Jackson's Knowledge Argument*, pp. 417–420. Cambridge, MA: MIT Press, 2004.

Jackson, F.C. (2003). Mind and illusion. In O'Hear, A. (ed.) *Minds and Persons* (Royal Institute of Philosophy Supplement 53), pp. 251–271. Cambridge: Cambridge University Press. Reprinted in Ludlow, P., Nagasawa, Y., and Stoljar, D. (eds) *There's Something About Mary: Essays on Phenomenal Consciousness and Frank Jackson's Knowledge Argument*, pp. 421–442. Cambridge, MA: MIT Press, 2004.

Jackson, F.C. (2004). Foreword. In Ludlow, P., Nagasawa, Y., and Stoljar, D. (eds) *There's Something About Mary: Essays on Phenomenal Consciousness and Frank Jackson's Knowledge Argument*, pp. xv–xix. Cambridge, MA: MIT Press.

Jackson, F.C. (2005a). Consciousness. In Jackson, F.C. and Smith, M. (eds) *The Oxford Handbook of Contemporary Philosophy*, pp. 310–333. Oxford: Oxford University Press.

Jackson, F.C. (2005b). The case for *a priori* physicalism. In Nimtz, C. and Beckermann, A. (eds) *Philosophy—Science—Scientific Philosophy: Main Lectures and Colloquia of GAP 5, Fifth International Congress of the Society for Analytical Philosophy, Bielefeld, 22–26 September* 2003, pp. 251–265. Paderborn: Mentis.

Jackson, F.C. (2006). On ensuring that physicalism is not a dual attribute theory in sheep's clothing. *Philosophical Studies* **131**, 227–49.

Jackson, F.C. (2007). The knowledge argument, diaphanousness, representationalism. Alter, T. and Walter, S. (eds) *Phenomenal Concepts and Phenomenal Knowledge: New Essays on Consciousness and Physicalism*, pp. 52–64. Oxford: Oxford University Press.

Kirk, R. (2006). Zombies. In Zalta, E.N. (ed.) *The Stanford Encyclopedia of Philosophy* (Winter 2006 edition), http://plato.stanford.edu/archives/win2006/entries/zombies/.

Kripke, S.A. (1980). *Naming and Necessity*. Cambridge, MA: Harvard University Press.

LeDoux, J.E. (1996). *The Emotional Brain: The Mysterious Underpinnings of Emotional Life*. New York: Simon and Schuster.

Levine, J. (1983). Materialism and qualia: The explanatory gap. *Pacific Philosophical Quarterly* **64**, 354–361.

Levine, J. (1993). On leaving out what it's like. In Davies, M. and Humphreys, G.W. (eds) *Consciousness: Psychological and Philosophical Essays*, pp. 121–136. Oxford: Blackwell.

Levine, J. (2007). Phenomenal concepts and the materialist constraint. In Alter, T. and Walter, S. (eds) *Phenomenal Concepts and Phenomenal Knowledge: New Essays on Consciousness and Physicalism*, pp. 145–166. Oxford: Oxford University Press.

Lewis, D. (1988). What experience teaches. *Proceedings of the Russellian Society*. Sydney: University of Sydney. Reprinted in Lycan, W.G. (ed.) *Mind and Cognition: A Reader*, pp. 499–518. Oxford: Blackwell, 1990.

Loar, B. (1997). Phenomenal states (revised version). In Block, N. Flanagan, O., and Güzeldere, G. (eds). *The Nature of Consciousness: Philosophical Debates*, pp. 597–616. Cambridge, MA: MIT Press. Reprinted in Ludlow, P., Nagasawa, Y. and Stoljar, D. (eds) *There's Something About Mary: Essays on Phenomenal Consciousness and Frank Jackson's Knowledge Argument*, pp. 219–239. Cambridge, MA: MIT Press, 2004.

Lodge, D. (2001). *Thinks ...* London: Secker and Warburg.

Ludlow, P., Nagasawa, Y. and Stoljar, D. (eds) (2004). *There's Something About Mary: Essays on Phenomenal Consciousness and Frank Jackson's Knowledge Argument*. Cambridge, MA: MIT Press.

McGinn, C. (2004). *Consciousness and Its Objects*. Oxford: Oxford University Press.

McLaughlin, B.P. (2005). *A priori* versus *a posteriori* physicalism. In Nimtz, C. and Beckermann, A. (eds) *Philosophy—Science—Scientific Philosophy: Main Lectures and Colloquia of GAP 5, Fifth International Congress of the Society for Analytical Philosophy, Bielefeld, 22–26 September* 2003, pp. 267–285. Paderborn: Mentis.

Merikle, P.M., Smilek, D., and Eastwood, J.D. (2001). Perception without awareness: Perspectives from cognitive psychology. *Cognition* **79**, 115–134.

Nagel, T. (1974). What is it like to be a bat? *Philosophical Review* 83, 435–450. Reprinted in Nagel, T. *Mortal Questions*, pp. 165–180. Cambridge: Cambridge University Press, 1979. Also reprinted in Block, N. Flanagan, O., and Güzeldere, G. (eds) *The Nature of Consciousness: Philosophical Debates*, pp. 519–527. Cambridge, MA: MIT Press, 1997.

Nemirow, L. (1980). Review of *Mortal Questions* by Thomas Nagel. *Philosophical Review* **89**, 473–477.

Nemirow, L. (2007). So *this* is what it's like: A defense of the ability hypothesis. In Alter, T. and Walter, S. (eds) *Phenomenal Concepts and Phenomenal Knowledge: New Essays on Consciousness and Physicalism*, pp. 32–51. Oxford: Oxford University Press.

Nida-Rümelin, M. (1995). What Mary couldn't know: Belief about phenomenal states. In Metzinger, T. (ed.) *Conscious Experience*, pp. 219–441. Paderborn: Mentis. Reprinted in Ludlow, P., Nagasawa, Y. and Stoljar, D. (eds) *There's Something About Mary: Essays on Phenomenal Consciousness and Frank Jackson's Knowledge Argument*, pp. 241–267. Cambridge, MA: MIT Press, 2004.

Nida-Rümelin, M. (2002). Qualia: The knowledge argument. In Zalta, E.N. (ed.) *The Stanford Encyclopedia of Philosophy* (Fall 2002 edition), http://plato.stanford.edu/archives/fall2002/entries/qualia-knowledge/.

Papineau, D. (2002). *Thinking About Consciousness*. Oxford: Oxford University Press.

Papineau, D. (2007). Phenomenal and perceptual concepts. In Alter, T. and Walter, S. (eds) *Phenomenal Concepts and Phenomenal Knowledge: New Essays on Consciousness and Physicalism*, pp. 111–144. Oxford: Oxford University Press.

Quine, W.V.O. (1953). *From a Logical Point of View*. Cambridge, MA: Harvard University Press.

Rosenthal, D.M. (1991). The independence of consciousness and sensory quality. In Villanueva, E. (ed.) *Philosophical Issues, Volume 1: Consciousness*, pp.15–36. Atascadero, CA: Ridgeview. Reprinted in Rosenthal, D.M. *Consciousness and Mind*, pp. 135–148. Oxford: Oxford University Press, 2005.

Rosenthal, D.M. (2005). *Consciousness and Mind*. Oxford: Oxford University Press.

Russell, B. (1927). *The Analysis of Matter*. London: Kegan Paul.

Ryle, G. (1949). *The Concept of Mind*. London: Hutchinson.

Ryle, G. (1954). *Dilemmas: The Tarner Lectures* 1953. Cambridge: Cambridge University Press.

Seager, W. and Bourget, D. (2007). Representationalism about consciousness. In Velmans, M. and Schneider, S. (eds) *The Blackwell Companion to Consciousness*, pp. 261–276. Oxford: Blackwell.

Smart, J.C.C. (1959). Sensations and brain processes. *Philosophical Review* **68**, 141–56.

Stoljar, D. (2001). Two conceptions of the physical. *Philosophy and Phenomenological Research* **62**, 253–281. Reprinted in Ludlow, P., Nagasawa, Y. and Stoljar, D. (eds) *There's Something About Mary: Essays on Phenomenal Consciousness and Frank Jackson's Knowledge Argument*, pp. 309–331. Cambridge, MA: MIT Press, 2004.

Stoljar, D. (2004). The argument from diaphanousness. In Ezcurdia, M., Stainton, R. and Viger, C. (eds) *New Essays in the Philosophy of Language and Mind (Supplementary Volume of the Canadian Journal of Philosophy)*, pp. 341–390. Calgary: University of Calgary Press.

Stoljar, D. (2005). Physicalism and phenomenal concepts. *Mind and Language* **20**, 469–494.

Stoljar, D. (2006). *Ignorance and Imagination: The Epistemic Origin of the Problem of Consciousness.* Oxford: Oxford University Press.

Stoljar, D. and Nagasawa, Y. (2004). Introduction. In Ludlow, P., Nagasawa, Y. and Stoljar, D. (eds). *There's Something About Mary: Essays on Phenomenal Consciousness and Frank Jackson's Knowledge Argument*, pp. 1–36. Cambridge, MA: MIT Press.

Sturgeon, S. (1994). The epistemic view of subjectivity. *Journal of Philosophy* **91**, 221–235.

Tye, M. (1995). *Ten Problems of Consciousness: A Representational Theory of the Phenomenal Mind.* Cambridge, MA: MIT Press.

Tye, M. (1999). Phenomenal consciousness: The explanatory gap as a cognitive illusion. *Mind* **108**, 705–725. Reprinted in Tye, M. *Consciousness, Color, and Content*, pp. 21–42. Cambridge, MA: MIT Press, 2000.

Tye, M. (2000). *Consciousness, Color, and Content.* Cambridge, MA: MIT Press.

Velmans, M. and Schneider, S. (eds) (2007). *The Blackwell Companion to Consciousness.* Oxford: Blackwell.

Watkins, M. (1989). The knowledge argument against 'the knowledge argument'. *Analysis* **49**, 158–160.

Weiskrantz, L. (1986). *Blindsight: A Case Study and Implications.* Oxford: Oxford University Press.

Weiskrantz, L. (1997). *Consciousness Lost and Found.* Oxford: Oxford University Press.

Chapter 2

Explanatory gaps and dualist intuitions

David Papineau

2.1 Introduction

I agree with nearly everything Martin Davies says. He has written an elegant and highly informative analysis of recent philosophical debates about the mind–brain relation. I particularly enjoyed Davies' discussion of B.A. Farrell, his precursor in the Oxford Wilde Readership (now Professorship) in Mental Philosophy. It is intriguing to see how closely Farrell anticipated many of the moves made by more recent 'type-A' physicalists who seek to show that, upon analysis, claims about conscious states turn out to be nothing more than complex third-personal claims about internal and external behaviour. Davies is also exemplary in his even-handed treatment of those contemporary 'type-B' physicalists who have turned away from the neo-logical-behaviourism of Farrell and his ilk. Davies explains how type-B physicalists recognize distinctive subjective 'phenomenal concepts' for thinking about conscious states and so deny that phenomenal claims can be deduced *a priori* from behavioural or other third-personal claims. However, type-B physicalists do not accept that these subjective phenomenal concepts refer to any distinct non-material reality. In their view, phenomenal concepts and third-personal scientific concepts are simply another example of the familiar circumstance where we have two different ways of referring to a single reality.

Since I am persuaded by pretty much all Davies says about these matters, I shall not comment substantially on the dialectical points he covers. Instead I want to raise two rather wider issues. The first is the set of ideas associated with the phrase 'the explanatory gap'. Davies specifies that he is using this phrase in a specific technical sense. But the phrase has further connotations, and this can lead to a distorted appreciation of the philosophical issues. The second issue is the methodological implications of the philosophical debate. I shall argue that the philosophical issues addressed by Davies suggest that there are unexpected limitations on what empirical brain research can achieve.

In what follows, sections 2.2–2.5 will be devoted to 'the explanatory gap', while the rest of my remarks will address the methodological issue.

2.2 **Terminology**

Along with many other contemporary philosophers of consciousness, Davies takes the phrase 'the explanatory gap' to refer to the absence of any *a priori* route from the physical facts to the conscious facts. Davies agrees that, however detailed a physical description of some system we are given, it will never follow *a priori* that the system must have certain conscious experiences. Even if it is specified that some being is physically exactly like one of us humans, it will remain conceivable that this being is a 'zombie' with no experiences at all. As Davies uses the phrase, 'the explanatory gap' thus refers to the lack of any conceptual tie between physical descriptions and claims about consciousness.

Given this terminology, type-A and type-B physicalists can be distinguished by their differing attitudes to this 'explanatory gap'. Type-A physicalists will insist that at bottom there is no 'explanatory gap' and that, despite appearances to the contrary, claims about consciousness can after all be shown to follow *a priori* from the physical facts. Type-B physicalists, by contrast, accept the 'explanatory gap', but maintain that no ontological conclusions about the immateriality of consciousness follow.

At one point (his section 1.2.4) Davies considers Michael Tye's objection to using the phrase 'explanatory gap' for nothing more than the lack of a conceptual tie between brain and mind. Tye complains that, if there is no conceptual connection between the physical and mental realms, then this isn't the kind of lacuna that can be filled by future empirical discoveries, and that therefore it is odd to call it an 'explanatory gap', as if it were something that will go away once we get hold of the right empirical information.

Davies immediately concedes to Tye that type-B physicalists who recognize distinctive phenomenal concepts of conscious states will hold that the conceptual lacuna is here to stay. Davies also agrees with Tye that this isn't necessarily fatal to the type-B position. But he opts to keep the term 'explanatory gap' for the lack of a conceptual tie between the physical and mental realms.

Davies says that this is just a terminological matter, and of course he is entitled to specify how he is to be understood. But even so I think there is a real danger hiding behind the usage he endorses. The trouble is that there is something else that makes people posit an 'explanatory gap' between brain and mind, quite apart from the lack of conceptual brain–mind ties. Moreover, this something else threatens physicalism far more directly than any difficulties raised by the conceptual lacuna. Given this, Davies' preferred usage can easily create the impression that the conceptual issue is more important than it is.

2.3 The intuition of distinctness

I think that the real reason most people feel that there is an 'explanatory gap' between the physical and conscious realms is simply that they find physicalism incredible. Physicalism about consciousness is a strong claim. It isn't the relatively anodyne claim that conscious states are closely connected with what is going on in the brain, in the way that smoke is connected with fire. Rather physicalism says that conscious states are brain processes, in the way that water is H_2O. Smoke is caused by fire. But water isn't caused by H_2O—it is H_2O. Similarly, according to physicalism, pain isn't caused by a brain process—it is a brain process.

This is a very hard thing to believe. The more intuitive thought is surely that conscious states are extra to brain processes, even if they are closely correlated with them. Elsewhere I have called this natural dualist thought the 'intuition of distinctness' (Papineau 2002.) I would say that it strikes most people as obvious that the conscious mind is something more than the brain and that physicalism is therefore false.

Indeed I would say that there is a sense in which even professed philosophical physicalists, including myself, cannot fully free themselves from this intuition of distinctness. Of course, we deny dualism in our writings, and take the theoretical arguments against it to be compelling. But when we aren't concentrating, we slip back into thinking of conscious feelings as something extra to the brain, in the way that smoke is extra to fire, and not as one and the same as the brain, in the way that water is one and the same as H_2O.

I take this to be the real lesson of Saul Kripke's celebrated argument at the end of *Naming and Necessity* (1980). Kripke points out that even 'identity theorists' will admit that mind–brain identity claims appear contingent: even professed physicalists admit that it seems possible that pain could have failed to accompany the brain processes with which it is inevitably found in this world. Kripke then argues, and I think he is quite right, that this appearance can't be explained consistently with the professed physicalism. If you really believe that pain is one and the same as some brain process, then how can it so much as seem to you that pain and that selfsame brain process might have come apart in some other possible world? How can something come apart from itself? The moral of Kripke's argument, as I read it, is thus that even professed physicalists don't always fully believe their physicalism. They slip into thinking of pain as something that accompanies the relevant brain process, rather than as identical to it, and that's why it appears to them that the two things may come apart. (It should be said that this isn't the standard reading of Kripke. For more on this see my 2007.)

There is also some rather more immediate evidence that professed physicalists don't always believe their physicalism, even before we bring in Kripke's sophisticated modal considerations. The very language used by physicalists often gives the game away. Consider the phrases normally used to raise questions about the relation between the brain and the conscious mind. What are the 'neural correlates' of consciousness? Which brain states 'accompany' conscious states? Again, how do brain processes 'generate' conscious states, or 'give rise to' them? Such questions are commonly posed in discussions of consciousness by many people who would adamantly insist that they reject dualism. But their phraseology shows that they are not consistent in this denial. If they really thought that conscious states are one and the same as brain states, they wouldn't say that the one 'generates' or 'gives rise to' the other, nor that it 'accompanies' or 'is correlated with' it. H_2O doesn't 'generate' water, nor does it 'give rise to' or 'accompany' or 'become correlated with' it. H_2O simply is water. To speak of brain states as 'generating' consciousness, and so on, only makes sense if you are implicitly thinking of the consciousness as ontologically additional to the brain states.

The obvious question to ask at this point is why physicalism should strike us all as intuitively false, even if it is true. This is a very interesting issue, on which a number of writers have expressed views (Papineau 1993, 2002, 2006; Melnyk 2003; Bloom 2004).

One possibility is that the intuition is forced on us by some deep feature of our cognitive system. Alternatively, it may be a relatively superficial phenomenon, stemming from the theoretical presuppositions of contemporary culture, and so likely to disappear if those presuppositions change. However, this is not the place to pursue this topic. Let me content myself by pointing out that, even if the intuition of distinctness is here to stay, it is something that theoretical physicalists can happily live with.

After all, there are plenty of other cases where our best theoretical beliefs conflict with deep-seated and unmalleable intuitions. Consider the familiar Müller–Lyer illusion. At a theoretical level, we know that the lines are the same length. But at a more intuitive level of judgement they strike us as of different lengths. Nor is this kind of set-up restricted to cases involving perceptual illusion. At a theoretical level, I am entirely convinced that there is no moving present and nothing is left out by a 'block universe' description of reality which simply specifies the dates at which everything occurs. But at an intuitive level I can't stop myself thinking that I am moving through time. At a theoretical level, I am persuaded that reality splits into independent branches whenever a quantum chance is actualized. But at an intuitive level I can't shake off the belief that there will be a fact of the matter about whether the Geiger counter will click in the next two seconds or not.

I take it that in all these cases it is clear that, whatever the source of the contrary intuitions, they do not constitute serious obstacles to the theoretical beliefs. Similarly, I say, with consciousness. We may be unable to shake off the intuitive feeling that mind and brain are distinct. But that is no substantial reason to doubt the theoretical reasons for identifying them.

2.4 The feeling of a gap

In my view, the intuition of dualist distinctness is the real reason for the widespread feeling of an 'explanatory gap'. This feeling is nothing to do with the lack of conceptual ties between physical and conscious claims. When people feel that a 'why' question remains even after the 'what' question of consciousness has been answered, as Davies puts it in his introductory remarks, this isn't because they are worried that they can't deduce the conscious facts *a priori* from the physical facts. It's simply because they are assuming that conscious states are ontologically distinct from brain states. And this assumption of course gives rise to an obvious and urgent explanatory question—why do certain physical processes have the mysterious power of extruding a special mind-stuff?

Still, this question is generated directly by the intuition of dualism, and not by the relatively esoteric circumstance that conscious facts can't be deduced *a priori* from the physical facts. The best way to see this is to note that there are plenty of other cases where such *a priori* deductions are not available, yet we don't have any feeling of a worrying explanatory gap. Davies himself offers one kind of example: identities like 'location L is here' (where location L is somehow specified objectively). Clearly there isn't any question of *a priori* deducing the here-facts (there is water here, say) from the objective spatial facts (such as that there is water at location L) however many objective spatial facts we are given. Yet we feel no explanatory gap in this case (we don't ask why location L is here). Or, again, consider identities framed with the help of proper names, such as that Kripke is the greatest living philosopher. I take it that there is no question of deducing this claim *a priori* from the physical facts, or any other set of facts stated without using the name 'Kripke'. Yet surely we don't on that account feel that we lack some further explanation of why the greatest living philosopher is Kripke.

I infer that the real reason people want to say there is an 'explanatory gap' in the mind–brain case is that they can't help thinking in dualist terms—they want to understand why certain physical processes extrude the extra mind-stuff. If the only problem were that they can't deduce the conscious facts *a priori* from the physical facts, they wouldn't experience any worrying 'explanatory gap'. They would be no more puzzled about the mind–brain case than they are about why location L is here, or why Kripke is the greatest living philosopher.

2.5 **Dangers of confusion**

Davies makes it clear that he is using the phrase 'the explanatory gap' to refer specifically to the lack of any conceptual tie between brain and conscious mind and not to any dualist intuition. And, as I said, it is up to him to stipulate how he is going to use words. Still, I hope it is now clear why I think there is some danger that this usage may mislead unwary readers.

After all, talk of 'the explanatory gap' will inevitably be understood by most people as expressing the dualist intuition that the conscious mind is separate from the brain. Yet at the same time Davies, along with many others, specifies that this phrase is to be understood as referring to the conceptual separation between brain and conscious mind. This can scarcely fail to create the impression that this lack of a conceptual tie provides prima facie support for dualism. To my mind, of course, it does no such thing. As I have just explained, there are plenty of other cases apart from the mind–brain relation where a conceivability gap is manifestly compatible with identity. So from my perspective it is little more than a coincidence that the mind–brain case is associated with both a conceivability gap and an intuition of ontological distinctness. The two things really have nothing to do with each other. Still, it can be hard to see this, if we refer to the conceivability gap using a phrase that is standardly understood as expressing the dualist intuition.

In effect, my worry is that Davies' preferred terminology gives the Jackson–Chalmers line of argument an unwarranted rhetorical boost. Jackson and Chalmers hold that there is a sound argument from the lack of a conceptual tie to dualism (e.g. Chalmers and Jackson 2001). This is a serious and influential thesis, and there is no doubt that it deserves the careful yet critical attention that Davies gives it. But I wonder whether it would carry quite the same initial plausibility, were it not for the widespread practice of describing the premise with a phrase that is naturally understood as expressing the conclusion.

I can make a related point about explanation itself. In stipulating that the 'explanatory gap' signifies the absence of conceptual ties between physical and conscious realms, Davies and others who adopt this usage strongly suggest that something will be left unexplained if such conceptual links cannot be forged. The very phraseology implies that a type-B physicalism that tries to live without such ties will inevitably leave us with unfinished explanatory business. This view is certainly supported by Davies' opening remarks, where he urges that even an answer to the question of 'what' physical states correlate with consciousness will fail to resolve the 'intractable' question of 'why' they do so.

But this casts an unwarranted aspersion on type-B physicalism. From the perspective urged here, it is a mistake to think it leaves anything unexplained. Of course, as I said earlier, if you slip into hearing the 'explanatory gap' as expressing an intuition of dualist distinctness, then of course you will think it raises an urgent explanatory demand—the need to explain why some special physical processes should extrude a distinctive mind-stuff. But any clear-thinking type-B physicalist will resist this dualist intuition, and then it is by no means clear that there is anything left to explain.

We have already seen reason to doubt that a conceptual lacuna *per se* generates explanatory demands. There are no conceptual ties between objective spatial facts and egocentrically identified ones, yet we don't feel we need to explain 'why location L is here'. Again, there is no conceptual tie between descriptive and proper name facts, yet we don't feel called to explain why 'the greatest living philosopher is Kripke'. I say the same about the mind–brain case. There may be no conceptual route from brain descriptions to claims about consciousness, but that doesn't mean non-dualists need to explain why given brain states are the conscious states they are.

Claims about identity or constitution don't need explanation, even when they aren't conceptually guaranteed. If two things accompany each other, we can rightly ask why this is so. But if something is the same as something else, it makes no sense to continue asking for an explanation. If you think that Samuel Clemens and Mark Twain are two different people, you will rightly be puzzled why they are always in the same place at the same time. But once you realize the truth, you won't go on asking why Samuels Clemens is Mark Twain. Similarly, if you think that pains accompany C-fibre stimulation, you might well wonder why this is so. But once you accept that they are one and the same state, there is no remaining need to explain why that state is itself.

From my point of view, the so-called 'explanatory gap' does not indicate a failing in type-B physicalism, so much as a failing in many professed type-B physicalists. The trouble isn't that type-B physicalism is somehow incomplete, but rather that many type-B physicalists fail to believe it wholeheartedly. They slide back into dualist thinking, and then find themselves hankering for explanations.

So the right reaction to talk of the 'explanatory gap' isn't that type-B physicalism somehow needs fixing or supplementing. It is just fine as it is. The difficulty is that we don't commit ourselves properly. If only we fully believed our physicalism, our explanatory hankerings would die away.

2.6 **Methodological limitations**

I turn now to the methodological implications of type-B physicalism. As we have seen, type-B physicalists explain the relation between mind and brain by positing special 'phenomenal concepts'. These concepts have no *a priori* connections with any scientific third-personal concepts. Still, in reality they refer to nothing except brain processes that can also be referred to using scientific concepts. So type-B physicalists combine ontological monism with conceptual dualism. There is just one physical reality, but we can think about the conscious part of that reality in two ways, either with third-person scientific concepts or with subjective phenomenal concepts.

However, once we bring in these special phenomenal concepts, then it turns out that they allow us to pose certain questions about consciousness that are very difficult to answer. Indeed I don't think that they can be answered at all. However, I don't think that this points to any real epistemological limitation facing consciousness research. The trouble is rather that phenomenal concepts are not precise enough to frame all the questions that we intuitively expect them to frame.

2.7 **Subjects' reports**

We can bring out the problem by thinking about the methodology of consciousness research. How do we find out which brain processes constitute pain, or consciously seeing something, or indeed being conscious at all? The obvious technique is to appeal to subjects' reports. We take human subjects in given conditions, ask them whether or not they were in the relevant conscious state, and seek to find some brain state that is present whenever they say 'yes' and absent whenever they say 'no'.

I shall have more to say about the methodological significance of subjects' reports below. But we can bring out the way in which consciousness research normally hinges on such reports by considering subjects who are not capable of making reports, like vervet monkeys, say. You can submit vervet monkeys to various experimental conditions, you can get them to perform various tasks, and you can check on what is going on in their brains at such times. But since the monkeys can't tell you what they consciously experience, none of this will cast any immediate light on the monkeys' phenomenal consciousness. For all this research establishes, the monkeys might share very similar experiences to our own, or have no consciousness at all, or be somewhere in between. By contrast, humans can explicitly tell us whether or not they are feeling a pain, seeing something red, or experiencing something rather than nothing, and this offers a crucial handle by which to identify the material nature of phenomenal properties.

2.8 **Too many candidates**

Now, there is no doubt that research based on such reports by human subjects can tell us many interesting and indeed surprising things about consciousness. For example, recent investigations have shown that subjects report no conscious experience of many cognitive processes which we might pre-theoretically have expected to be conscious: these include the visual guidance of object-oriented hand movements (Weiskrantz 1986; Goodale and Milner 1992) and the instigation of voluntary actions (Libet 1993). There are also cases of the converse kind, where subjects report conscious experiences in cases where we might have expected none: for example, subjects who take morphine to allay an existing pain will report that they can still consciously feel the pain, even after it has ceased to agitate them (Dennett 1978).

However, such research can only take us so far. It can narrow down the candidates for the physical nature of given conscious states, but it can never identify the relevant physical state uniquely. For example, we can be sure that more is required for conscious vision than the processes guiding object-oriented hand movements, and that less is required for conscious pain than the processes present in unmorphiated sufferers. But even after we have done all the narrowing down that subjects' reports permit, there will inevitably remain more than one candidate physical state P for any given phenomenal state C.

2.9 Structure and substance

The easiest way to see this is to take up a point mentioned by Davies at various points. Not all contemporary type-B physicalists aim to identify conscious states with strictly physical states. For there is also the option of identifying conscious states with properties that supervene on such strictly physical states, such as functionalist states defined in terms of causal structure. Thus one possibility might be to identify pain with the strictly physical state P which distinguishes those humans who are in pain from those who aren't. In this case pain would be identified with the property of having such-and-such organic molecules arranged into neurons in such-and-such ways. But then there is the option of abstracting away from the specific way this arrangement is realized in humans, and identifying pain with the structural property S of having such and such a causal arrangement of 'neurons' without specifying what the 'neurons' are made of.

Now these are certainly different accounts of the nature of pain. To see this clearly, consider the possibility of a 'silicon doppelgänger'—a being whose brain is structured just like yours, down to a fine level of detail, but whose 'neurons' are made of silicon-based circuits rather than organic molecules. If pain is identical with the strictly physical human property P, then this being will not experience pain. On the other hand, if pain is identical with the structural property S, then the doppelgänger will have pain.

The problem should now be becoming clear. How are we to decide between the claim that pain is identical to the strictly physical state P and the claim that it is identical to the structural state S? The trouble is that both these states will be present in all humans who report pain, and both absent from all who deny it. After all, the two states are coextensive within humans, so there is no way that they can be distinguished by empirical research that relies on human reports of pain (cf. Levine 1983; Block 2002).

You might think that this is just a familiar problem of insufficiently varied data. Don't we simply need some cases where the two properties dissociate? A silicon doppelgänger, for example, would help. This would give us a case with the structural S but not the strictly physical P. So in principle shouldn't we be able to decide the issue by empirically testing whether a silicon doppelgänger is in pain or not?

However we can see that, even if we did have such a test case, it wouldn't really help. We already know that the doppelgänger will say 'I am in pain' when it has S. After all, it is structured just like us. But it would say this even if conscious pain did depend on the organic realization P and were absent in beings that have S without P. (In that case, we could think of the doppelgänger as using the term 'pain' for a different state from our conscious pain.)

The difficulty of deciding between structural and strictly physical properties as the material essence of conscious properties is just one of a number of analogous problems that arise for consciousness research (see Papineau 2002, section 7.7). Thus, to take just one further example, consider the difficulty of deciding whether pain consists of (a) some material state M that is present just in case humans report pain or (b) M-plus-the-brain-processes-underlying-reports-of-pain (call this latter state 'H'—for 'higher-order'). Here too it seems clear that empirical research will be unable to adjudicate between M and H, since the two will be coextensive in humans: by definition, H will always be accompanied by M; and M must always be accompanied by H (brain processes underlying pain reports), given that M is picked out in the first instance as a state that is present whenever humans report pain (cf. Block 2007).

In this case too it is clear that the problem won't simply be resolved by gathering more varied data. We can of course seek out beings who have M but who lack the higher-order mental processes required for H. Vervet monkeys may well fit this bill. But such cases won't decide whether M without H suffices for pain. Since the subjects involved will lack the mental wherewithal to report on whether or not they feel pain, they will cast no light on the issue.

2.10 **Epistemology or semantics?**

What are we to make of these methodological conundrums? One possible response is that there is some kind of epistemological barrier preventing us from finding out about the material nature of conscious states—there really is a fact of the matter about whether silicon doppelgängers or vervet monkeys feel pain, but we human investigators are for some reason prevented from ascertaining it.

I'm not happy with this conclusion. What exactly is supposed to be stopping us from finding out these putative facts? It's not as if the relevant data are far away, or too small for our instruments to detect, or only manifest at high energies that are very expensive to engineer. But in the absence of any such specific difficulty it seems mysterious that there should be some barrier to our finding out about the material nature of conscious states.

I prefer a different kind of explanation. The difficulty isn't that we can't access the facts, but that there aren't any real facts here to start with. The phenomenal concepts we use to pose the relevant questions—for example, do silicon doppelgängers or vervet monkeys have pain?—aren't sharp enough to determine definite answers. The phenomenal concept pain is fine for distinguishing those humans who are in pain from those who aren't. But it goes fuzzy when we try to extend it to further cases. Asking whether a quite different kind of being feels pain is like asking whether it is five o'clock on the sun right now.

This suggestion might seem ad hoc. It is not in general a good idea to conclude that there is no fact of the matter just because we can't find out about something. However in this case there are good independent reasons for thinking that phenomenal concepts are insufficiently precise to determine the relevant facts. Phenomenal concepts are peculiar, imagistic, introspective ways of thinking about different kinds of conscious states. They serve a perfectly good purpose in enabling us to keep track of the mental lives of other human beings. But there is no reason to suppose that they draw a sharp line right through nature.

2.11 The intuition of distinctness again

At first sight it might seem absurd to deny that there is a fact of the matter about whether different kinds of being feel pain or other conscious states. Surely it is either like this for them, or it isn't. How can it be indeterminate whether the experience I am undergoing is also present in them?

But I wonder whether this intuitive reaction to the suggestion of indeterminacy is not just another manifestation of the intuition of distinctness I discussed earlier. Suppose that you think that certain brain processes give rise to the presence of some extra non-physical conscious mind-stuff. Then you will certainly think that there is some fact of the matter as to whether or not this same mind-stuff is present in different kinds of being. On the other hand, it is not so clear that this conviction of determinacy will remain if you fully believe that conscious states are nothing but physical processes.

In support of this diagnosis, note that when we think of mental categories in non-phenomenal terms, there seems nothing intuitively objectionable about the idea that their application might sometimes be indeterminate. For example, the human visual system uses non-conscious stereopsis to help judge distance: it matches features across the left and right eye images to estimate the distance of their common source. Suppose some other beings also judge distance by matching features, but use different features from humans—relative light intensity rather than edges and lines, say. Or suppose they match the same features, but the processing is realized in silicon-based circuits rather than organic neurons. Is this still stereopsis? The natural reaction here is—who cares? Why should the concept of stereopsis be refined enough to determine a definite answer to these questions? If we do need a definite answer, we can refine the concept to determine one, but that's clearly not the same as ascertaining whether the divergent beings really instantiate our prior notion of stereopsis.

I infer from this comparison that our resistance to suggestions of indeterminacy arises specifically in those cases where we are thinking of mental states in phenomenal terms—in terms of what they are like, so to speak. When we do think of our mental life in this phenomenal way, then we are hard-pressed to avoid the intuitive thought that it involves some ontological addition to the physical processes of the brain. And once we slip into this dualist frame of mind, we will of course think there are definite facts about where the extra mind-stuff appears. But all this is driven by the dualist intuition that is incompatible with genuine physicalism.

Consider this analogy (cf. Papineau 2003). Eskimos make extensive use of whale oil, for heating, lubrication and many other purposes. At some point in Eskimo history, commercial manufacturers introduced a substitute petroleum product which behaved just like natural whale oil. Did the original Eskimo notion (let's call it 'whale oil') apply to this new substance or not? If 'whale oil' referred to a biologically or chemically identified type, it did not; but if 'whale oil' referred to anything with the requisite appearance, then it did.

Of course there was likely to be no answer here. Nothing in previous Eskimo thinking need have determined which way 'whale oil' should be understood when this question arose. Of course, the question could have been resolved by refining the Eskimo concept. The Eskimos could have decided henceforth to understand 'whale oil' as referring to a biological or chemical substance, in which case the new stuff wouldn't have counted as 'whale oil'. Alternatively, they could have decided to understand it as referring to anything with the appropriate appearance, in which case the new stuff would have counted as 'whale oil'. But clearly neither option need have been determined by the prior Eskimo notion of 'whale oil'.

But now imagine people who were unhappy about the idea that this was just a matter for decision. They said, 'Yes, I can see that this new manufactured stuff is like the old stuff in appearance, while different in chemical constitution and biological origin. But that doesn't yet answer the important question, which is whether it is like the old stuff in being made of whale oil.'

Now clearly these people would have been assuming that being made of whale oil is some extra property, over and above constitution and behaviour. They would have been dualists about the property of being made of whale oil. Because of this, they would have thought that there must be some fact of the matter as to whether the new product involved this extra dualist whale oil or not.

I think the same about people who are convinced that there must be some fact of the matter as to whether the doppelgänger is really in pain. They are assuming that pain is some extra property, over and above physical and structural properties, and so of course they think that there is a substantial matter as to whether this extra property is present in the doppelganger. But in truth there is no such extra property. There is nothing to refer to except the relevant physical and structural properties, and our phenomenal concepts leave it indeterminate which of these is to be identified with pain.

2.12 Conclusion

I am not aiming to belittle empirical research into consciousness. As I said, it can tell us many interesting and surprising things about the material processes present in human beings in different conscious states. However, there are certain apparent questions about consciousness that empirical research won't be able to answer. For it will always leave us with a plurality of candidates for the material essence of any conscious property. And in the face of this plurality empirical science will be impotent, for the methodology of consciousness research offers no handle by which to separate material properties that are perfectly correlated in normal humans.

To this extent, scientific research into consciousness is fated to deliver less than we might have hoped. This does not mean, however, that we would do better to turn to some alternative mode of investigation. It is not as if conscious properties have true material essences, yet science is unable to discover them. Rather the whole idea of precisely identifying such essences is a chimera, fostered by the impression that our phenomenal concepts of conscious states are more precise than they are.

References

Block, N. (2002). The harder problem of consciousness. *Journal of Philosophy* **99**, 391–425.

Block, N. (2007). Consciousness, accessibility, and the mesh between psychology and neuroscience. *Behavioral and Brain Sciences* **30**, 481–499.

Bloom, P. (2004). *Descartes' Baby*. New York: Basic Books.

Chalmers, D. and Jackson, F. (2001). Conceptual analysis and reductive explanation. *Philosophical Review* **110**, 315–360.

Dennett, D. (1978). Why you can't make a computer that feels pain. In *Brainstorms*, pp. 190–229. Cambridge, MA: Bradford Books.

Goodale, M. and Milner, A. (1992). Separate visual pathways for perception and action. *Trends in Neurosciences* **15**, 20–25.

Kripke, S. (1980). *Naming and Necessity*. Oxford: Blackwell.

Levine, J. (1983). Materialism and qualia: the explanatory gap. *Pacific Philosophical Quarterly* **64**, 354–361.

Libet, B. (1993). The neural time factor in conscious and unconscious events. In Bock, G. and Marsh, J. (eds) *Experimental and Theoretical Studies of Consciousness*, pp. 122–137. London: Wiley.

Melnyk, A. (2003). Papineau on the Intuition of Distinctness. SWIF Forum on *Thinking about Consciousness*, http://lgxserver.uniba.it/lei/mind/forums/004_0003.htm.

Papineau, D. (1993). Physicalism, consciousness, and the antipathetic fallacy. *Australasian Journal of Philosophy* **71**, 169–183.

Papineau, D. (2002). *Thinking about Consciousness*. Oxford: Oxford University Press.

Papineau, D. (2003). Could there be a science of consciousness? *Philosophical Issues* **13**, 205–220.

Papineau, D. (2006). Comments on Strawson's 'Realistic monism: why physicalism entails panpsychism'. *Journal of Consciousness Studies* **13**, 100–109.

Papineau, D. (2007). Kripke's proof is *ad hominem* not two-dimensional. *Philosophical Perspectives* **21**, 475–494.

Weiskrantz, L. (1986). *Blindsight*. Oxford: Oxford University Press.

Chapter 3

Emotional colouration of consciousness: how feelings come about

Joseph LeDoux

3.1 Emotions, feelings, and consciousness: an overview

When you are emotionally aroused, you usually feel something—you feel the emotion. That is, the emotion comes to be what you are conscious of at the moment. This implies that emotions and feelings are not the same thing—that to have a feeling requires that you be conscious of the emotion. The emotion, in other words, exists independent of the conscious experience of the emotion: emotions are processed unconsciously, while feelings are conscious emotional experiences (LeDoux 1984, 1996; Damasio 1994).

In this chapter I will develop an account of how the brain consciously feels emotions. I will argue that feelings are, in most cases, emotionally coloured cognitions, and that emotional consciousness is just a particular instance of cognitive awareness rather than a separate form of awareness mediated by a distinct neural system. I have presented this general view previously (LeDoux 1984, 1992, 1996, 2002) but will develop a considerably more detailed account here.

In order to set the stage for the discussion that follows, I lay out the basic ideas in this introductory section. Specifically, I will define emotions, justify a focus on fear, explain my idea that a fearful feeling is an emotionally coloured cognition, argue that working memory provides a useful framework for accounting for the emotional colouration of cognition, and consider the relation of language to consciousness and feelings.

3.1.1 Defining emotions

Emotion theorists often distinguish basic or primary emotions and higher-order emotions (Tomkins 1962; Izard 1971; Plutchik 1980; Scherer 1984; Frijda 1986; Ekman 1993; Panksepp 1998). Basic emotions (fear, anger, joy, disgust, etc.) are believed to be innately organized in humans, conserved to some degree across mammalian species, automatically or unconsciously elicited, expressed in characteristic ways in the body of all humans independent of culture, and mediated by distinct neural systems. Higher-order emotions (e.g. empathy, jealousy, guilt) are often considered to be less conserved across species, cognitively mediated rather than automatically elicited, less rigidly expressed in the body, mainly due to learning and social factors, and possibly different in different cultures. Because some higher-order emotions may be unique to humans or our close relatives (Scherer 1984; Frijda 1986; Ortony *et al.* 1988), they might be thought of as more interesting from the point of view of human nature. However, higher-order emotions are less amenable to experimental research, especially brain research, and thus give us less traction when trying to explore the relation of emotions to feelings. In the effort to explore the relation between emotions and feelings in the brain, I will focus on the general area of basic emotions.

Basic emotions have been delineated in a number of ways, including facial expressions (Tomkins 1962; Izard 1971; Ekman 1992) linguistic categories (Scherer *et al.* 1986), and brain systems (Panksepp 1998). A key problem with most approaches to basic emotions is that they are top heavy. That is, they are based on human introspection and on the linguistic labels we attribute to our experiences. An alternative bottom-up approach first identifies fundamental behavioural adaptations that allow organisms to survive and then determines the brain systems involved in these adaptations (LeDoux 1996).

3.1.2 Fear as a model system

An important behavioural adaptation is the capacity to detect and respond to danger. This capacity is conserved behaviourally and neurobiologically throughout mammalian evolutionary history (LeDoux 1996). By convention, the brain system mediating these capacities is called the fear system. This terminology potentially raises the problem mentioned above—that is, the use of introspection and linguistic labels to identify emotions. However, fear in this case refers not to the subjective experience of fear but to the network that detects and responds to danger. A more appropriate term might be the defence system. For consistency with the literature, I will use the terms *fear* and *fear system* to refer to the behavioural adaptation that allows organisms to detect and respond to threats. When I refer to the subjective conscious experiences of fear I will use the expression *fearful feelings*.

I focus on fear because much has been learned about the organization and function of the fear system, and there is good agreement amongst brain scientists about how this system works in animals and humans (LeDoux 1996, 2000; Maren 2001; Dolan and Vuilleumier 2003; Fanselow and Gale 2003;

Phelps and LeDoux 2005). The fear system thus provides a useful context in which to consider the relation of emotions to feelings.

The fear system is programmed by evolution to respond to species-typical threats and by individual learning to respond to stimuli that, through experience, are determined to be dangerous or predictive of danger (Blanchard and Blanchard 1972; Bolles and Fanselow 1980; Marks and Tobena 1990; Öhman and Mineka 2001). This system can operate independent of conscious awareness and thus without first having a conscious feeling (LeDoux 1996, 2002; Robinson 1998; Dolan and Vuilleumier 2003; Öhman 2005; Phelps and LeDoux 2005). When activated to a sufficient degree it produces a complex constellation of coordinated responses in the brain and body that help the organism cope with the threat (see LeDoux 1987, 1996, 2002; Davis 1992; Kapp et al. 1992; Fendt and Fanselow 1999; Lang et al. 2000; Davis and Whalen 2001; Maren 2001; Phelps and LeDoux 2005; Phelps 2006).

Within the brain, output connections of the fear system to widespread areas result in the mobilization of brain resources and redirection of attention to the threat. Direct connections to areas involved in sensory and cognitive processing are important, as are connections are to neuronal groups that regulate brain arousal levels that then, in turn, affect sensory and cognitive processing. Within the body, behavioural, autonomic, and endocrine responses are expressed. Feedback from these responses can then influence further processing, but in ways that are not fully understood (Damasio 1994, 1996; LeDoux 1996, 2002; Critchley et al. 2001, 2004; North and O'Carroll 2001; Hinson et al. 2002; Heims et al. 2004; Maia and McClelland 2004; Bechera et al. 2005; Nicotra et al. 2006; Dunn et al. 2007).

As noted above, the fear system responds to innate and learned threats (Blanchard and Blanchard 1972; Bolles and Fanselow 1980; LeDoux 1987, 1996; Marks and Tobena 1990; Davis 1992; Maren 2001; Öhman and Mineka 2001). Stimuli that activate the fear system without prior learning are unconditioned stimuli. When novel stimuli occur in conjunction with unconditioned stimuli they acquire the capacity to activate the fear system and elicit protective responses. These are conditioned stimuli. Both unconditioned and conditioned fear stimuli, through processing in the fear system, can reinforce the learning of new instrumental responses that reduce or terminate exposure to the conditioned or unconditioned stimulus (Mowrer 1939; Miller 1948; McAllister and McAllister 1971; Bolles and Fanselow 1980; Killcross et al. 1997; Nader and LeDoux 1997; Everitt et al. 1999; Amorapanth et al. 2000).

Given that the fear system processes reinforcing properties of unconditioned and conditioned stimuli, one might be tempted to define the fear system, or other emotion systems, in terms of behavioural or neural responses to reinforcing stimuli (Millenson 1967; Gray 1987; Rolls 1999, 2005). However, many brain regions contain neurons that process reinforcing properties of stimuli while at the same time those neurons do not directly participate in the emotional reactions to those stimuli. As defined here, an emotion system is one that both (a) detects the presence of unconditioned or conditioned stimuli and (b) orchestrates and coordinates brain and body responses to specifically deal with the consequences that result from the presence or anticipation of such stimuli.

3.1.3 Fearful feelings

It is common when discussing basic emotions to use the same label for the conscious feeling and for the unconscious processing and control of emotional responses. For example, the term fear is used interchangeably to refer to conscious feelings of fear and to behavioural and physiological responses elicited when threatening stimuli are unconsciously processed by the fear system. The use of the same term for feelings and responses, or experiences and expressions of emotions, causes confusion when we are discussing human emotions. But this practice is even more problematic for the study of basic emotions in other animals since we are severely limited in our ability to know what other creatures experience, if anything, when a basic emotion system is active. I will argue that we cannot simply reason by analogy (that if humans and rats both freeze in the face of a sudden danger, and humans feel fear, then rats also feel the same fear) about other animals (see also Heyes, this volume). To minimize confusion, I will refer to conscious experiences of fear as fearful feelings and to expressions of fear in the brain and body as fear responses.

3.1.4 Emotional colouration of cognition

As noted above, my proposal is that fearful feelings are emotionally coloured cognitions. In this view, the capacity of a particular species to feel an emotion is directly tied to its cognitive capacities. Because of the problems in assessing consciousness in other animals, I will focus on the features that characterize feelings in humans. This will allow us to then ask which of the features are present in other animals, and by inference, what kinds of feelings they might have.

I will argue that feelings occur (cognitions become emotionally coloured) through several capacities of the human brain. These include (1) binding of cognitive and emotional information into a unified representation, (2) temporary maintenance of the representation, (3) attention to the representation.

The first capacity underlying a feeling is cognitive-emotional binding in which three diverse classes of information are integrated to create a multidimensional representation of an emotional experience: sensory properties of an immediately present emotional stimulus, long-term memories that are activated by the stimulus, and the emotional arousal being elicited at the same time. The sensory stimulus will be processed by the relevant sensory system in terms of its physical properties (in the case of a visual stimulus, its shape, colour, movement, location, etc.). Two classes of long-term memories activated in cortical memory systems by the stimulus will allow the encoding of the perceptual meaning of the stimulus, giving it an identity: semantic memories (factual memories about what the stimulus is and what other kinds of stimuli it relates to) and episodic memories (memories of personal experiences you've had with that stimulus or related stimuli) (Tulving 1972, 1983). A special form of episodic memory, called autonoetic memory, or memory of the self, is likely to also be important (Levine *et al.* 1998; Stuss *et al.* 1998; Tulving 2002). Autonoetic memory allows us to relate immediate and past experiences and facts to our sense of who we are. The stimulus, being an emotional stimulus, will also activate an emotion system. Whether a particular stimulus activates an emotion system is due to information stored, from past encounters, in the emotion system itself. In contrast to semantic and episodic memories, which are called explicit memories because they are encoded in a way that allows retrieval into conscious awareness, memory in an emotion system is a form of implicit memory since it cannot be directly explicitly retrieved and experienced consciously and can only affect consciousness indirectly through other processing in the brain or through feedback from bodily responses. Through brain processing and feedback, the emotional arousal is integrated with the sensory and memory representations, leading to an emotional colouration of the representation.

Fleeting representations do not have long-term consequences. In order for cognitive-emotional representations to have consequences they have to be maintained in an active state (Schneider and Shiffrin 1977; Norman and Shallice 1980; Goldman-Rakic 1987; Fuster 1997; Schneider *et al.* 1984; Baddeley 1986, 1993, 2000; Dehaene and Naccache 2001; Baars, 1993, 2005). This is achieved in part by the actions of arousal systems which release neuromodulator chemicals (e.g. norepinephrine, acetylcholine, dopamine) throughout the cortex and alter the level of arousal and vigilance during information processing (e.g. Arnsten and Li 2005; Aston-Jones and Cohen 2005; Robbins 2005; Sarter *et al.* 2005). Modulatory systems are responsive to novel, attention-demanding, and emotionally significant stimuli. Outputs of emotion processing systems, such as the fear system, play an important role in activating modulatory systems and thus in regulating the ability of cortical areas to process information from external sensory systems or internal sources (thoughts and memories), and in maintaining information processing in an active form for further processing (Durstewitz et al. 2000; Tanaka 2001). Through active maintenance, information processing can be used in thought and action.

Creation and active maintenance of sensory–memory–emotion representations is not sufficient to be conscious of the representation. It is commonly assumed in cognitive science that we are conscious of those things that attention is focused on (James 1890; Broadbent 1962; Posner and Snyder 1975; Schneider and Shiffrin 1977; Norman and Shallice 1980; Schneider et al. 1984; Mack et al. 1992; Posner 1994; Dehaene and Naccache 2001). Unless processing is selected by attention it remains outside our conscious awareness, and thus is unconscious.

Much work on attention has been about attention to a sensory event in one modality. Thus, when attending to an object presented in one part of the visual field you are less likely to notice objects presented in other parts of the visual field (Posner 1994). In such situations, attention involves selection of which object to attend to and suppression of processing of other stimuli in the same or other modalities (Desimone and Duncan 1995; Maunsell 1995). When this occurs we are consciously aware of the item we are attending to.

Something similar probably occurs for multimodal representations that involve integration of processing across modalities. Cortical areas that receive and integrate inputs from multiple unimodal areas are called convergence zones (Geschwind 1965a, 1965b; Jones and Powell 1970; Mesulam et al. 1977; Damasio 1989). Several such zones exist in the cortex. Very likely these various regions continuously create and maintain multimodal representations.

I propose that just as we are not aware of sensory processing that goes unattended, we are unaware of processing that take place in convergence zones unless that processing is attended to. Attention allows a multidimensional representation in a convergence zone to be the object of active operations that then make the representation available for use in though and action.

When we are attending to one thing, ongoing processing of other things is actively prevented from entering awareness. So how do we ever stop attending to one thing and start attending to something else? This is an important contribution of emotional arousal—it interrupts the current focus of attention (Simon 1967; Armony et al. 1997, 1998). This allows the attention system to redirect its resources to the most salient immediate event, the emotionally arousing event. Sensations and memories relevant to the emotional arousal are selected and amplified (Posner 1994). The net result of emotional arousal is an increase in the likelihood that the representation being created during an emotional episode occupies the spotlight of attention. In order for this sequence of events to play out, it is necessary that emotion systems operate pre-attentively and be capable of being activated without the need for attentional resources. As we'll see, evidence exists for this (e.g. Vuilleumier *et al.* 2001).

3.1.5 Working memory as a functional framework of cognition, consciousness, and feeling

The various capacities just described fall within the framework of cognition called *working memory* (Baddeley 1986, 1993, 2000). Working memory includes a mental workspace in which diverse kinds of information can be integrated to form episodic representations that are held temporarily for use in thought and action. It crucially involves attention for the selection of the information to be held in the workspace for use in the control of thought and action. Working memory does not usually involve emotional processing but is easily extended to do so. Related models are the supervisory attention system (Norman and Shallice 1980), operating systems (Johnson-Laird 1988), attentional amplification network (Posner 1994), and global workspace (Baars 1988, 2005; Dehaene *et al.* 1998, 2006; Dehaene and Naccache 2001). These models attempt to explain roughly the same kinds of data as the working memory model, often invoke overlapping brain systems, and lead to similar conclusions. Although there are some subtle differences, for simplicity, I will use the term working memory when referring to these models and their component processes.

The idea that we are not conscious of a sensory, memory, or emotional representation until it is further represented in working memory and attended to relates to the notion that consciousness requires that sensory or memory representations be processed by an interpreter system (Gazzaniga 1985, 1988; Baynes and Gazzaniga 2000) or commentary system (Weiskrantz 1997). It also relates to Rosenthal's (1990, 1993) higher-order theory of consciousness, which argues that in order for a representation to be conscious it has to be the subject of a higher-order thought, and to Weiskrantz's (1997) and Rolls' (2005) ideas about the brain mechanisms of consciousness that are built upon Rosenthal's higher-order thought theory. The main difference with these other proposals is that the working memory proposal is considerably more specific about the underlying processes that make consciousness possible. Also, it builds upon these well-characterized cognitive processes in explaining how representations are made available for use in thought and action, and thus become conscious. The working memory account would thus seem to be stated in ways more amenable to empirical research.

3.1.6 Language and consciousness

Many philosophers and scientists have noted that there is a special relation between language and consciousness. Much debate exists about whether this is simply because language is a convenient way of assessing conscious states, or whether it contributes in more profound ways. I will propose that human emotional experience is greatly affected by our capacity for language. This does not mean that language is necessary for conscious experience of emotions (or other states), but instead that language influences how we experience emotions and what we can do with those experiences. Language allows us to categorize and discriminate experiences, thus extending the range of emotions that can be differentiated in episodic representations. And through syntax, language allow executive functions to parse situations online in real time to determine who is feeling what when, what they are likely to do on the basis of those feelings, and increases our flexibility in deciding how to respond. The idea that language influences thought was long in poor standing, but has been revived in recent years. As discussed later, the evidence for a role for language in the experience of emotion (in feelings) is strong.

3.1.7 Looking forward

Having provided an extensive overview of what is to come, I will now turn to specific aspects of the proposal and expand on them. I begin by discussing the nature of unconscious fear processing by circuits in the brain. I then turn to the question of how fearful feelings are created.

3.2 Unconscious processing of fear-arousing events

Responding to threats is one of the most essential things that organisms do. Unlike eating, sleep, sex or other activities, responding to threats cannot be postponed. There is an urgency to threat processing that is absent in most other emotions (LeDoux 1996; Robinson 1998).

In mammals, threatening stimuli elicit hard-wired responses protective of the brain and body (Blanchard and Blanchard 1972; Bolles and Fanselow 1980; LeDoux 1987; 1996, 2002; Davis 1992; Kapp et al. 1992; Lang et al. 2000). Body responses include behavioural reactions (i.e. freezing, facial postures), autonomic nervous system responses (i.e. changes in blood flow distribution accompanied by changes in blood pressure and heart rate; increased respiration; sweat gland activity), and endocrine responses (release of pituitary–adrenal hormones). Reactive brain responses include activation of arousal systems that release modulatory chemical signals throughout the brain and regulate the level of activation and receptivity to stimulation of neurons in target regions. The net result is the mobilization, in fact monopolization, of cerebral resources so that attention is focused on the threat and other competing external and internal demands are suppressed. These reactive brain and body responses occur naturally (without learning) to unconditioned stimuli (intense stimuli, especially those causing pain, or to predators or other species-specific threats) and can be conditioned to occur in response to otherwise meaningless stimuli when they are associated with unconditioned stimuli.

In this section I will first argue that threat processing occurs unconsciously in humans. I will then describe findings from animal research that show that the amygdala is an essential component of the threat-processing circuitry in the brain. The last part of this section shows that amygdala processing occurs unconsciously in humans.

3.2.1 Unconscious processing of threats in humans

Everyone has in some situation found their heart pounding before recognizing the nature of the fear-eliciting event. This occurs because the threatening nature of emotional stimuli is determined unconsciously and initiates emotional responses before we are consciously aware of what the stimulus is.

Much scientific evidence supports the idea that unconscious stimulus processing occurs in humans (Erdelyi 1985; Kihlstrom 1987; Zajonc 1980, 1984; Jacoby 1991; Bargh 1992; Greenwald 1992; Merikle 1992; Wilson 2002) and that unconsciously processed threatening stimuli can control emotional responses (Lazarus and McCleary 1951; Moray 1959; Erdelyi 1985; Öhman and Soares 1994; LeDoux 1996; Robinson 1998; Whalen *et al.* 1998, 2004; Dolan and Vuilleumier 2003; Olsson and Phelps 2004; Öhman 2005). In the older literature, conscious processing of an emotional stimulus (such as a stimulus previously paired with painful electric shock) was prevented by using very brief stimulus presentations (e.g. Lazarus and McCleary 1951). Although subjects were unable to report the stimulus, autonomic nervous system responses occurred. These studies were criticized on various grounds and led to the adoption of more sophisticated stimulus presentation and analytic techniques (see Erdelyi 1985; Merikle 1992). Much of the recent work has used a masking procedure where conscious processing of an emotional stimulus is prevented both by brief presentation of the emotional stimulus and by presenting an emotionally neutral second stimulus (the mask) a few milliseconds after the emotional stimulus (e.g. Morris *et al.* 1998; Öhman and Soares 1998; Whalen *et al.* 1998; Olsson and Phelps 2004). In addition, subjects are probed more carefully to ensure that their failure to be conscious of the stimulus is not due to a shift in their tendency to report or not. Typically, pictures of humans expressing fear or anger are used as emotional stimuli, with some studies using these as unconditioned stimuli and others pairing the face with an electric shock for additional emotional amplification. As with subliminal presentation alone, masking prevents awareness but does not disrupt the expression of autonomic responses.

Studies in neurological patients also provide strong evidence that emotional processing can occur unconsciously. For example, patients with unilateral damage to primary visual cortex have no conscious awareness of stimuli presented in the area of visual space normally processed by that part of the brain (Milner A.D. and Goodale 1995; Weiskrantz 1997; Milner A.D., this volume). However, when conditioned fear stimuli (faces previously paired with electric shock) were presented to the blind area of space, the patient expressed conditioned fear responses (Hamm et al. 2003).

These various studies clearly show that emotional stimuli do not have to be consciously processed in order to elicit emotional responses. This does not mean that conscious emotional processing never occurs. It simply means that conscious processing of emotional stimuli is not necessary for the elicitation of emotional responses.

3.2.2 Brain mechanisms mediating threat processing in animals

Early approaches to the emotional brain led to an emphasis on the hypothalamus as the seat of emotional control by the brain (e.g. Cannon 1929; Hess and Brugger 1943). This emphasis shifted somewhat following publication of the Papez (1937) circuit theory, which emphasized the role of higher forebrain areas in regulating the hypothalamus, and Klüver and Bucy's (1937) discovery that removal of the temporal lobes in monkeys led to a loss of fear and other emotional reactions to external stimuli. Building on the Papez model and the Klüver–Bucy syndrome, MacLean (1949, 1952) proposed that a set of forebrain areas centred on the temporal lobe, and especially involving the hippocampus, formed an emotion network called the the limbic system. Subsequently, Scoville and Milner (1957) implicated the hippocampus in cognitive memory, and Weiskrantz (1956) showed that damage to the amygdala accounted for the emotional changes in the Klüver–Bucy syndrome produced by larger lesions of the temporal lobe. These observations should have been problematic for the limbic theory of emotion. However, in part because the amygdala was part of the limbic system, and data on the role of the amygdala in emotional functions continued to pile up (Kaada 1960; Goddard 1964; Mishkin and Aggleton 1981; Sarter and Markowitsch 1985), the limbic system theory persisted. Over the years, though, questions about the anatomical validity and lack of predictive power of the limbic system concept led to its demise as a general explanation of emotions in the brain (Swanson 1983; LeDoux 1987, 1991, 1996; Kotter and Meyer 1992).

Today, researchers are less focused on the search for a universal mechanism to explain all emotions and instead are attempting to understand how specific classes of emotion are represented in the brain. Research on fear has been especially fruitful. Much of this work has focused on fear learning in rats and other small mammals using a procedure called fear conditioning, a variant of Pavlov's celebrated conditioning procedure (Figure 3.1). Fear conditioning has been especially useful in understanding how neutral stimuli, through learning, come to elicit fear reactions. But in addition to passively and automatically responding to danger, organisms also perform instrumental actions that cope with the treat. The brain mechanisms underlying instrumental coping responses have been studied using an extension of the fear conditioning paradigm. Fear conditioning has also been used to study fear regulation. Finally, some progress has also been made in exploring the neural basis of unconditioned threat processing.

3.2.2.1 Learning and storing information about threats

Most of the work on fear learning and memory has focused on how neutral stimuli acquire threat value through fear conditioning. In a typical study a conditioned stimulus (CS), usually a tone, is paired with an unconditioned stimulus (US), such as footshock. After conditioning the CS elicits a coordinated set of bodily reactions involving behavioural, autonomic, and endocrine responses, as well as brain arousal. These are hard-wired responses that are automatically elicited by the CS.

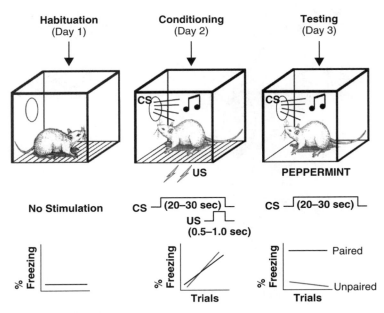

Fig. 3.1 Fear conditioning

Weiskrantz's (1956) study implicating the amygdala in the fear component of the Kluver–Bucy syndrome involved a task called avoidance conditioning in which fear conditioning to a shock occurs and is then followed by the learning of an instrumental response that avoids the shock. Much subsequent research in rats and other small mammals used avoidance conditioning to study the neural basis of emotion (Goddard 1964; Gold *et al.* 1975; Sarter and Markowitsch 1985). However, with avoidance conditioning the ability to separate the effects of brain manipulations on the Pavlovian and instrumental components of learning is difficult, and much confusion resulted.

Subsequent work focused on the use of pure Pavlovian fear conditioning tasks and made much progress (Davis 1992; Kapp *et al.* 1992; LeDoux 1992, 1996, 2000; Davis and Whalen 2001; Maren 2001; Fanselow and Poulos 2005). This research showed that two regions of the amygdala are especially important in fear conditioning (Figure 3.2). The lateral nucleus is the sensory gateway to the amygdala. It receives information about external stimuli from thalamic and cortical processing regions in various sensory systems and forms associations between neural and painful stimuli. Either the thalamic or cortical input is sufficient for learning about or responding to simple stimuli but the cortical

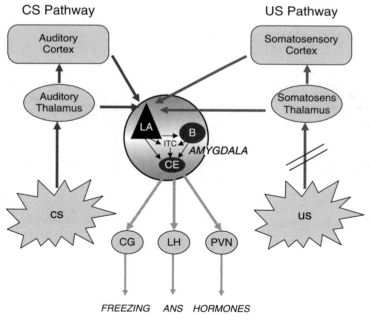

Fig. 3.2 Fear conditioning circuitry

pathway to the amygdala is required for processing complex stimuli. The lateral amygdala connects with the central amygdala both directly and by way of intra-amygdalar pathways (Smith and Paré 1994; Pitkänen et al. 1997). The central amygdala, in turn, connects with lower areas, including the hypothalamus and central grey, that then connect with specific motor neuron groups that control various emotional response modalities (behavioural, autonomic, endocrine). The central amygdala also connects with brainstem circuits that synthesize monoamines and release these in widespread areas of the forebrain. Thus, once the lateral amygdala detects a threat, the central amygdala can lead to the expression of bodily responses and to the activating of brain arousal circuits (for an alternative view see Balleine and Killcross 2006). Below we will discuss the role of brain arousal and feedback from bodily responses in conscious feelings. Lesion and imaging studies have confirmed the role of the human amygdala in fear conditioning (for review see Buchel and Dolan 2000; Critchley et al. 2002; Labar 2003; Phelps and LeDoux 2005; Phelps 2006). However, it is not possible at this point to explore subregions of the human amygdala.

3.2.2.2 Emotional control over instrumental actions

In fear-arousing situations we not only react to the danger by expressing hard-wired responses but can also perform instrumental actions that help cope with the situation. Avoidance conditioning, mentioned above, is a means of studying these emotional actions. Although early work on avoidance led to confusing results (see Sarter and Markowitsch 1985; LeDoux 1996), recent work, building on the progress made in studies of fear conditioning, has begun to elucidate how the brain makes the transition from reaction to action. The connection from the lateral amygdala to the central amygdala is involved in reaction but not action in response to a CS (Amorapanth et al. 2000). Actions, by contrast, involve connections from the lateral nucleus to the basal amygdala (Nader and LeDoux 1997). The basal amygdala then likely connects with the striatum in the control of the instrumental responses (Everitt et al. 1999; Cardinal et al. 2002). In pathological fear, active coping is more effective than passive coping (Cloitre, personal communication) and the action circuitry within the amygdala may play a key role (LeDoux and Gorman 2001).

Instrumental responses that help cope with an emotional situation are learned because they reduce the emotional arousal being elicited by the situation. In the studies described above, the instrumental response is reinforced by the termination of the CS (Balleine and Dickinson 1998; Everitt et al. 1999). In this case the CS is a secondary reinforcer that acquires its reinforcing properties by being associated with a primary reinforcer (unconditioned stimulus).

3.2.2.3 Emotional regulation

A prominent idea in recent years has been that the prefrontal cortex, especially the medial prefrontal cortex, also contributes to emotional processing (Damasio 1994, 1999; Rolls 1999, 2005; Davidson 2002; Ochsner *et al.* 2004; Sander *et al.* 2005b; Kalisch *et al.* 2006). For example, interactions between the medial prefrontal cortex and the amygdala contribute to aspects of emotional regulation learned through exposure/extinction processes (for review see Damasio 1994; Quirk *et al.* 2000; Phelps *et al.* 2004; Phelps and LeDoux 2005; Phelps 2006; Sotres-Bayon *et al.* 2006). These circuits may confer the benefits of exposure in cognitive-behavioural treatment of fear and anxiety (Foa 2006). Medial prefrontal cortex also plays a role in decision-making processes in which emotional information is used to guide mental activity and instrumental behaviour (Damasio 1994; Rolls 2005). Some studies also suggest a role for the lateral prefrontal cortex in emotional regulation in some situations (Ochsner *et al.* 2004).

3.2.2.4 Processing unlearned threats

With the rise of the amygdala as an essential component of the threat-processing circuitry, the hypothalamus came to be viewed simply as an output of the amygdala in the expression of emotional responses (see above). However, recent evidence suggests that the hypothalamus may play a role in processing unconditioned threats (Rosen and Donley 2006; Canteras *et al.* 1995; Petrovich *et al.* 1996). The exact manner in which sensory information reaches the hypothalamus in processing unconditioned threats is less clearly worked out than the manner in which the amygdala participates in conditioned threats. Also, the relative role of the hypothalamus and amygdala in unconditioned threat processing is still uncertain. The amygdala is clearly responsive to the threatening value of unconditioned stimuli both in animals (Blair *et al.* 2005) and in humans (i.e. faces expressing fear or anger; Whalen *et al.* 1998), otherwise fear learning could not take place through amygdala plasticity. However, the ability of unconditioned stimuli to activate emotional responses may also depend on the hypothalamic processing.

3.2.3 Unconscious threat processing in the human amygdala

Above we saw that unconscious emotional processing occurs. Then we examined how emotional processing, especially fear processing, occurs in brain circuits involving the amygdala in both animals and humans. In this section we will first see that amygdala processing in humans occurs unconsciously. Then I will discuss neural pathways involved in unconscious processing by the amygdala. The last part of this section considers some of the functional consequences of unconscious processing by the amygdala.

3.2.3.1 Unconscious amygdala processing in humans

Fearful and angry faces are unconditioned threatening stimuli for humans and produce strong activation of the amygdala both when used as an unconditioned stimulus (the face is presented without any prior training) and when used as conditioned stimulus paired with shock (e.g. Morris et al. 1998, 2002; Whalen et al. 1998, 2001; Thomas et al. 2001; Vuilleumier et al. 2001, 2002; Armony and Dolan 2002; Olsson and Phelps 2004; Critchley et al. 2005). When masked stimulus presentations of conditioned or unconditioned faces are used, and the stimuli are blocked from entering conscious awareness and are not reported as having been seen by the subjects, amygdala activity still occurs (Whalen et al. 1998; Morris et al. 1998, 1999; Critchley et al. 2002; Olsson and Phelps 2004). The eyes alone are sufficient to activate the amygdala in unmasked or masked situations (Morris et al. 2002; Whalen et al. 2004). There is even evidence that amygdala activity is stronger when masked presentations are used than when the stimuli are freely visible (Whalen et al. 1998). Particularly important is the fact that masked presentations not only produce amygdala activation but also autonomic nervous system responses (Morris et al. 1998, 1999), indicating that unconscious emotional processing by the amygdala is capable of controlling the bodily expression of emotion. Implicit processing of faces of people from a race different from the observer also elicits amygdala activation (Phelps et al. 2000).

3.2.3.2 Sensory pathways mediating unconscious processing in humans

As noted above, there are two routes by which sensory information can reach the amygdala (LeDoux 1996). One involves a direct pathway from the thalamus to the amygdala and the other involves transmission from the sensory thalamus to the sensory cortex and from there to the amygdala. Might the thalamic pathway be the source of unconscious processing in the amygdala and the cortical pathway the source of conscious processing?

Functional imaging results support the idea that thalamic processing is unconscious. Thus, in studies using masked presentation of emotional faces to elicit functional activity in the amygdala and autonomic responses, amygdala activity is correlated with subcortical but not cortical sensory processing areas (de Gelder et al. 1999; Morris et al. 1999, 2001a). Does this mean that when information is transmitted to the amygdala from sensory cortex the amygdala receives a consciously processed signal?

Studies examining the cortical activation patterns that occur when human subjects are not aware of a stimulus give important clues. For example, a region of visual cortex called the fusiform face area is specialized for face processing (Kanwisher *et al.* 1997). Face stimuli activate this region when masked or consciously perceived, but consciously perceived faces additionally activate areas of prefrontal cortex and parietal cortex (Rees *et al.* 2002; Vuilleumier *et al.* 2002), regions implicated in attention (see below). When emotional faces are used, both masked and unmasked stimuli produce activation in the face cortex and amygdala (e.g. Morris *et al.* 1998, 1999, 2001b; Whalen *et al.* 1998, 2004; Etkin *et al.* 2004; de Gelder *et al.* 2005), but unmasked presentations leading to conscious awareness additionally lead to activation of other cortical areas, including prefrontal, insular, and parietal regions (Critchley *et al.* 2002, 2004; Vuilleumier *et al.* 2002; Etkin *et al.* 2004; Carter *et al.* 2006). These findings suggest that sensory cortex activation alone is not sufficient for conscious awareness of the stimulus and thus that activity in sensory cortex that activates the amygdala is unlikely involve conscious processing. Sensory transmission from both the thalamus and cortex to the amygdala is thus likely to involve unconscious processing.

In sum, emotional processing can occur unconsciously in the amygdala on the basis of sensory processing in both thalamic and cortical pathways. This indicates that conscious processing of an emotional stimulus is not necessary for emotional processing. The fact that amygdala activity is weaker under unmasked than masked conditions suggests that conscious processing is also insufficient for emotional processing by the amygdala. Conscious processing of emotional stimuli obviously occurs, but this is unrelated to amygdala processing and the control of emotional responses by such processing. At the same time, once a conscious emotional state is present we can, from that vantage point, exercise unique control or regulation of our emotional responses (Ochsner and Gross 2005). Functional imaging stuides indeed show that explicit cognitive control can influence amygdala reactivity, such as asking subjects to re-evaluate the meaning of an emotionally arousing stimulus (Ochsner *et al.* 2004).

3.2.3.3 Functional implications of unconscious processing by the amygdala in humans

The amygdala, once activated, is capable of recruiting a host of brain systems in the service of dealing the with immediate threat. Thus, the brain's resources are diverted from their current focus and directed towards the goal of protection and survival. The net result is that the brain's resources are monopolized until the threat has subsided, or exhaustion occurs. Here, I will describe how amygdala activity affects attention. The role of the amygdala in cognition has been considered in detail in several recent reviews (Whalen *et al.* 1998; McGaugh 2000; Dolan and Vuilleumier 2003; Phelps and LeDoux 2005; LaBar and Cabeza 2006; Phelps 2006).

When danger occurs, the brain has to be redirected towards the threat. In other words, the mechanisms of attention have to be recruited. Although the job of attention is to prevent distraction, attention is also subject to priorities (Norman and Shallice 1980). As a result, if a more important event occurs then the current focus of attention can be shifted. One hypothesis about how this occurs is that when a threatening stimulus is detected by the amygdala, outputs of the amygdala lead to an interruption of attention, allowing the organism to reassess the focus of attention (Armony *et al.* 1997; Whalen *et al.* 1998; Armony and LeDoux 1999).

Two kinds of outputs of the amygdala are particularly important. On the one hand the amygdala has direct connections to widespread cortical areas. Thus, once the amygdala is activated it can in turn influence what cortical areas are processing. When the amygdala activation is driven by thalamic inputs, the amygdala may be able to influence sensory cortex just as or shortly after thalamic inputs arrive. On the other hand, as already noted, the amygdala is connected with arousal systems that release modulator chemicals (norepinephrine/ noradrenaline, acetylcholine, serotonin, dopamine) in widespread cortical areas. These serve to regulate the excitability of cortical neurons, determining the extent to which and efficiency with which incoming signals will be processed, such as sensory processing in posterior cortical areas and cognitive processing in cortical areas. Thus, the amygdala's transitory regulation of cortical regions through direct or modulatory pathways might result in an increased attentional focus (selective attention) and an enhanced level of attention (vigilance) (Armony *et al.* 1997; Whalen *et al.* 1998; Armony and LeDoux 1999; Davis and Whalen 2001).

Evidence from a range of techniques in humans is consistent with the idea that emotional stimuli, via the amygdala, can influence attention and perception. For example, brain imaging studies show that amygdala activation to fearful (vs neutral) faces does not depend on whether or not the faces are the focus of attention (Vuilleumier *et al.* 2001, Anderson *et al.* 2003, Williams *et al.* 2004). These studies indicate that the amygdala responds to a fear stimulus automatically and prior to awareness (Dolan and Vuilleumier 2003). This preattentive amygdala response, early in stimulus processing, may enable modulation of subsequent attention and perception (Whalen *et al.* 1998). Further, patients with amygdala damage fail to show the normal facilitation of attention for emotional stimuli (Anderson and Phelps 2001). Additional support comes from the fact that the degree of activation of visual cortex for fearful vs neutral faces is correlated with the magnitude of amygdala activation

(Morris *et al.* 1998). This enhanced activation in visual regions to fearful faces is absent in patients with amygdala damage (Vuilleumier *et al.* 2004).

The idea that the amygdala always functions pre-attentively has been questioned by studies showing that if the attention demands are especially high amygdala activation in the absence of attention does not occur (Pessoa *et al.* 2002). However, in more routine situations it seems that amygdala functions pre-attentively.

Together, the findings indicate that amygdala processing influences cortical processing, often in the absence attention. Indeed, amygdala processing meets most of the principles of automaticity—that is, it is independent of attention and awareness, with certain limitations for highly demanding attention tasks. By influencing attention, the amygdala alters the gateway of information processing into consciousness, the topic to which we now turn.

3.3 Feeling afraid: awareness of being in danger

In this section I will first discuss what I think a feeling is, and then turn to what I believe are the basic requirements necessary to create a feeling. I will then argue that these are met by the set of cognitive processes collectively called working memory. The contribution of language to feelings will be considered, as will the nature of feelings in non-human organisms.

3.3.1 What is a feeling?

The modern debate about conscious feelings was started by William James' article entitled 'What is an emotion?' (James 1884). James distinguished the expression of emotion in the body from the conscious experience of emotion. His goal was to explain the conscious experience of emotion, the feeling that occurs when we are emotionally aroused. As part of this exercise, he asked his famous question: do we run from a bear because we are afraid, or are we afraid because we run? James rejected the common-sense answer—that we run because we are afraid—and instead promoted the idea that the bodily responses that occur during an emotion determine the feeling: that we are afraid because we run. James's model is called a peripheral feedback theory since it proposed that sensory feedback from the responses determines what we feel.

James's theory ran into trouble in the hands of the physiologist Walter Cannon who argued that body responses are too slow and non-specific to account for feelings (Cannon 1927, 1929). The issue was fairly dormant in psychology during the behaviourist reign since feelings, like other forms of consciousness, were not viewed as appropriate topics for scientific psychology (for a thorough review of early theories of emotion see Plutchik 1980).

By the 1960s, the cognitive movement had overthrown behaviourism and research on mental processes had returned to psychology.

The cognitive revolution in psychology was the backdrop against which Stanley Schachter and Jerome Singer (Schachter and Singer 1962; Schachter 1975) revisited the James–Cannon debate. They proposed that although body feedback is non-specific it is essential as a trigger of cognitive processes that interpret the arousal in light of the environmental context. Ever since, cognitive theories have dominated the psychology of emotion. But unlike the Schachter–Singer theory, most subsequent cognitive theories argued that body feedback is not essential to feelings and that what defines a feeling is simply one's cognitive assessment (appraisal) of the situation (Arnold 1960; Lazarus 1966, 1984, 1991; Scherer 1984; Smith and Ellsworth 1985; Frijda 1986; Leventhal and Scherer 1987; Ortony et al. 1988; Clore 1994; Clore and Ketelaar 1997; Sander et al. 2005a). Through cognitive processing we determine the valence (positive or negative), category (fearful, frustrating, joyful), intensity (high or low arousal), cause, degree of control, and so on of an emotional situation. The net result of these assessments is a subjective feeling, which we then use in the control of thought and behaviour. When we appraise a situation as dangerous, we feel afraid, and when we appraise it as joyful we feel happy. Appraisal theorists often allow for low-level appraisals (which involve automatic, preattentive, unconscious processes) and higher-level appraisals that involve controlled processing and attention and that lead to conscious feelings. Appraisals leading to feelings are in the high-level category.

Central theories are another explanation of feelings. After arguing why peripheral feedback could not explain feelings, Cannon (1929) proposed an alternative. In James's theory a feeling occurred when sensations resulting from body responses were perceived in the cerebral cortex. Cannon held on to the idea that feelings occur in cortex but argued that feelings occur when sensory information about the stimulus is integrated in the cerebral cortex with information from within the brain about emotional arousal. So in contrast to James's model, the emotional information was not fed back from the body but was instead entirely in the brain. And the key brain region responsible for the emotional signal was the hypothalamus. According to Cannon, the hypothalamus detects the emotional significance of external stimuli. By way of descending connections to the body, the hypothalmalus the controls emotional responses and by way of ascending connections to the cortex the hypothalamus provides the signal that gives emotional colouration to sensory stimuli and thus gives rise to a feeling.

Most subsequent theories of the emotional brain (Papez 1937; MacLean 1949, 1952; LeDoux 1996) have continued to emphasize that feelings involve the emotional colouration of sensory processing. The theories differ in their assumptions about which brain regions are involved in processing the emotional stimulus, in producing emotional responses, and in cortical integration, but the basic structure of the hypothesis is the same— that a feeling occurs when the cerebral cortex integrates sensory information about an external stimulus with emotional information processed subcortically within the brain.

Central and cognitive theories are not as far apart as is sometimes assumed. Central theories propose that feelings result when cortical circuits integrate cortical sensory processing with subcortical emotional processing. The subcortical processing essentially overlaps with what the appraisal theorists refer to as low-level appraisal processes that occur automatically, unconsciously, and without attention. The nature of the cortical integration, though, is usually left unexplained. Cognitive theories propose that feelings are the result of high-level appraisals that take into account the external situation, the way the situation is encoded through memory retrieval, the nature of the emotional arousal, and ways of behaving to deal with the situation. High-level appraisals involve controlled processing, short-term memory, long-term memory, and attention. Because the various cognitive processes subsumed under the concept of high-level appraisal are likely to be cortically mediated, they begin to fill in the gap that central theories leave open about what the cortex does when it integrates sensory and subcortical emotional information.

Below, I develop a model that further pursues the cortical mechanisms through which feelings are created. I build on the common themes of central and cognitive theories. However, my model is somewhat more specific in the identification of the cognitive processes involved. I argue that the requirements of a feeling are the creation of an episodic representation of an emotional event and the direction of attention to that representation. I then propose that these requirements are met by the set of cognitive functions called working memory.

3.3.2 Emotional representation: integration of sensory, memory, and emotional information

Let's start with an example of a situation in which a strong emotional (fear) response is automatically elicited. If while walking down a path you suddenly encounter a large snake slithering towards you, you will probably feel somewhat afraid (at least I would). A basic component of this feeling is the sensory processing of the object before you—if your visual system doesn't detect the snake you can't react to it. But a sensory stimulus itself has no meaning until it has been linked up to memories, which are a second contributor to a feeling. In this way the perception is encoded in light of past experience. Both semantic and episodic memories (Tulving 1972, 1983) are important. Semantic memory is memory about facts—in this case, things you know about snakes. Episodic memories are memories of personal experiences you've had—in this case experiences with snakes or other similar creatures, and/or similar environmental conditions. Recent research suggests that semantic and episodic memories of emotional experiences are processed differently (Robinson and Clore 2002). A third component of a feeling is the fact that your brain and body are emotionally aroused. When these three factors (sensory, memory, emotional) are bound together they form an episodic representation of the emotional experience, an experience coloured by emotional arousal.

Most experiences are not unimodal in nature. Your experience with the snake occurs in a context involving information from multiple sensory modalities. The episodic memory you form of the event will therefore consist of the entire situation, not just the isolated visual image of the snake. Thus, diverse forms of sensory processing, forming stimulus configurations and other kinds of relational representations, are often linked to memory information and emotional information in the process of creating emotional representations.

Another kind of episodic memory is also important. This is memory of the self. The self can be defined as the collection of genetic and individual memories that constitute your personality (LeDoux 2002). Self memories are complex multimodal representations not about experiences you've had so much as representations about who you are, both historically and futuristically (Fuster 1997; Levine et al. 1998; Stuss et al. 1998; LeDoux 2002; Tulving 2002). Whenever you encode an immediate experience, you not only retrieve semantic and episodic memories related to the stimulus, but also episodic memories of the self (Kihlstrom 1987; LeDoux 2002; Tulving 2002). Episodic memories of the self are sometimes called autonoetic memories and conscious awareness of the self through such memories is called autonoetic awareness (Tulving 2002).

How does the brain integrate all this information? Sensory systems are separate functional modules that operate more or less independent of one another. Memory and emotional systems depend on sensory processing. Sensory-elicited memory retrieval and emotional arousal can affect subsequent sensory processing. However, in order to create a representation of an experience that integrates sensory information from multiple modalities with memory and emotional information, the outputs of the various systems have to be transferred to brain regions called convergence zones (Damasio 1989). These are regions that receive inputs from various systems and that integrate the information from the separate sources into a unified represenation that can be used in thought and action. We will discuss how this takes place below.

A key question, though, is whether an emotional episodic representation created in convergence zones is itself a feeling or whether something else has to happen. It seems likely that, while necessary, the creation of such a representation is not sufficient. The additional factor that is required is that the representation be attended to, and thus made available for thought and action.

3.3.3 Attention to emotional representations allows their use in thought and behaviour

Emotional representations are fairly passive constructions that are due to the integration of information from diverse sources in one brain region, a convergence zone. Actually, it is unlikely that they are completely passive since a certain degree of selection and processing may be required in each system in order for the information to make it through the system and reach a modality-independent convergence zone where the unified representation is created. Nevertheless, integration of information in a convergence zone is probably not sufficient for it to be consciously experienced.

It is well known that we are not consciously aware of all the processing that takes place in our sensory systems and that conscious awareness only occurs to those things we attend to (James 1890; Broadbent 1962; Posner and Snyder 1975). In early cognitive science it became common to distinguish between automatic and controlled processing (Posner and Snyder 1975; Schneider and Shiffrin 1977; Schneider *et al.* 1984). Automatic processing occurs unconsciously, without effort, and is not subject to interference by increasing cognitive demands by other tasks and is not limited by the capacity of short-term or working memory. Controlled processing, by contrast, requires attention, involves conscious awareness, is limited by short-term memory capacity, and is subject to distraction. These features are consistent with the idea that automatic processing occurs in parallel in multiple processing systems (which is why it is not subject to interference and short-term memory limits) whereas controlled processing occurs serially (because attention and short-term memory are limited in their capacity we can only focus on one thing at time and shifting to a new focus eliminates the past one) (Schneider and Shiffrin 1977; Schneider *et al.* 1984). Controlled processing can be thought of as occurring when attention is directed to one of the various parallel sources that are otherwise processing automatically and unconsciously. In general, the assumption is that what we are conscious of is what, through controlled processing (especially attention), is occupying our short-term memory. As we'll see below, the notion of working memory subsumes all of these processes in a fairly comprehensive model of cognitive function.

Just as we are not aware of all the processing that takes place in sensory systems, we are unlikely to be aware of all of the processing that take place in various convergence zones. Presumably there are multiple such higher-order representations available in convergence zones and attention is used to select which one to focus on and which ones to suppress.

In sum, the binding of sensory, memory, and emotional information is not sufficient to give rise to a conscious experience (Wolfe and Bennett 1997). Conscious experiences, including feelings, come about when we attend to an episodic emotional representation that includes sensory, memory, and emotional information, thus making the representation available for use in thought and action.

3.3.4 Working memory underlies feelings

The requirements outlined above for emotional experience, for feelings, overlap with the functions of what is called working memory (Baddeley and Hitch 1974; Baddeley 1986, 1993, 2000; Repovs and Baddeley 2006). Working memory integrates information from diverse sources, including sensory systems and long-term memory, into episodic representations. Through executive functions, working memory controls attention and allows us to use bound or integrated episodic representations to guide thought and action. Although working memory is not usually thought of as being involved in emotional information processing, a simple extension of the standard model shows how this can occur. Let's look at working memory and its neural basis before turning to a consideration of how to extend working memory to include emotional information.

3.3.4.1 What is working memory?

Since the 1950s it has been widely accepted that what we are conscious of at any one moment is the information we are attending to and occupying our limited-capacity short-term memory (Miller 1956; Atkinson and Shiffrin 1968; Posner and Snyder 1975; Schneider and Shiffrin 1977; Schneider et al. 1984). Later research showed that this limited-capacity temporary storage device is better viewed as a component of a more elaborate cognitive architecture that not only holds onto information temporarily but also is involved in the selection and control of inputs by attention and the use of selected information in thought and action. This view of cognition is called the working memory model (Baddeley and Hitch 1974; Baddeley 1986, 1993). It incorporates the notions of automatic and controlled processing but in a more elaborate and comprehensive framework. Related models are the supervisory attention system (Norman and Shallice 1980), attentional amplification network

(Posner and Dehaene 1994), operating system (Johnson-Laird 1988), and global workspace (Baars 1988, 2005; Dehaene *et al.* 1998, 2006; Dehaene and Naccache 2001). As already noted, since these attempts to explain roughly the same kinds of data and lead to similar conclusions as the working memory model, I concentrate on the working memory model here.

Today, working memory is generally viewed as consisting of several components. These include (a) input processing systems that operate automatically and in parallel, (b) long-term memory, (c) an episodic buffer, and (d) an executive control system involved in attention and other cognitive control functions (Baddeley 2000; Repovs and Baddeley 2006). Each of these will be described.

3.3.4.1.1 **Input processing systems** Two input processing systems were originally identified as components of working memory: the articulatory loop and the visuospatial scratchpad (Baddeley and Hitch 1974; Baddeley 1986, 1993). The former was an auditory processing channel devoted to language, and the latter a non-verbal visuospatial channel. The visuospatial channel has recently been shown to consist of at least two components, one visual and one spatial (Della Sala *et al.* 1999). It seems reasonable that each sensory system provides inputs to working memory. It is now known that there are multiple processing streams within sensory systems (Ungerleider and Mishkin 1982; Milner A.D. and Goodale 1995; Barrett and Hall 2006), and working memory appears to receives separate inputs from the various steams (Goldman-Rakic 1996; Courtney *et al.* 1998; Romanski *et al.* 1999; Fuster 2000; Miller and Cohen 2001; Constantinidis and Procyk 2004; Romanski 2004; Pasternak and Greenlee 2005; Curtis *et al.* 2005).

3.3.4.1.2 **Long-term memory** Long-term memory, especially declarative or explicit memory, plays important roles in working memory. Through the integration of sensory processing with semantic memory sensations are given meaning (sensations + semantic memories = perceptions). However, as we've seen, these perceptions are not conscious perceptions when processed solely in the sensory system. Conscious perception requires that the information also be represented in other cortical areas. As argued below, representation in these additional cortical areas places the perceptual information in the episodic buffer.

3.3.4.1.3 **Episodic buffer** An important role of the episodic buffer is to integrate information from various independent processors and bind them into unified representations or episodes (Baddeley 2000; Repovs and Baddeley 2006). We don't usually separately experience all the sensations that make up an event. We have unified experiences in which the component sensations are fused. The episodic buffer not only receives perceptual (sensory + semantic memory) information from multiple sensory systems, but also direct information from declarative memory systems so that both semantic and episodic memories can be integrated with online processing. This is important since perceptual information from each sensory modality is modality-specific, whereas the episodic buffer has multimodal representations. The meaning of an individual stimulus when contextualized by other stimuli may be quite different from its meaning alone. These multimodal (conceptual) memories can then be used in online processing. Further, as noted above, a particularly important form of episodic memory is memory about the self. Retrieval of these memories is important in further contextualizing encoding that is taking place about a multimodal stimulus situation. In these ways, the episodic buffer is able to create representations that consist of multiple diverse stimuli and semantic and episodic memories that are bundled into a single episode or experience.

3.3.4.1.4 **Executive control** A crucial aspect of working memory is the ability to select which of the sensory processors (both within and between systems) should be allowed to pass information in to the episodic buffer, which semantic and episodic long-term memories to retrieve, and whether to store a representation in long-term memory. These are jobs of the executive control system of working memory. It manages attention to input sources and also regulates the retrieval of information from long-term memory during online processing of the situation and controls storage of new memories.

3.3.4.2 **Brain mechanisms of working memory** The brain region most commonly associated with working memory is the dorsolateral prefrontal cortex (Goldman-Rakic 1987; D'Esposito et al. 1995; Smith and Jonides 1999; Fuster 2000; Rowe et al. 2000; Miller and Cohen 2001; Ranganath et al. 2003; Muller and Knight 2006). This region is a classic convergence zone, as it receives inputs from all of the posterior sensory processing systems (visual, auditory, and somatosensory corticies) and can thus perform multimodal integration. It also receives information from medial temporal lobe systems involved in storage and retrieval of semantic and episodic memory. It has output connections with motor and premotor areas involved in the planning and control of voluntary behaviour. Damage to this region disrupts the ability to use information held in temporary storage in problem solving, and also interferes with executive functions, including attention.

If working memory truly does involve the dorsolateral prefrontal cortex, and if working memory is an important part of conscious experience, it would seem natural to expect that damage to this region should eliminate the ability to be conscious of one's episodic experiences. Such lesions do alter conscious experience in various ways but do not eliminate the ability to have conscious experiences (Knight and Grabowecky 2000).

The fact that dorsolateral prefrontal cortex lesions do not prevent consciousness has sometimes been used to challenge the idea that working memory is a key to understanding consciousness. However, it is now apparent that the various functions ascribed to working memory are not solely mediated by dorsolateral prefrontal cortex. For example, additional areas of prefrontal cortex (including anterior cingulate cortex, Broca's area, orbital cortex, insular cortex, and premotor cortex) as well as areas of parietal cortex appear to contribute to certain aspects of working memory, including temporary storage and executive functions such as attention and volitional behavioural control (see Posner and Dehaene 1994; Mesulam 1999; Colby and Goldberg 1999; Platt and Glimcher 1999; Rolls 1999; Smith and Jonides 1999; LeDoux 2002; Ranganath et al. 2003; Constantinidis and Procyk 2004; Pasternak and Greenlee 2005; Curtis 2006; Muller and Knight 2006). These cortical areas are highly interconnected with one another, leading to the idea that working memory involves distributed frontoparietal circuits rather than a strictly localized area of prefrontal cortex. The distributed nature of cortical areas involved in working memory likely explains why circumscribed prefrontal cortical lesions do not completely disrupt consciousness.

That frontal and parietal areas are involved in consciousness is indicated by studies that have compared differences in brain activation patterns when subjects are explicitly aware and unaware of stimuli (Frith and Dolan 1996; Frith et al. 1999; Beck et al. 2001; Rees et al. 2002; Feinstein et al. 2004; Frith, this volume). Similar effects, reviewed above, occur for emotional stimuli that are explicity and implicitly processed (Critchley et al. 2002, 2004; Vuilleumier et al. 2002; Cunningham et al. 2004; Carter et al. 2006). Awareness engages cortical areas, especially frontal and parietal areas, in ways that implicit processing does not. Alternatively, engagement of cortical areas makes awareness possible.

Several times we've emphasized the role of episodic memories both of past experiences and of the self. Considerable evidence implicates medial temporal lobe areas such as the hippocampus in the encoding and retrieval of episodic memories of past expereinces (Tulving and Markowitsch 1998; Squire et al. 2004; Eichenbaum and Fortin 2005; Suzuki 2006). However, frontal areas are also involved (Tulving et al. 1994; Nyberg et al. 1996; Fletcher et al. 1997; Yancey and Phelps 2001). Conscious experience of an episodic memory probably involves the representation of episodic memory in the episodic buffer of working memory via storage and attention functions of the prefrontal cortex. Episodic memory of the self has been studied far less. A report of one case, patient M.L., suggests that autonoetic memory (Tulving 2002) crucially involves the prefrontal cortex (Levine *et al.* 1998). Following damage to the ventral prefrontal cortex in the right hemisphere this patient lost the capacity to re-experience personal events and form coherent plans for the future, and lacking this information was unable to self-regulate his behaviour in adaptive productive ways. Recent imaging studies have also emphasized the role of prefrontal cortex, especially right prefrontal cortex, in self-representations, though the exact areas vary from study to study (Keenan *et al.* 2000; Gusnard *et al.* 2001; Johnson *et al.* 2002; Kelley *et al.* 2002; Fossati *et al.* 2003; Schmitz *et al.* 2004; Goldberg *et al.* 2006).

An important challenge for the future is to untangle the exact contribution of the various components of these circuits to different aspects of working memory, specifically to temporary storage and executive functions. Some progress has been made (Smith and Jonides 1999; Miller and Cohen 2001) but much remains unknown. Some key questions are these. Does attention transfer information from posterior cortical areas to frontal areas where it is held in temporary storage for use in the control of thought and action, or does attention simply determine which posterior cortical area is focused on? The latter might work for attention to unimodal events but would be less satisfactory for multimodal representations. Is there an anatomical episodic buffer, a convergence zone, that integrates information across modalities for use in thought and action, or is the episodic representation mediated by synchronized firing across brain regions? Is there a single key convergence zone or do several participate in integration and temporary storage of multimodal information, and what defines the involvement of each? How does attention work, and what is the role of different brain circuits that have been implicated in attention (lateral and medial frontal areas and posterior parietal cortex)? What do neuromodulators contribute, and how are they engaged? What triggers attention shifts as the environmental situation changes?

3.3.4.3 Representation of emotional information in working memory

As noted above, working memory is not usually considered to be the answer to how feelings come about. However, to the extent that working memory is used to create episodic representations, and episodic representations of emotional events are the most important ones we create, it seems natural that working memory would be involved in feelings. The key requirement would be some way for emotional information to be represented in working memory and integrated with on-line processing of sensory inputs and long-term memories (episodic and semantic) activated by the sensory inputs (Figure 3.3). We'll consider this by exploring ways that the consequences of amygdala activity might influence working memory circuits in prefrontal cortex (Figure 3.4).

96 | EMOTIONAL COLOURATION OF CONSCIOUSNESS

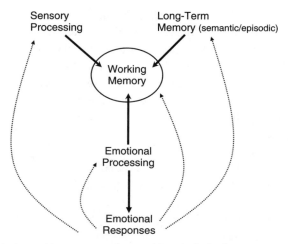

Fig. 3.3 Inputs to working memory that could underlie feelings

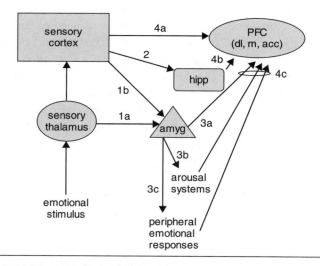

1a,b: sensory inputs to the amygdala
2: sensory inputs to hippocampus
3a,b,c: amygdala outputs to central (a,b) and peripheral (c) systems
4a: allows awareness of a stimulus
4a + 4b: allows awareness of the memory of a stimulus
4c: allows awareness of emotional arousal
4a+4b+4c: conscious feeling (see Fig 3.3)

Fig. 3.4 Neural circuits through which amygdala outputs contribute to feelings

We'll consider three kinds of outputs: direct connections from amygdala to cortical areas, indirect influences involving arousal systems, and indirect influences involving feedback from peripheral responses.

3.3.4.3.1 **Direct connections to cortical areas** The amygdala sends widespread connections to cortical areas. It connects with both early and late stages of sensory processing, as well as with a number of higher-order cortical areas, such as the insular, orbital, and anterior cingulate cortices (Amaral 1987; McDonald 1998; Ghashghaei and Barbas 2002). Importantly, though, the amygdala does not connect with the dorsolateral prefrontal cortex (Ghashghaei and Barbas 2002). In spite of its lack of connectivity with dorsolateral prefrontal cortex, amygdala activity seems to influence it (Dolcos et al. 2004; Dolcos and McCarthy 2006). For example, prefrontal activity increases when human subjects performed working memory tasks and amygdala activity increased when processing emotional stimuli. When the emotional stimuli were presented during the working memory task behavioural performance and frontal activity were both disrupted. This suggests that emotional stimuli can redirect attention by altering prefrontal processing. This could be achieved by the fact several other frontal areas (orbital frontal, anterior cingulate, insular) that are connected with the amygdala are also interconnected with dorsolateral prefrontal cortex (Amaral 1987; McDonald 1998; Ghashghaei and Barbas 2002) or by connections from modulatory systems and by body feedback, as discussed next.

3.3.4.3.2 **Indirect influence via modulatory systems** The amygdala can also influence cortical areas through output connections to brainstem and forebrain arousal systems. These are called modulatory systems because they typically do not cause cells to fire action potentials so much as to change the propensity of cells to respond to other inputs. For example, noradrenergic, dopaminergic, serotonergic, and cholinergic neurons are stimulated to release their chemicals throughout the cortex, including dorsolateral prefrontal cortex, in response to amygdala activity. These chemicals alter the excitability of cortical neurons and can thus enhance or inhibit their responsiveness to incoming inputs form sensory or memory areas. These indirect routes could also participate in the amgyala's ability to alter processing in dorsolateral other frontal areas.

3.3.4.3.3 **Body feedback** Activity in the amygdala also leads to the expression of bodily responses that provide feedback that can influence the brain in multiple ways (see Davis 1992; Kapp et al. 1992; LeDoux 1992, 2000; Maren 2001). The three main classes of responses that could provide feedback are (1) skeletal muscle responses that lead to proprioceptive signals; (2) autonomic nervous system responses that lead to sensory activity in the vagus nerve, the main visceral afferent pathway to the brain; and (3) hormonal responses. I'll discuss each of these, and then consider Antonio Damasio's modern feedback theory.

Amygdala activation leads to the expression of a variety of autonomic nervous system responses (Davis 1992; LeDoux 1996). Autonomic activity, in turn, is sensed by the vagus nerve. Vagus nerve activity has been shown to influence cognitive functions. For example, it is believed to regulate brainstem arousal systems, and thus the degree to which these modulatory networks influence forebrain processing (McGaugh 2000). This has been especially well documented in terms of memory processing. Thus, vagus activity appears to be capable of enhancing or weakening memory storage in the hippocampus and striatum, possibly via the amygdala. However, these effects are relative non-specific in that they occur for both memory in both appetitively (positively) and aversively (negatively) motivated tasks. Visceral afferent activity can thus modulate emotion but is probably not sufficient to determine the particular emotional state that one will experience. Nevertheless, there is much more specificity to autonomic control than previously thought (see de Groat *et al.* 1996). The specificity, though, is largely determined by the output to the autonomic nervous system from the brain (Cabot 1996). Although specificity associated with different emotions is thus possible, as Cannon noted, the actual response of visceral organs is sluggish (relative to striated muscles) and probably too slow to directly determine feelings.

Amygdala activation also leads to the release of hormones from the pituitary–adrenal axis (Sullivan *et al.* 2004). Glucocorticoid hormones released from the adrenal cortex travel in the blood stream and affect widespread brain areas, including frontal areas (Diorio *et al.* 1993; Radley and Morrison 2005). These can, like visceral afferent activity, affect cognitive functions such as memory and do so for both appetitively and aversively motivated tasks. Catecholamines released from the adrenal medulla do not cross the blood–brain barrier but do affect the vagus nerve, which as we saw can stimulate brainstem arousal systems and influence their release of modulatory chemicals that influence other areas (McGaugh 2000). Hormonal effects, in addition to being relatively non-specific, are even slower than direct visceral afferent activity and thus are also unlikely to determine exactly what we feel in a given situation.

Obviously, amygdala activity also leads to behavioural responses. Cannon criticized James's theory as if it were based on autonomic/endocrine feedback. This criticism mainly applied to Lange's version of the James–Lange theory. James actually emphasized the entire act of responding, including behavioural responses (after all, his famous question was about running from the bear). Proprioceptive sensations occur when skeletal muscles contract and relax.

These have the speed and specificity needed to provide feedback that could influence feelings. On the basis of this conclusion, several theories have emphasized that idea that proprioceptive signals from facial muscles help define emotional experiences (Tomkins 1962; Izard 1971; Buck 1980). Although facial muscles do indeed distinguish between emotions (Darwin 1872; Ekman 1993), compelling evidence in support of the facial feedback theory has not emerged.

In recent years, Antonio Damasio and colleagues (Damasio 1994, 1999; Damasio *et al.* 1996; Bechera *et al.* 2000, 2005) have developed a comprehensive theory of the role of body feedback in feelings. Like James, Damasio argues that we use body feedback to figure out what we are feeling. He goes on to propose that this information plays an important role in the way we make decisions. For example, when trying to decide between two courses of action, we play these out in our minds and the body sensations that result in one vs the other, according to Damasio, help us decide which to choose (which one 'feels' better). Also, like James, Damasio emphasizes feedback from all possible body sources, including proprioceptive feedback from the body as well as the face, and all the visceral afferent and endocrine activity as well. Damasio proposes that the key brain areas involved in feelings are those that receive all these body signals, including the orbitofrontal cortex and the body-sensing regions of the cortex (somatosensory and insular regions). A parallel idea has been propsoed by Craig (2003). His view is that the insular cortex is a site of interoceptive integration. In this region various viscerosensory signals are integrated and 'meta-representations' are formed. Craig views these representations as feelings.

Damasio and colleagues have performed imaging studies that support his hypothesis (Damasio *et al.* 2000). For example, subjects were asked to recall certain personal emotional experineces of a particular type (joyous, fearful, sad, etc.). Different patterns of brain activity were found in each case. These findings clearly show that different recalled emotional episodes lead to different patterns of cortical and subcortical activation correlated with the different states. However, the fact that activity in body sensing areas correlates with emotional experiences does not show that such activity plays a causal role in the experience.

In light of what we have been discussing, I propose a slightly different view of Damasio's findings. As we've seen, activity in visual cortex is necessary but not sufficient for visual awareness. The same is probably true for body sensing regions such as the somatosensory cortex. Thus, feedback from the body may activate these cortical and subcortical areas in relatively specific ways. However, until this information is represented in the episodic buffer and attended to it is unlikely to be felt. Notably prefrontal, including insular, areas were also activated. These could be the gateway into the working memory circuitry. In this view, proprioceptive and visceral feedback is not a special determinant of feelings. It influences feelings same way that any other sensory system does—as a source of sensory inputs to working memory (also see Hinson et al. 2002).

One possible resolution of the Damasio/Craig view with my view comes from recent findings suggesting that insular and parietal cortex areas contribute to working memory kinds of functions and are sometimes active when subjects consciously process stimuli in contrast to situations where stimuli are masked (Critchley et al. 2002, 2004; Vuilleumier et al. 2002; Etkin et al. 2004; Carter et al. 2006). Perhaps these additional areas belong to a broader working memory network, sometimes referred to as a global workspace (Baars 1997, 2005; Dehaene et al. 1998, 2006; Dehaene and Naccache 2001).

What then can we say about feedback? First, it clearly affects the brain and also the way we feel. Whether visceral inputs are as non-specific as once thought is less clear. However, visceral inputs are still generally believed to be too slow to directly determine what we feel. The main contributions may be to increase intensity and duration of an emotional experience by altering other systems such as arousal systems. Proprioceptive feedback has greater potential to contribute directly as one of the various sensory inputs to working memory.

3.3.4.3.4 **Monitoring behaviour** Working memory also has access to another kind of information that results from amygdala activation of behavioural responses: when we observe or monitor our mental or behavioural activities. Monitoring is an important executive function of working memory (Smith and Jonides 1999; Knight and Grabowecky 2000; Fuster 2003; Mitchell et al. 2004; Muller and Knight 2006). When we see ourselves acting a certain way we can use that information to update what we are experiencing.

This is dramatically illustrated by observations from split-brain patients (Gazzaniga and LeDoux 1978). In a typical patient the left hemisphere but not the right hemisphere can speak. If the right hemisphere is provoked to produce a bodily response (such as a wave or laugh) and the patient is asked why he did that, an answer is usually confabulated by the left hemisphere. For example, when the right hemisphere waves, the left hemisphere explains the response as, 'Oh, I thought I saw a friend.' Time after time, these kinds of narrative stories are told, as if the left hemisphere was trying to make sense of life (Gazzaniga et al. 1977; LeDoux et al. 1977; Gazzaniga and LeDoux 1978).

In social psychology it is commonplace to assume that people produce explanations when faced with uncertain information about their behaviour or the behaviour of others (Nisbett and Wilson 1977; Ross et al. 1977; Wilson 2002;

Johansson *et al.* 2006). This is called attribution theory (Kelley and Michela 1980). Gazzaniga and LeDoux fused the split-brain findings just described with attribution theory to propose that a key function of consciousness is the ongoing construction of narratives about who we are and that allows consciousness to have a continuity over time, giving rise to a unified sense of self (Gazzaniga *et al.* 1977; Gazzaniga and LeDoux 1978). On the basis of these and other findings Gazzaniga developed the interpreter theory of consciousness (Gazzaniga 1985, 1988; Baynes and Gazzaniga 2000), which will be discussed further below when we discuss language and consciousness.

The interpretation of behaviour on the basis of self-observation is readily viewed as a function of working memory. When we attend to visual and other sensory signals produced by behaviour, those signals enter working memory, where they can be evaluated in terms of past experiences. If the current behaviour contradicts expectations stored in memory, executive functions initiate thoughts and actions to cope with this discrepancy. We'll return to the role of language in working memory, consciousness and feelings below.

3.3.4.4 **Working memory and feeling**

Most modern theories view feelings as being cognitively constructed. Schachter and Singer's theory was an attribution theory. They proposed that unexplained arousal elicits dissonance, which demands reduction. By attending to the social environment the emotional situation could be encoded and labelled in certain way. This attribution of a cause to the emotional arousal (blaming the social environment) reduces the dissonance and frees the mind to think of other things. Appraisal theories also placed great emphasis on attention and encoding of external conditions (stimuli) and internal states (arousal, valence). All of the cognitive processes invoked by attribution and appraisal theories are easily accounted for by working memory. The working memory account of feelings thus has the advantage of allowing the same cognitive model to be used to study the processing functions that underlie conscious cognitions and emotions.

That working memory circuits can perform the kinds of integrations that could give rise to a feeling seems pretty clear. Somewhat more mysterious is the manner in which feelings are given their unique signatures. Why does fear feel different from anger or joy?

My proposal is that the uniqueness of feelings in different emotion states is due to two factors. The first is that the inputs to the episodic buffer of working memory are distinct in different states. They are distinct because the environmental conditions that elicit them are different, and they are encoded differently through memory retrieval about the environment and the self. They are also different because activation of different emotion networks results in different patterns of brain arousal, body responses, and peripheral feedback. The net result is that the episodic representation is experienced differently for different kinds of emotions. The second factor that contributes to the uniqueness of different emotion states is that once the state is represented in the episodic buffer it can, through language, be labelled in different ways that reflect subtle differences in the way the inputs to working memory are processed and how these are evaluated in relation to one's past experiences and self concept. Let's turn to language.

3.3.5 The role of language in feelings and other forms of consciousness

Most scientists would surely agree that organisms with language process information differently than do organisms without language. All sorts of evolutionary modifications of the brain affect the way information is processed so there is every reason to think that the evolution of language would have such effects (Arbib 2001).

The precise role of language in cognition and consciousness is heavily debated (see Fodor 1975, 1983; Dennett 1991; Lucy 1992; Landau and Jackendoff 1993; Pinker 1994; Clark A. 1998; Carruthers 2002). The traditional view, called the Sapir–Whorf hypothesis, proposed that language shapes thought (Sapir 1921; Whorf 1956). This view lost favour relative to the idea that thinking is mediated by an innate, universal language that is translated into specific languages when used (Fodor 1975, 1983; Pinker 1994). According to the latter view we don't think in English or Chinese, but in a universal 'language of thought' nicknamed 'mentalese.' More recently the Sapir–Whorf hypothesis has been revived (Bowerman and Levinson 2001; Gentner and Goldin-Meadow 2003). For example, it has been shown that spatial reasoning is influenced by the spatial lexicon (Levinson *et al.* 2002). Also, and particularly relevant here, there is a long tradition of research showing that culture, and the language used by a culture, influences the emotions that are commonly experienced in that culture (e.g. Harré 1986; Kitayama and Markus 1994; Wierzbicka 1994).

Before discussing the role of language and culture in emotion it is important to re-emphasize the difference between the expression of emotional responses and conscious emotional feelings. Language and culture can influence both. In the case of emotional expression, facial and other bodily responses can be inhibited to some extent in light of social constraints—people in different cultures have different tendencies to express anger, joy, fear, or other emotions in different situations. These constraints are called display rules and they specify who can do what in the presence of whom (Ekman 1980). Display rules modulate rather than completely replace emotional expressions, especially for basic emotions. Regardless of the effects on emotional responses, though, what we are most concerned with here is the role of culture, and especially language, in feelings.

Clearly, different cultures have different words and concepts to describe emotional experiences (Harré 1986; Wierzbicka 1986; Lutz 1988; Russell 1991; Mesquita and Frijda 1992; Shweder 1993; Scherer and Wallbott 1994). When members of a given culture label emotions in themselves or others they are relating those emotion labels to culturally shared concepts that define how they encode experiences and respond to them (Mesquita and Frijda 1992). Labelling of an experience in a certain culturally defined way can influence the degree to which that experience is viewed as manageable or not and may guide behaviour in a certain direction. There are emotion words and concepts in English that do not appear in other European languages and that have to be translated using combinations of other words. The emotion concepts that a culture has can influence the way members of that culture experience situations and make judgments about how to act in those situations, and can be perplexing to members of other cultures who do not have those words and concepts.

In general, the semantic aspect of language allow us to classify and categorize experience in ways that extend the range of emotions that can be distinguished and represented episodically, and thus experienced. Without words it would be difficult to easily distinguish different categories of emotion. For example, because we have labels for different emotions we can easily distinguish when we are feeling afraid, angry, or sad. Words also allow us to divide up each category of emotion. There are more than 30 words in English related to fear (Marks 1987). One person seeing a snake in front of them might feel terror, another trepidation, and another concern. Semantic categories give us ways of conveniently labelling the degree to which an emotion system is active. Words are especially useful in communicating with others about our emotions.

Cultures with different languages also have different syntactic structures that influence the way emotions are communicated and experienced (Wierzbicka 1994). Syntax allows us to construct sentences that, on the basis of their structure, convey information about action. Thus, depending on the particular construction, we can determine who did what to whom when. Using syntactic information in thought processes gives us a way of parsing social situations so that we can, online and with facility, tell ourselves who is feeling what and what they are likely to do on the basis of those feelings. This helps us decide what we should do, to whom we should do it, and when we should do it.

Where does all this leave the mentalese hypothesis? The fact that there are linguistically influenced, culturally defined ways of experiencing and emotions does not rule out a role for culturally independent universals that reflect the underlying structure of an innate language of thought. Indeed, to the extent that the evolution of mentalese resulted in an innate wiring plan that has altered the cognitive processing capacities of the primate brain, mentalese helps explain why people with language impairments can think even though they cannot speak or comprehend speech.

The importance of language for cognition and consciousness is dramatically illustrated by studies of split-brain patients. In most of the early split-brain patients, language is weakly present or largely absent in the right hemisphere (Gazzaniga 1970). The right hemisphere in these patients was cognitively dull, more like a higher primate than like a human. But in some rare cases that emerged as the population of patients expanded, language was found to be robustly present in both hemispheres (Gazzaniga *et al.* 1977; LeDoux *et al.* 1977, 1985; Gazzaniga and LeDoux 1978; Baynes and Gazzaniga 2000), probably because of early damage to the left hemisphere. When this happened, the right hemisphere had clear signs of advanced cognitive capacities and conscious awareness: self-concept and plans for the future. It seemed unlikely that the presence of language simply made it easier to communicate. Instead, it seemed that the extensive invasion of language gave the right hemisphere cognitive prowess that it could not possess otherwise.

As noted above, these and other findings led to the idea that the unity of consciousness is made possible by the ongoing construction of a narrative that in part depends on the attribution of causes to mental states (Gazzaniga *et al.* 1977; LeDoux *et al.* 1977; Gazzaniga and LeDoux 1978). This was then developed into the interpreter theory of consciousness by Gazzaniga (Gazzaniga 1985, 1988; Baynes and Gazzaniga 2000).

The basic idea underlying the interpreter theory goes like this. The human brain evolved from brains in which various modules are poorly integrated and that operate unconsciously (Rozin 1976). The addition of language to the brain allows integration across and executive control over these modules, especially in the left hemisphere (Baynes and Gazzaniga 2000). Still, most of what the brain does it does unconsciously (Wegner and Wheatley 1999; Velmans 2000; Wegner 2002; Hassin *et al.* 2005). Normally, we don't pay much attention to these unconscious responses, such as gestures during conversation. However, when they violate some expectation, they create a state of cognitive dissonance (Festinger 1957). The dissonant cognition in this case is the thought that I am normally in control of my behaviour and the realization that I just did something I did not realize I was doing (and more dramatically, that I don't like or approve of). To resolve the dissonance, an explanation is produced. If you got angry with someone you thought you liked, you might say, well they weren't really that nice after all, or, I didn't get that angry. Language, in short, provides a means of integrating information processing in and executive control over modules that were previously independent.

As implied by the last sentence, language and working memory are highly interlocked in humans (Aboitiz 1995; Baddeley 2003). The articulatory loop is a speech-processing system. It feeds its outputs into working memory, allowing the monitoring of inner speech (Vygotsky 1934). Working memory is essential to speech processing, as we have to keep the first part of a sentence in mind until we hear the end in order to comprehend the entire sentence. Working memory becomes even more important for processing long stretches of speech or written language. Planning speech acts can be thought of as an executive function. Many have argued that thought is a kind of inner speech. To that extent, it is the organization and manipulation of verbal information in working memory. Finally, as explained above, just as language enhances how we can process emotional information it also enhances what we do in cognitive processing.

The basic ideas about language, working memory, consciousness, and feelings developed here are consistent with Weiskranz's notion of a commentary system (Weiskrantz 1997). Weiskrantz makes the useful point that it is not sufficient to simply monitor what is going on in order to be conscious of it. It is necessary for the information processing to reach an additional state where it can be acknowledged by the subject in some form of commentary or report, whether verbal or non-verbal. The commentary system thus interprets the activity of other systems, such as sensory processing and memory systems, thus making that information either conscious or, in a milder form, available for conscious awareness. Like the present proposal focused on working memory, Weiskrantz proposes that the commentary system depends on prefrontal cortex. As pointed out by Weiskrantz, ideas such as these overlap with Rosenthal's philosophical theory that consciousness depends on higher-order thought: that we are not aware of a sensation or memory until we have a thought about that sensation or memory (Rosenthal 1990, 1993). Rolls (2005) has also recently adopted Rosenthal's approach in the context of understanding feelings as higher-order thoughts about one's affective experinces. In a similar vein, in the present proposal one is not aware of a sensation or memory until it is represented in a higher-order system, namely working memory. There is an important advantage to explicitly hypothesizing that working memory is the higher-order system that is responsible for the creation of feelings via higher-order representations. Although the neural mechanisms of higher-order thought, in the general sense, are poorly understood, the neural mechanisms of working memory are understood in great detail. The abundant research paradigms that exist for studying the processes involved in working memory can therefore be readily used to explore the contribution of working memory to emotional awareness, including both the awarenes of emotional stimuli and the awareness of the expereinces elicited by such stimuli (i.e. conscious feelings).

3.3.6 Neuro-evolutionary aspects of feelings and consciousness

All humans have pretty much the same kinds of brains. And there are fundamental similarities in organization of all mammalian brains, in fact, all vertebrate brains (Nauta and Karten 1970). But there are also important differences between the brains of humans and primates and between other primates and other mammals, and between mammals and other vertebrates. If the circuits that mediate consciousness are circuits that are conserved across species, we would probably feel pretty comfortable in assuming some similarity in consciousness. But if the circuits involved in consciousness in humans are different from circuits present in other species, we would have to be very cautious in attributing consciousness to those animals.

It turns out that the areas of the brain that are most different between humans and other primates are mostly located in the prefrontal cortex. The term prefrontal cortex is used fairly loosely in the literature. For more precision I will follow the distinction made by Reep (1984) and Braak (1980). Based on a review of comparative studies of cortical organization, they argue that the prefrontal cortex has two components, one mesocortical and the other isocortical. Mesocortical areas are connected with limbic regions, notably the amygdala, but isocortical regions are not. Further, the isocortical prefrontal cortex has a well-defined layer IV. Included in mesocortical prefrontal cortex are the various regions such as the cingulate, insular, medial orbital, and parahippocampal cortex, among others. Isocortical prefrontal cortex is comprised of various lateral prefrontal areas we have been discussing such as dorsal lateral prefrontal cortex, Broca's area, and parts of orbitofrontal cortex, among others. Primates have isocortical and mesocortical prefrontal cortex. Other mammals mainly have mesocortical prefrontal cortex. Further, isocortical prefrontal cortex occupies a considerably larger proportion of cortical tissue in humans than in other primates. Because working memory in humans depends on isocortical prefrontal cortical areas, we have good reasons to question the nature of working memory-dependent consciousness in other animals.

Although isocortical prefrontal cortex is smaller in other primates than in humans, it has multimodal input connections and participates in temporary storage and executive functions, including attention (Goldman-Rakic 1987; Tomita et al. 1999; Miller and Cohen 2001; Fuster 2003). Thus, isocortical prefrontal cortex contributes to working memory functions in monkeys in much the same way it does in humans. However, because other primates lack natural language, and because natural language is important in human consciousness, we have to conclude that although other primates may be conscious in some ways made possible by the working memory functions of isocortical lateral prefrontal cortex, they are not likely to be conscious in the same way as humans.

That primates are consciously aware of visual stimuli is strongly suggested by studies using the binocular rivalry paradigm in which presenting two stimuli simultaneously, one to each eye, leads to conscious awareness of one stimulus at a time, with the stimulus that is consciously perceived changing over time (Logothetis 1998). Monkeys were first trained to perform separate responses to the two stimuli. The stimuli were then presented in such a way to create binocular rivalry. The monkeys' responses indicated which stimulus they were seeing at a given point in time. Recordings in the visual cortex showed that cells in the earlier stages of cortical processing (V1) tended to respond to both stimuli whereas cells in later stages mainly responded to the stimulus that the monkey indicated, by its response, that it was seeing. These basic results have been shown in the human brain as well (Moutoussis et al. 2005). Studies in humans additionally show that activity also increases in frontal areas during the processing of the 'seen' stimulus (Srinivasan and Petrovic 2006). This is consistent with the literature discussed above indicating that a key difference between conscious and unconscious processing is the activation of frontal areas. It would be interesting to determine whether prefrontal cortex is selectively active when the monkey brain processes the 'seen' stimulus.

Given that sub-primate mammals lack both isocortical prefrontal cortex and natural language, whatever conscious experience they have is likely to be even more different from what we experience than what other primates experience. Sub-primate mammals do have the capacity to selectively attend to external stimuli (Mackintosh 1965; Hall and Pearce 1979; Holland and Gallagher 1999) and they have important convergence zones, including the hippocampus and related areas of the medial temporal lobe. The hippocampus, in fact, has been proposed to play an important role in conscious experience (O'Keefe 1985; Gray 2004; Kandel 2006).

Several lines of evidence are consistent with the idea that the hippocampus plays a some role in conscious experience. First, the hippocampus has long been implicated in what is called declarative or explicit memory in humans (e.g. Mishkin 1982; Squire 1987; Cohen and Eichenbaum 1993; Milner B. *et al.* 1998; Mishkin *et al.* 1998; Squire and Zola 1998; Eichenbaum 2002; Milner B. 2005), the defining feature of which is that the memory is consciously accessible. The fact that tasks that are meant to be analogues of declarative memory are disrupted by damage to the hippocampus in monkeys and rats (see Squire and Zola 1998; Eichenbaum 2002) suggests that maybe they are consciously aware of their hippocampal-dependent memories. Second, damage to the hippocampus disrupts trace conditioning (where there is a temporal gap between the conditioned and unconditioned stimuli) in humans (Clark and Squire 1998; Clark *et al.* 2002; Carter *et al.* 2003) and in rats (Moyer *et al.* 1990; Huerta *et al.* 2000; Tseng *et al.* 2004). In humans (Clark and Squire 1998; Clark *et al.* 2002; Carter *et al.* 2003, 2006), the hippocampus is only involved in trace conditioning if the subject is consciously aware that the conditioned stimulus reliably predicts the unconditioned stimulus (but see LaBar and Disterhoft 1998; Shanks and Lovibond 2002), again suggesting that maybe the hippocampus allows for conscious awareness in rats. Third, Eichenbaum and colleagues have taken advantage of a particular feature of human conscious recollection and shown that this feature not only occurs in rats but also depends on the hippocampus in rats (Fortin *et al.* 2004).

In spite of these compelling findings, it is important to note that evidence that the hippocampus contributes to conscious memory in humans does not show that the hippocampus mediates the conscious component of the memory. Just as the face area of visual cortex is required to represent a face and other areas (frontoparietal areas) are necessary for conscious perception of the face, conscious recollection of a memory requires that hippocampal representations be processed in frontal areas. For example, episodic memory retrieval, and hence conscious experience of the episodic memory, involves not just hippocampus (Vargha-Khadem 1997, *et al.* 2003; Squire and Zola 1998; Baddeley *et al.* 2001; Tulving 2002; *et al.* Fortin *et al.* 2004; Eichenbaum and Fortin 2005; Kennedy and Shapiro 2004; Suzuki 2006) but also frontal areas (Levine *et al.* 1998; Squire and Zola 1998; Allan *et al.* 2000; Knight and Grabowecky 2000; Maratos *et al.* 2001; Greenberg *et al.* 2005). In this sense hippocampus is like sensory cortex—it is a processing zone that provides inputs to prefrontal areas that make information available in conscious awareness, possibly by serving as or contributing to an episodic buffer.

As noted, rats and other sub-primate mammals do not have the isocortical prefrontal areas present in humans and other primates. However, in both rats and primates the hippocampus connects with mesocortical prefrontal areas that have also been implicated in multimodal processing, attention, and working memory in humans (anterior cingulate cortex, insular cortex, medial orbitofrontal cortex). Whether these regions can subserve the functions that isocortical prefrontal areas mediate in primates is not known. At this point we might therefore remain open to the possibility of some form of awareness of processing in sub-primate mammals. Nevertheless, because of significant difference in the organization of key brain areas and the lack of language, any experience they have is likely to be considerably different from what humans consciously experience. Even strong proponents of animal awareness tend to argue for only 'simple levels of consciousness' (Griffin and Speck 2004).

What might these simple levels of consciousness be? Given that sub-primate mammals have control systems that can select which sensory modality to attend to, might they at least have unimodal conscious experiences? It seems very likely (though not easily proven) that when a rat sees or smells food or its skin is cut, it is consciously aware of the sensation—in other words that it has a conscious perception of the food or pain. These kinds of primitive feelings are sometimes called 'raw feels,' which we'll discuss again in the next section. However, lacking an elaborate lateral prefrontal cortex, the animal may lack the capacity to fully integrate such perceptions with past memories formed through other modalities, and may lack the capacity to flexibly use such perceptions in the guidance of behaviour to other modalities. This does not mean that the rat lacks the ability to use multimodal representations to flexibly guide behaviour to multiple sensory modalities, but only that it lacks the capacity to use its limited conscious perceptions to do so.

The existence of modality-specific consciousness under the control of attention in sub-primate mammals fits with the idea that humans can be conscious of the activity of specific cortical modules (such as the colour module in visual cortex) (Zeki and Bartels 1998). Even though we can have much more complex conscious experiences, we might also have simpler experiences that involve the selection of which posterior cortical area to attend to. This capacity might also account for Block's phenomenal consciousness, a perception that one is conscious of but not accessing in awareness (working memory) (Block 2005). However, it is also possible that what Block calls phenomenal consciousness involves the ability of executive functions to attend to an unconscious memory trace in sensory cortex and, for a limited time after the experience, retrieve it into the episodic buffer and thus become aware of it.

The ability to attend to information integrated across multiple modalities and use that kind of representation in thought and action would seem to have important advantages for information processing. Organisms with such a capacity would have survival advantages over those without the capacity. Such advantages may have been driving pressures for the evolution of highly integrated convergence zones in both posterior and prefrontal isocortex, which in turn may have contributed to the evolution of language (Geschwind 1965). The combination of language and convergence, which typifies the human brain, may account for what appears, at least to humans, to be our unique cognitive capacities, including our capacity for being aware of our own brain states.

In sum, just as primates can have multimodal conscious experiences without language, sub-primates may have simpler (unimodal or limited multimodal) experiences without isocortical prefrontal cortex. And while much of human conscious experience may depend on language and prefrontal cortex, we still have the capacity for non-verbal unimodal conscious experiences. However, integrated representations combined with language offer special advantages and may underlie the unique features of human cognition and emotion.

3.3.7 Sorting out the consciousness network

When I first proposed the idea that feelings might involve the representation of a current emotional state in working memory (LeDoux 1992, 1996), the dorsolateral prefrontal cortex was the main brain region believed to mediate the temporary storage and executive functions of working memory. However, as noted above, a number of other areas of prefrontal cortex, including isocortical (orbitofrontal, Broca's area, premotor) and mesocortical (anterior cingulate, insular) areas, and parietal (inferior parietal) cortex have additionally been implicated in various aspects of working memory, attention and other aspects of executive control. The proliferation of brain areas might suggest that something like a global workspace (Baars 1997, 2005; Dehaene et al. 1998, 2006; Dehaene and Naccache 2001) is an appropriate model. However, because the key areas are restricted to limited areas of cortex (mostly prefrontal regions, but including both isocortical and mesocortical prefrontal areas), the designation global seems inappropriate. Nevertheless, the idea that some kind of multimodal workspace that interacts with attention to control thought and action exists seems undeniable, and is in fact a leading hypothesis about consciousness.

A number of regions have been postulated to mediate the multimodal integration underlying consciousness. Most are areas that are part the working memory circuitry, including areas of isocortical prefrontal cortex (e.g. Weiskrantz 1997; Wheeler et al. 1997; Courtney et al. 1998; Knight and Grabowecky 2000; Maia and Cleeremans 2005), anterior cingulate cortex (e.g. Oakley 1999; Reinders et al. 2003; Ochsner et al. 2004), a region of insular cortex called the claustrum (Crick and Koch 2005) and the hippocampus (e.g. O'Keefe 1985; Gray 2004;

Kandel 2006). Each has the connectivity required to perform multimodal integration. Some (especially isocortical prefrontal areas and hippocampus) are known from imaging studies to be activated by multimodal inputs. Some (isocortical prefrontal cortex, anterior cingulate) also participate in attention and executive function. Some are mainly activated during conscious experience (especially isocortical prefrontal areas and insular). The isocortical prefrontal cortex pops up in all of the above categories and would thus seem to be the most likely culprit. However, as noted, damage to this region interferes with but does not eliminate conscious awareness. Perhaps the isocortical prefrontal cortex is required for consciousness in all its glory but there is some redundancy such that undamaged isocortical, neocortical, and/or parietal areas may be able to cobble together less elaborate experiences.

The existence of a workspace made up of multiple areas distributed throughout prefrontal cortex clearly makes it harder to study the functions mediated by such a network. However, there are probably fundamental principles that underlie the involvement of each of the various areas. The challenge is to figure out what the essential areas are, what each does, and what principles underlie their combined involvement. With this information about human consciousness we could then turn to animal studies and ask two kinds of questions. First, which of the key brain areas are present in a given animal? Second, which of the component capacities are they capable of? Dissociations would clarify the link between the function and its neural basis and also shed light on consciousness in other animals.

3.3.8 Minimal requirements of a feeling: raw feels

The conditions outlined above are meant as a description of a full-blown feeling, one that includes representations of stimuli, explicit memories evoked by the stimuli (including semantic, episodic, and self memories), and emotional arousal (including arousal within the brain and feedback from the body). A key question is whether all of these are required, or whether there might be conditions under which a more limited set of requirements might give rise to a feeling.[1] If the latter were true, the study of emotional consciousness might be facilitated since some of the complexities described above might be eliminated. I suspect that very intense sensory stimuli might give rise to feeling without the involvement of explicit semantic or episodic memories of the stimulus memories of the self. For example, an intense painful stimulus might be able

---

[1] This issue was raised by Rafi Malach of the Weizmann Institute in a conversation in August 2006.

to elicit emotional responses that produce strong brainstem arousal and body (proprioceptive) feedback that come to be represented in working memory independent of cognitive awareness. Such signals could create content-empty working memory representations, and might constitute what philosophers have called *raw feels*. In their classic studies, Schachter and Singer (1962) used drugs to produce strong arousal in the absence of content—these feelings were subsequently given meaning through cognitive attributions.

Raw feels might be very similar in diverse species. In species lacking prefrontal cortex and working memory, a raw feel might simply be a strong sensation that is attended to using whatever attentional mechanism the organism has available to it. Obviously, in species lacking language and working memory, attribution of cause to raw feels would be limited or non-existent.

3.4 Conclusion: feeling our way through consciousness

The original title of this chapter included the phrase, 'thinking our way through feelings.' That phrase was meant to suggest that feelings are made possible by processes that create complex representations and that allow attention to be directed to those representations. The inclusion of the phrase 'feeling our way through consciousness' in the title of this concluding section is meant to imply that the study of emotion is a rich area through which consciousness itself might be profitably explored. Most work on the brain mechanisms of consciousness has so far focused on sensory awareness. This has been a useful starting point. However, because sensory awareness is a component of emotional awareness, we can build on the former to study the latter. The reason to take the additional step is that so much of our conscious life is occupied by our feelings, not all of them welcome. The conscious lives of people suffering from anxiety disorders, depression, and related conditions are dominated by negative feelings. Most of what we know about the neural basis of mental problems is based on the unconscious emotional functions of the brain. Research aimed at treating these disorders, especially through drug therapy, is also mostly based on studies of processes that work unconsciously. Studies of emotional consciousness are thus not just of immense intellectual interest but may also offer new opportunities for understanding mental problems and improving the quality of life for people suffering from these conditions.

Acknowledgements

This chapter was written while I was a Visiting Fellow at All Souls College, Oxford, May–June 2006. It was a beautiful spring, and a wonderful experience.

Along with other contributors to this volume, we were the Chichele Fellows of this year, and were otherwise known as the "consciousness crowd." Thanks to Warden John Davis and All Souls College for hosting us, and to Julie Edwards for coordinating all the details. Special thanks to Edmund Rolls and Chris Frith for comments on the text. And thanks also to all the other consciousness fellows (Chris Frith, Cecilia Heyes, Adam Zeman, David Milner, and Martin Davies), the Oxford folks who also participated in the consciousness discussions (Larry Weiskrantz, Nick Shea, and the late Susan Hurley), the real All Souls Fellows, and other Oxford faculty. I still have many pleasant memories of stimulating discussions over dinner, in the garden, and around Oxford.

References

Aboitiz, F. (1995). Working memory networks and the origin of language areas in the human brain. *Medical Hypotheses* **44**, 504–506.

Allan, K., Dolan, R.J., Fletcher, P.C., and Rugg, M.D. (2000). The role of the right anterior prefrontal cortex in episodic retrieval. *NeuroImage* **11**, 217–227.

Amaral, D.G. (1987). Memory: anatomical organization of candidate brain regions. In Plum, F. (ed.) *Handbook of Physiology. Section 1: The Nervous System. Vol. V, Higher Functions of the Brain*, pp. 211–294. Bethesda, MD: American Physiological Society.

Amorapanth, P., LeDoux, J.E., and Nader, K. (2000). Different lateral amygdala outputs mediate reactions and actions elicited by a fear-arousing stimulus. *Nature Neuroscience* **3**, 74–79.

Anderson, A.K. and Phelps, E.A. (2001). Lesions of the human amygdala impair enhanced perception of emotionally salient events. *Nature* **411**, 305–309.

Anderson, A.K., Christoff, K., Panitz, D., De Rosa, E., and Gabrieli, J.D. (2003). Neural correlates of the automatic processing of threat facial signals. *Journal of Neuroscience* **23**, 5627–5633.

Arbib, M.A. (2001). Co-evolution of human consciousness and language. *Annals of the New York Academy of Science* **929**, 195–220.

Armony, J.L. and LeDoux, J.E. (1999). How danger is encoded: Towards a systems, cellular, and computational understanding of cognitive-emotional interactions in fear circuits. In Gazzaniga, M.S. (ed.) *The Cognitive Neurosciences*. Cambridge: MIT Press.

Armony, J.L. and Dolan, R.J. (2002). Modulation of spatial attention by fear-conditioned stimuli: an event-related fMRI study. *Neuropsychologia* **40**, 817–826.

Armony, J.L., Quirk, G.J., and LeDoux, J.E. (1998). Differential effects of amygdala lesions on early and late plastic components of auditory cortex spike trains during fear conditioning. *Journal of Neuroscience* **18**, 2592–2601.

Armony, J.L., Servan-Schreiber, D., Cohen, J.D., and LeDoux, J.E. (1997). Computational modeling of emotion: Explorations through the anatomy and physiology of fear conditioning. *Trends in Cognitive Sciences* **1**, 28–34.

Arnold, M.B. (1960). *Emotion and Personality*. New York: Columbia University Press.

Arnsten, A.F. and Li, B.M. (2005). Neurobiology of executive functions: catecholamine influences on prefrontal cortical functions. *Biological Psychiatry* **57**, 1377–1384.

Aston-Jones, G. and Cohen, J.D. (2005). An integrative theory of locus coeruleus-norepinephrine function: adaptive gain and optimal performance. *Annual Review of Neuroscience* **28**, 403–450.

Atkinson, R.C. and Shiffrin, R.M. (1968). Human memory: a proposed system and its control processes. In Spence, K.W. and Spence, J.T. (eds) *The Psychology of Learning and Motivation*. New York: Academic Press.

Baars, B.J. (1988). *A Cognitive Theory of Consciousness*. New York: Cambridge University Press.

Baars, B.J. (1993). How does a serial, integrated and very limited stream of consciousness emerge from a nervous system that is mostly unconscious, distributed, parallel and of enormous capacity? *Ciba Foundation Symposium* **174**, 282–290; discussion 291–303.

Baars, B.J. (1997). In the theatre of consciousness. *Journal of Consciousness Studies* **4**, 292–309.

Baars, B.J. (2005). Global workspace theory of consciousness: toward a cognitive neuroscience of human experience. *Progress in Brain Research* **150**, 45–53.

Baddeley, A.D. (1986). *Working Memory*. Oxford: Oxford University Press.

Baddeley, A.D. (1993). Verbal and visual subsystems of working memory. *Current Biology* **3**, 563–565.

Baddeley, A. (2000). The episodic buffer: a new component of working memory? *Trends in Cognitive Science* **4**, 417–423.

Baddeley, A. (2003). Working memory and language: an overview. *Journal of Communication Disorders* **36**, 189–208.

Baddeley, A. and Hitch, G.J. (1974). Working memory. In Bower, G. (ed.) *The Psychology of Learning and Motivation, vol.*8. New York: Academic Press.

Baddeley, A., Vargha-Khadem, F., and Mishkin M (2001). Preserved recognition in a case of developmental amnesia: implications for the acquisition of semantic memory? *Journal of Cognitive Neuroscience* **13**, 357–369.

Balleine, B.W. and Dickinson A (1998). Goal-directed instrumental action: contingency and incentive learning and their cortical substrates. *Neuropharmacology* **37**, 407–419.

Balleine, B.W. and Killcross, S. (2006). Parallel incentive processing: an integrated view of amygdala function. *Trends in Neurosciences* **29**, 272–279.

Bargh, J.A. (1992). Being unaware of the stimulus vs. unaware of its interpretation: Why subliminality *per se* does matter to social psychology. In Bornstien, R. and Pittman, T. (eds) *Perception without Awareness*. New York: Guilford.

Barrett, D.J. and Hall, D.A. (2006). Response preferences for 'what' and 'where' in human non-primary auditory cortex. *NeuroImage* **32**, 968–977.

Baynes, K. and Gazzaniga, M.S. (2000). Consciousness, introspection, and the split-brain the two minds/one body problem. In Gazzaniga, M.S. (ed) *The New Cognitive Neurosciences*, 2nd edn, pp. 1355–1364. Cambridge, MA: MIT Press.

Bechara, A., Damasio, H., and Damasio, A.R. (2000). Emotion, decision making and the orbitofrontal cortex. *Cerebral Cortex* **10**, 295–307.

Bechara, A., Damasio, H., Tranel, D., and Damasio, A.R. (2005). The Iowa Gambling Task and the somatic marker hypothesis: some questions and answers. *Trends in Cognitive Science* **9**, 159–162; discussion 162–154.

Beck, D.M., Rees, G., Frith, C.D., and Lavie N (2001). Neural correlates of change detection and change blindness. *Nature Neuroscience* **4**, 645–650.

Blair, H.T., Sotres-Bayon, F., Moita, M.A., and LeDoux, J.E. (2005). The lateral amygdala processes the value of conditioned and unconditioned aversive stimuli. *Neuroscience* **133**, 561–569.

Blanchard, C.D. and Blanchard, R.J. (1972). Innate and conditioned reactions to threat in rats with amygdaloid lesions. *Journal of Comparative Physiological Psychology* **81**, 281–290.

Block, N. (2005). Two neural correlates of consciousness. *Trends in Cognitive Science* **9**, 46–52.

Bolles, R.C. and Fanselow, M.S. (1980). A perceptual-defensive-recuperative model of fear and pain. *Behavioral and Brain Sciences* **3**, 291–323.

Bowerman, M. and Levinson, S.C. (eds) (2001). *Language Acquisition and Conceptual Development*. Cambridge: Cambridge University Press.

Braak, E (1980). On the structure of IIIab-pyramidal cells in the human isocortex. A Golgi and electron microscopical study with special emphasis on the proximal axon segment. *Journal für Hirnforschung* **21**, 437–442.

Broadbent, D.E. (1962). Attention and the perception of speech. *Scientific American* 206(4), 143–151.

Buchel, C. and Dolan, R.J. (2000). Classical fear conditioning in functional neuroimaging. *Current Opinions in Neurobiology* **10**, 219–223.

Buck, R (1980). Nonverbal behavior and the theory of emotion: the facial feedback hypothesis. *Journal of Personal and Social Psychology* **38**, 811–824.

Cabot, J.B. (1996). Some principles of the spinal organization of the sympathetic preganglionic outflow. *Progress in Brain Research* **107**, 29–42.

Cannon, W.B. (1927). The James–Lange theory of emotions: a critical examination and an alternative theory. *American Journal of Psychology* **39**, 106–124.

Cannon, W.B. (1929). *Bodily Changes in Pain, Hunger, Fear, and Rage*. New York: Appleton.

Canteras, N.S., Simerly, R.B., and Swanson, L.W. (1995). Organization of projections from the medial nucleus of the amygdala: a PHAL study in the rat. *Journal of Comparative Neurology* **360**, 213–245.

Cardinal, R.N., Parkinson, J.A., Hall, J., and Everitt, B.J. (2002). Emotion and motivation: the role of the amygdala, ventral striatum, and prefrontal cortex. *Neuroscience and Biobehavioral Reviews* **26**, 321–352.

Carruthers, P. (2002). The cognitive functions of language. *Behavioral and Brain Sciences* **25**, 657–674; discussion 674–725.

Carter, R.M., Hofstotter, C., Tsuchiya, N., and Koch, C. (2003). Working memory and fear conditioning. *Proceedings of the National Academy of Sciences of the USA* **100**, 1399–1404.

Carter, R.M., O'Doherty, J.P., Seymour, B., Koch, C., and Dolan, R.J. (2006). Contingency awareness in human aversive conditioning involves the middle frontal gyrus. *NeuroImage* **29**, 1007–1012.

Clark, A. (1998). *Being There*. Cambridge, MA: MIT Press.

Clark, R.E. and Squire, L.R. (1998). Classical conditioning and brain systems: the role of awareness. *Science* **280**, 77–81.

Clark, R.E., Manns, J.R., and Squire, L.R. (2002). Classical conditioning, awareness, and brain systems. *Trends in Cognitive Science* **6**, 524–531.

Clore, G. (1994). Why emotions are never unconscious. In Ekman, P. and Davidson, R.J. (eds) *The Nature of Emotion: Fundamental Questions*, pp. 285–290. New York: Oxford University Press.

Clore, G. and Ketelaar, T. (1997). Minding our emotions. On the role of automatic unconscious affect. In Wyer, R.S. (ed.) *Advances in Social Cognition*, pp. 105–120. Hillsdale, NJ: Erlbaum.

Cohen, N.J. and Eichenbaum, H. (1993). *Memory, Amnesia, and the Hippocampal System.* Cambridge, MA: MIT Press.

Colby, C.L. and Goldberg, M.E. (1999). Space and attention in parietal cortex. *Annual Review of Neuroscience* **22**, 319–349.

Constantinidis, C. and Procyk, E. (2004). The primate working memory networks. *Cognitive, Affective and Behavioral Neuroscience* **4**, 444–465.

Courtney, S.M., Petit, L., Haxby, J.V., and Ungerleider, L.G. (1998). The role of prefrontal cortex in working memory: examining the contents of consciousness. *Philosophical Transactions of the Royal Society of London Series B Biological Sciences* **353**, 1819–1828.

Craig, A.D. (2003). Interoception: the sense of the physiological condition of the body. *Current Opinions in Neurobiology* **13**, 500–505.

Crick, F.C. and Koch, C. (2005). What is the function of the claustrum? *Philosophical Transactions Royal Society of London Series B Biological Sciences* **360**, 1271–1279.

Critchley, H.D., Mathias, C.J., and Dolan, R.J. (2001). Neuroanatomical basis for first- and second-order representations of bodily states. *Nature Neuroscience* **4**, 207–212.

Critchley, H.D., Mathias, C.J., and Dolan, R.J. (2002). Fear conditioning in humans: the influence of awareness and autonomic arousal on functional neuroanatomy. *Neuron* **33**, 653–663.

Critchley, H.D., Wiens, S., Rotshtein, P., Ohman, A., and Dolan, R.J. (2004). Neural systems supporting interoceptive awareness. *Nature Neuroscience* **7**, 189–195.

Critchley, H.D., Rotshtein, P., Nagai, Y., O'Doherty, J., Mathias, C.J., and Dolan, R.J. (2005). Activity in the human brain predicting differential heart rate responses to emotional facial expressions. *NeuroImage* **24**, 751–762.

Cunningham, W.A., Raye, C.L., and Johnson, M.K. (2004). Implicit and explicit evaluation: FMRI correlates of valence, emotional intensity, and control in the processing of attitudes. *Journal of Cognitive Neuroscience* **16**, 1717–1729.

Curtis, C.E. (2006). Prefrontal and parietal contributions to spatial working memory. *Neuroscience* **139**, 173–180.

Curtis, C.E., Sun, F.T., Miller, L.M., and D'Esposito M (2005). Coherence between fMRI time-series distinguishes two spatial working memory networks. *NeuroImage* **26**, 177–183.

D'Esposito, M., Detre, J., Alsop, D., Shin, R., Atlas, S., and Grossman M (1995). The neural basis of the central executive sytem of working memory. *Nature* **378**, 279–281.

Damasio, A. (1994). *Descartes' Error: Emotion, Reason, and the Human Brain.* New York: Gosset/Putnam.

Damasio, A.R. (1989). The brain binds entities and events by multiregional activation from convergence zones. *Neural Computation* **1**, 123–132.

Damasio, A.R. (1996). The somatic marker hypothesis and the possible functions of the prefrontal cortex. *Philosophical Transactions of the Royal Society of London Series B Biological Sciences* **351**, 1413–1420.

Damasio, A.R. (1999). *The Feeling of What Happens: Body and Emotion in the Making of Consciousness*. New York: Harcourt Brace.

Damasio, A.R., Tranel, D., and Damasio H (1996). Somatic markers and the guidance of behavior: theory and preliminary testing. In Levin, H., Eisenberg, H., and Benton, A. (eds) *Frontal Lobe Function and Injury*, pp. 2–23. New York: Oxford University Press.

Damasio, A.R., Grabowski, T.J., Bechara, A., Damasio, H., Ponto, L.L., Parvizi, J., and Hichwa, R.D. (2000). Subcortical and cortical brain activity during the feeling of self-generated emotions. *Nature Neuroscience* **3**, 1049–1056.

Darwin, C. (1872) *The Expression of the Emotions in Man and Animals*. London: John Murray.

Davidson, R.J. (2002). Anxiety and affective style: role of prefrontal cortex and amygdala. *Biological Psychiatry* **51**, 68–80.

Davis, M. (1992). The role of the amygdala in conditioned fear. In Aggleton, J.P. (ed.) *The Amygdala: Neurobiological Aspects of Emotion, Memory, and Mental Dysfunction*, pp. 255–306. New York: Wiley-Liss

Davis, M. and Whalen, P.J. (2001). The amygdala: vigilance and emotion. *Molecular Psychiatry* **6**, 13–34.

de Gelder, B., Morris, J.S., and Dolan, R.J. (2005). Unconscious fear influences emotional awareness of faces and voices. *Proceedings of the National Academy of Sciences of the USA* **102**, 18682–18687.

de Gelder, B., Vroomen, J., Pourtois, G., and Weiskrantz, L. (1999). Non-conscious recognition of affect in the absence of striate cortex. *Neuroreport* **10**, 3759–3763.

deGroat, W.C., Vizzard, M.A., Araki, I., and Roppolo J (1996). Spinal interneurons and preganglionic neurons in sacral autonomic reflex pathways. *Progress in Brain Research* **107**, 97–111.

Dehaene, S. and Naccache L (2001). Towards a cognitive neuroscience of consciousness: basic evidence and a workspace framework. *Cognition* **79**, 1–37.

Dehaene, S., Kerszberg, M., and Changeux, J.P. (1998). A neuronal model of a global workspace in effortful cognitive tasks. *Proceedings of the National Academy of Sciences of the USA* **95**, 14529–14534.

Dehaene, S., Changeux, J.P., Naccache, L., Sackur, J., and Sergent, C. (2006). Conscious, preconscious, and subliminal processing: a testable taxonomy. *Trends in Cognitive Science* **10**, 204–211.

Della Sala, S., Gray, C., Baddeley, A., Allamano, N., and Wilson, L. (1999). Pattern span: a tool for unwelding visuo-spatial memory. *Neuropsychologia* **37**, 1189–1199.

Dennett, D.C. (1991). *Consciousness Explained*. Boston: Little, Brown.

Desimone, R. and Duncan, J. (1995). Neural mechanisms of selective visual attention. *Annual Review of Neuroscience* **18**, 193–222.

Diorio, D., Viau, V., and Meaney, M.J. (1993). The role of the medial prefrontal cortex (cingulate gyrus) in the regulation of hypothalamic-pituitary-adrenal responses to stress. *Journal of Neuroscience* **13**, 3839–3847.

Dolan, R.J. and Vuilleumier P (2003). Amygdala automaticity in emotional processing. *Annals of the New York Academy of Science* **985**, 348–355.

Dolcos, F. and McCarthy, G. (2006). Brain systems mediating cognitive interference by emotional distraction. *Journal of Neuroscience* **26**, 2072–2079.

Dolcos, F., LaBar, K.S., and Cabeza, R. (2004). Dissociable effects of arousal and valence on prefrontal activity indexing emotional evaluation and subsequent memory: an event-related fMRI study. *NeuroImage* **23**, 64–74.

Dunn, B.D., Dalgleish, T., Ogilvie, A.D., and Lawrence, A.D. (2007). Heartbeat perception in depression. *Behaviour Research and Therapy* **45**(8), 1921–1930.

Durstewitz, D., Seamans, J.K., and Sejnowski, T.J. (2000). Dopamine-mediated stabilization of delay-period activity in a network model of prefrontal cortex. *Journal of Neurophysiology* **83**, 1733–1750.

Eichenbaum, H. (2002). *The Cognitive Neuroscience of Memory*. New York: Oxford University Press.

Eichenbaum, H. and Fortin, N.J. (2005). Bridging the gap between brain and behavior: cognitive and neural mechanisms of episodic memory. *Journal of the Experimental Analysis of Behavior* **84**, 619–629.

Ekman, P. (1980). Biological and cultural contributions to body and facial movement in the expression of emotions. In Rorty, A.O. (ed.) *Explaining Emotions*. Berkeley, CA: University of California Press.

Ekman, P. (1992). An argument for basic emotions. *Cognition and Emotion* **6**, 169–200.

Ekman, P. (1993). Facial expression and emotion. *American Psychologist* **48**, 4384–4392.

Erdelyi, M.H. (1985). *Psychoanalysis: Freud's Cognitive Psychology*. New York: W.H. Freeman.

Etkin, A., Klemenhagen, K.C., Dudman, J.T., Rogan, M.T., Hen, R., Kandel, E.R., and Hirsch, J. (2004). Individual differences in trait anxiety predict the response of the basolateral amygdala to unconsciously processed fearful faces. *Neuron* **44**, 1043–1055.

Everitt, B.J., Parkinson, J.A., Olmstead, M.C., Arroyo, M., Robledo, P., and Robbins, T.W. (1999). Associative processes in addiction and reward. The role of amygdala–ventral striatal subsystems. In McGintry, J. (ed.) *Advancing from the Ventral Striatum to the Extended Amygdala*, pp. 412–438. New York: New York Academy of Sciences.

Fanselow, M.S. and Gale, G.D. (2003). The amygdala, fear, and memory. *Annals of the New York Academy of Science* **985**, 125–134.

Fanselow, M.S. and Poulos, A.M. (2005). The neuroscience of mammalian associative learning. *Annual Review of Psychology* **56**, 207–234.

Feinstein, J.S., Stein, M.B., Castillo, G.N., and Paulus, M.P. (2004). From sensory processes to conscious perception. *Conscious Cognition* **13**, 323–335.

Fendt, M. and Fanselow, M.S. (1999). The neuroanatomical and neurochemical basis of conditioned fear. *Neuroscience and Biobehavior Review* **23**, 743–760.

Festinger, L. (1957). *A Theory of Cognitive Dissonance*. Evanston, IL: Row Peterson.

Fletcher, P.C., Frith, C.D., and Rugg, M.D. (1997). The functional neuroanatomy of episodic memory. *Trends in Neuroscience* **20**, 213–218.

Foa, E.B. (2006). Psychosocial therapy for posttraumatic stress disorder. *Journal of Clinical Psychiatry* **67** Suppl **2**, 40–45.

Fodor, J. (1975). *The Language of Thought*. Cambridge, MA: Harvard University Press.

Fodor, J. (1983). *The Modularity of Mind*. Cambridge, MA: MIT Press.

Fortin, N.J., Wright, S.P., and Eichenbaum H (2004). Recollection-like memory retrieval in rats is dependent on the hippocampus. *Nature* **431**, 188–191.

Fossati, P., Hevenor, S.J., Graham, S.J., Grady, C., Keightley, M.L., Craik, F., and Mayberg, H. (2003). In search of the emotional self: an, F.M.RI study using positive and negative emotional words. *American Journal of Psychiatry* **160**, 1938–1945.

Frijda, N. (1986). *The Emotions*. Cambridge: Cambridge University Press.

Frith, C. and Dolan, R. (1996). The role of the prefrontal cortex in higher cognitive functions. *Brain Research: Cognitive Brain Research* **5**, 175–181.

Frith, C., Perry, R., and Lumer E (1999). The neural correlates of conscious experience: an experimental framework. *Trends in Cognitive Sciences* **3**, 105–114.

Fuster, J. (1997). *The Prefrontal Cortex: Anatomy, Physiology, and Neuropsychology of the Frontal Lobe*, 3rd edn. Philadelphia: Lippincott-Raven.

Fuster, J.M. (2000). The prefrontal cortex of the primate: a synopsis. *Psychobiology* **28**, 125–131.

Fuster, J.M. (2003). *Cortex and Mind: Unifying Cognition*. Oxford: Oxford University Press.

Gazzaniga, M.S. (1970). *The Bisected Brain*. New York: Appleton-Century-Crofts.

Gazzaniga, M.S. (1985). *The Social Brain*. New York: Basic Books.

Gazzaniga, M.S. (1988). Brain modularity: Towards a philosophy of conscious experience. In Marcel, A.J. and Bisiach, E. (eds) *Consciousness in Contemporary Science*. Oxford: Clarendon Press.

Gazzaniga, M.S. and LeDoux, J.E. (1978). *The Integrated Mind*. New York: Plenum.

Gazzaniga, M.S., Wilson, D.H., and LeDoux, J.E. (1977). Language praxis, and the right hemisphere: clues to some mechanisms of consciousness. *Neurology* **27**, 1144–1147.

Gentner, D. and Goldin-Meadow, S. (eds) (2003). *Language in Mind: Advances in the Study of Language and Thought*. Cambridge: MIT Press.

Geschwind, N. (1965a). The disconnexion syndromes in animals and man. Part I. *Brain* **88**, 237–294.

Geschwind, N. (1965b). The disconnexion syndromes in animals and man. Part II. *Brain* **88**, 585–644.

Ghashghaei, H.T. and Barbas H (2002). Pathways for emotion: interactions of prefrontal and anterior temporal pathways in the amygdala of the rhesus monkey. *Neuroscience* **115**, 1261–1279.

Goddard, G. (1964). Functions of the amygdala. *Psychological Reviews* **62**, 89–109.

Gold, P.E., van Buskirk, R.B., and McGaugh, J.L. (1975). Effects of hormones on time-dependent memory storage processes. *Progress in Brain Research* **42**, 210–211.

Goldberg, I.I., Harel, M., and Malach, R, (2006). When the brain loses its self: prefrontal inactivation during sensorimotor processing. *Neuron* **50**, 329–339.

Goldman-Rakic, P.S. (1987). Circuitry of primate prefrontal cortex and regulation of behavior by representational memory. In Plum, F. (ed.) *Handbook of Physiology. Section 1: The Nervous System. Vol. V, Higher Functions of the Brain*, pp. 373–418. Bethesda, MD: American Physiological Society.

Goldman-Rakic, P.S. (1996). Regional and cellular fractionation of working memory. *Proceedings of the National Academy of Sciences of the USA* **93**, 13473–13480.

Gray, J.A. (1987). *The Psychology of Fear and Stress*. New York: Cambridge University Press.

Gray, J.A. (2004). *Consciousness: Creeping Up on the Hard Problem*. Oxford: Oxford University Press.

Greenberg, D.L., Rice, H.J., Cooper, J.J., Cabeza, R., Rubin, D.C., and Labar, K.S. (2005). Co-activation of the amygdala, hippocampus and inferior frontal gyrus during autobiographical memory retrieval. *Neuropsychologia* **43**, 659–674.

Greenwald, A.G. (1992). New look 3: Unconscious cognition reclaimed. American Psychologist **47**, 766–779.

Griffin, D.R. and Speck, G.B. (2004). New evidence of animal consciousness. *Animal Cognition* **7**, 5–18.

Gusnard, D.A., Akbudak, E., Shulman, G.L., and Raichle, M.E. (2001). Medial prefrontal cortex and self-referential mental activity: relation to a default mode of brain function. *Proceedings of the National Academy of Sciences of the USA* **98**, 4259–4264.

Hall, G. and Pearce, J.M. (1979). Latent inhibition of a CS during CS–US pairings. *Journal of Experimental Psychology. Animal Behavior Processes* **5**, 31–42.

Hamm, A.O., Weike, A.I., Schupp, H.T., Treig, T., Dressel, A., and Kessler C (2003). Affective blindsight: intact fear conditioning to a visual cue in a cortically blind patient. *Brain* **126**, 267–275.

Harré, R. (1986). *The Social Construction of Emotions.* New York: Blackwell.

Hassin, R.R., Uleman, J.S., and Bargh, J.A. (eds) (2005). *The New Unconscious.* New York: Oxford University Press.

Heims, H.C., Critchley, H.D., Dolan, R., Mathias, C.J., and Cipolotti, L. (2004). Social and motivational functioning is not critically dependent on feedback of autonomic responses: neuropsychological evidence from patients with pure autonomic failure. *Neuropsychologia* **42**, 1979–1988.

Hess, W.R. and Brugger, M. (1943). Das subkortikale Zentrum der affektiven Abwehrreaktion. Helvetica Physiologica Pharmacologica Acta **1**, 35–52.

Hinson, J.M., Jameson, T.L., and Whitney P (2002). Somatic markers, working memory, and decision making. *Cognitive, Affective and Behavioral Neuroscience* **2**, 341–353.

Holland, P.C. and Gallagher, M. (1999). Amygdala circuitry in attentional and representational processes. *Trends in Cognitive Sciences* **3**, 65–73.

Huerta, P.T., Sun, L.D., Wilson, M.A., and Tonegawa S (2000). Formation of temporal memory requires NMDA receptors within CA1 pyramidal neurons. Neuron **25**, 473–480.

Izard, C.E. (1971). *The Face of Emotion.* New York: Appleton-Century-Crofts.

Jacoby, L.L. (1991). A process dissociation framework: separating automatic from intentional uses of memory. *Journal of Memory and Learning* **30**, 513–541.

James, W. (1884) What is an emotion? *Mind* **9**, 188–205.

James, W. (1890) *Principles of Psychology.* New York: Holt.

Johansson, P., Hall, L., Skistrom, S., Tarning, B., and Lind, A. (2006). How something can be said about telling more than we know: on choice blindness and introspection. *Consciousness and Cognition* **15**, 673–692.

Johnson, S.C., Baxter, L.C., Wilder, L.S., Pipe, J.G., Heiserman, J.E., and Prigatano, G.P. (2002). Neural correlates of self-reflection. *Brain* **125**, 1808–1814.

Johnson-Laird, P.N. (1988). *The Computer and the Mind: An Introduction to Cognitive Science.* Cambridge, MA: Harvard University Press.

Jones, E.G. and Powell, T.P.S (1970). An anatomical study of converging sensory pathways within the cerebral cortex of the monkey. *Brain* **93**, 793–820.

Kaada, B.R. (1960). Cingulate, posterior orbital, anterior insular and temporal pole cortex. In Field, J., Magoun, H.J., and Hall, V.E. (eds) *Handbook of Physiology: Neurophysiology II*, pp. 1345–1372. Washington, DC: American Physiological Society.

Kalisch, R., Wiech, K., Critchley, H.D., and Dolan, R.J. (2006). Levels of appraisal: a medial prefrontal role in high-level appraisal of emotional material. *NeuroImage* 30, 1458–1466.

Kandel, E.R. (2006). *In Search of Memory: The Emergence of a New Science of Mind*. New York: W.W. Norton.

Kanwisher, N., McDermott, J., and Chun, M.M. (1997). The fusiform face area: a module in human extrastriate cortex specialized for face perception. *Journal of Neuroscience* 17, 4302–4311.

Kapp, B.S., Whalen, P.J., Supple, W.F., and Pascoe, J.P. (1992). Amygdaloid contributions to conditioned arousal and sensory information processing. In Aggleton, J.P. (ed) *The Amygdala: Neurobiological Aspects of Emotion, Memory, and Mental Dysfunction*, pp. 229–254. New York: Wiley-Liss.

Keenan, J.P., Wheeler, M.A., Gallup, G.G., Jr., and Pascual-Leone, A. (2000). Self-recognition and the right prefrontal cortex. *Trends in Cognitive Science* 4, 338–344.

Kelley, H.H. and Michela, J.L. (1980). Attribution theory and research. *Annual Review of Psychology* 31, 457–501.

Kelley, W.M., Macrae, C.N., Wyland, C.L., Caglar, S., Inati, S., and Heatherton, T.F. (2002). Finding the self? An event-related fMRI study. *Journal of Cognitive Neuroscience* 14, 785–794.

Kennedy, P.J. and Shapiro, M.L. (2004). Retrieving memories via internal context requires the hippocampus. *Journal of Neuroscience* 24, 6979–6985.

Kihlstrom, J.F. (1987). The cognitive unconscious. *Science* 237, 1445–1452.

Killcross, S., Robbins, T.W., and Everitt, B.J. (1997). Different types of fear-conditioned behaviour mediated by separate nuclei within amygdala. *Nature* 388, 377–380.

Kitayama, S. and Markus, H.R. (eds) (1994). *Emotion and Culture: Empirical Studies of Mutual Influence*. Washington, DC: American Psychological Association.

Klüver, H. and Bucy, P.C. (1937). 'Psychic blindness' and other symptoms following bilateral temporal lobectomy in rhesus monkeys. *American Journal of Physiology* 119, 352–353.

Knight, R.T. and Grabowecky, M. (2000). Prefrontal cortex, time and consciousness. In Gazzaniga, M.S. (ed.) *The New Cognitive Neurosciences*. Cambridge, MA: MIT Press.

Kotter, R. and Meyer, N. (1992). The limbic system: a review of its empirical foundation. *Behavioural Brain Research* 52, 105–127.

LaBar, K.S. (2003). Emotional memory functions of the human amygdala. *Current Neurology and Neuroscience Reports* 3, 363–364.

LaBar, K.S. and Cabeza R (2006). Cognitive neuroscience of emotional memory. *Nature Reviews. Neuroscience* 7, 54–64.

LaBar, K.S. and Disterhoft, J.F. (1998). Conditioning, awareness, and the hippocampus. *Hippocampus* 8, 620–626.

Landau, B. and Jackendoff, R. (1993). 'What' and 'where' in spatial language and spatial cognition. *Behavioral and Brain Sciences* 16, 217–238, 255–265.

Lang, P.J., Davis, M., and Ohman A (2000). Fear and anxiety: animal models and human cognitive psychophysiology. *Journal of Affective Disorders* 61, 137–159.

Lazarus, R.S. (1966). *Psychological Stress and the Coping Process*. New York: McGraw Hill.

Lazarus, R.S. (1984). On the primacy of cognition. *American Psychologist* 39, 124–129.

Lazarus, R.S. (1991). Cognition and motivation in emotion. *American Psychologist* **46**, 352–367.

Lazarus, R. and McCleary, R. (1951). Autonomic discrimination without awareness: a study of subception. *Psychological Review* **58**, 113–122.

LeDoux, J.E. (1984). Cognition and emotion: processing functions and brain systems. In Gazzaniga, M.S. (ed.) *Handbook of Cognitive Neuroscience*, pp. 357–368. New York: Plenum.

LeDoux, J.E. (1985). Brain, mind, and language. In Oakley, D.A. (ed.) *Brain and Mind*. London: Methuen.

LeDoux, J.E. (1987). Emotion. In Plum, F. (ed.) *Handbook of Physiology*. 1: *The Nervous System. Vol. V, Higher Functions of the Brain*, pp. 419–460. Bethesda, MD: American Physiological Society.

LeDoux, J.E. (1991). Emotion and the limbic system concept. *Concepts in Neuroscience* **2**, 169–199.

LeDoux, J.E. (1992). Emotion and the amygdala. In Aggleton, J.P. (ed.) *The Amygdala: Neurobiological Aspects of Emotion, Memory, and Mental Dysfunction*, pp. 339–351. New York: Wiley-Liss.

LeDoux, J.E. (1996). *The Emotional Brain*. New York: Simon & Schuster.

LeDoux, J.E. (2000). Emotion circuits in the brain. *Annual Review of Neuroscience* **23**, 155–184.

LeDoux, J.E. (2002). *Synaptic Self: How Our Brains Become Who We Are*. New York: Viking.

LeDoux, J.E. and Gorman, J.M. (2001). A call to action: overcoming anxiety through active coping. *American Journal of Psychiatry* **158**, 1953–1955.

LeDoux, J.E., Wilson, D.H., and Gazzaniga, M.S. (1977). A divided mind: observations on the conscious properties of the separated hemispheres. *Annals of Neurology* **2**, 417–421.

Leventhal, H. and Scherer K (1987). The relationship of emotion to cognition: a functional approach to a semantic controversy. *Cognition and Emotion* **1**, 3–28.

Levine, B., Black, S.E., Cabeza, R., Sinden, M., McIntosh, A.R., Toth, J.P., Tulving, E., and Stuss, D.T. (1998). Episodic memory and the self in a case of isolated retrograde amnesia. *Brain* **121**(10), 1951–1973.

Levinson, S.C., Kita, S., Haun, D.B., and Rasch, B.H. (2002). Returning the tables: language affects spatial reasoning. *Cognition* **84**, 155–188.

Logothetis, N.K. (1998). Single units and conscious vision. *Philosophical Transactions Royal Society of London Series B Biological Sciences* **353**, 1801–1818.

Lucy, J.A. (1992). *Language Diversity and Thought: A Reformulation of the Linguistic Relativity Hypothesis*. Cambridge: Cambridge University Press.

Lutz, C.A. (1988). *Unnatural Emotions*. Chicago: University of Chicago Press.

Mack, A., Tang, B., Tuma, R., Kahn, S., and Rock, I. (1992). Perceptual organization and attention. *Cognitive Psychology* **24**, 475–501.

Mackintosh, N.J. (1965). Selective attention in animal discrimination learning. *Psychological Bulletin* **64**, 124–150.

MacLean, P.D. (1949). Psychosomatic disease and the 'visceral brain': recent developments bearing on the Papez theory of emotion. *Psychosomatic Medicine* **11**, 338–353.

MacLean, P.D. (1952). Some psychiatric implications of physiological studies on frontotemporal portion of limbic system (visceral brain). *Electroencephalography and Clinical Neurophysiology* **4**, 407–418.

Maia, T.V. and Cleeremans, A. (2005). Consciousness: converging insights from connectionist modeling and neuroscience. *Trends in Cognitive Science* **9**, 397–404.

Maia, T.V. and McClelland, J.L. (2004). A reexamination of the evidence for the somatic marker hypothesis: what participants really know in the Iowa gambling task. *Proceedings of the National Academy of Sciences of the USA* **101**, 16075–16080.

Maratos, E.J., Dolan, R.J., Morris, J.S., Henson, R.N., and Rugg, M.D. (2001). Neural activity associated with episodic memory for emotional context. *Neuropsychologia* **39**, 910–920.

Maren, S. (2001). Neurobiology of Pavlovian fear conditioning. *Annual Review of Neuroscience* **24**, 897–931.

Marks, I. (1987). *Fears, Phobias, and Rituals: Panic, Anxiety and Their Disorders*. New York: Oxford University Press.

Marks, I. and Tobena, A. (1990). Learning and unlearning fear: a clinical and evolutionary perspective. *Neuroscience and Biobehavioral Reviews* **14**, 365–384.

Maunsell, J.H.R (1995). The brain's visual world: Representation of visual targets in cerebral cortex. *Science* **270**, 764–769.

McAllister, W.R. and McAllister, D.E. (1971). Behavioral measurement of conditioned fear. In Brush, F.R. (ed.) *Aversive Conditioning and Learning*, pp. 105–179. New York: Academic Press.

McDonald, A.J. (1998). Cortical pathways to the mammalian amygdala. *Progress in Neurobiology* **55**, 257–332.

McGaugh, J.L. (2000). Memory—a century of consolidation. *Science* **287**, 248–251.

Merikle, P.M. (1992). Perception without awareness. Critical issues. *American Psychologist* **47**, 792–795.

Mesquita, B. and Frijda, N.H. (1992). Cultural variations in emotions: a review. *Psychological Bulletin* **112**, 179–204.

Mesulam, M.M. (1999). Spatial attention and neglect: parietal, frontal and cingulate contributions to the mental representation and attentional targeting of salient extrapersonal events. *Philosophical Transactions of the Royal Society of London Series B Biological Sciences* **354**, 1325–1346.

Mesulam, M.M., van Hoesen, G., Pandya, D.N., and Geschwind, N. (1977). Limbic and sensory connections of the inferior parietal lobule (area pg) in the rhesus monkey: a study with a new method for horseradish peroxidase histochemistry. *Brain Research* **136**, 393–414.

Millenson, J.R. (1967). *Principles of Behavioral Analysis*. New York: Macmillan.

Miller, E.K. and Cohen, J.D. (2001). An integrative theory of prefrontal cortex function. *Annual Review of Neuroscience* **24**, 167–202.

Miller, G. (1956). The magical number seven, plus or minus two: Some limits on our capacity for processing information. *Psychological Review* **63**, 81–97.

Miller, N.E. (1948). Studies of fear as an acquirable drive: I. Fear as motivation and fear reduction as reinforcement in the learning of new responses. *Journal of Experimental Psychology* **38**, 89–101.

Milner, A.D. and Goodale, M.A. (1995). *The Visual Brain in Action*. New York: Oxford University Press.

Milner, B. (2005). The medial temporal-lobe amnesic syndrome. *Psychiatric Clinics of North America* **28**, 599–611, 609.

Milner, B., Squire, L.R., and Kandel, E.R. (1998). Cognitive neuroscience and the study of memory. *Neuron* **20**, 445–468.

Mishkin, M. (1982). A memory system in the monkey. *Philosophical Transactions of the Royal Society of London Series B Biological Sciences* **298**, 85–95.

Mishkin, M. and Aggleton, J. (1981). Multiple functional contributions of the amygdala in the monkey. In Ben-Ari, Y. (ed.) *The Amygdaloid Complex*, pp. 409–420. Amsterdam: Elsevier/North-Holland.

Mishkin, M., Vargha-Khadem, F., and Gadian, D.G. (1998). Amnesia and the organization of the hippocampal system. *Hippocampus* **8**, 212–216.

Mitchell, K.J., Johnson, M.K., Raye, C.L., and Greene, E.J. (2004). Prefrontal cortex activity associated with source monitoring in a working memory task. *Journal of Cognitive Neuroscience* **16**, 921–934.

Moray, N. (1959). Attention in dichotic listening: affective cues and the influence of instructions. *Quarterly Journal of Experimental Psychology* **11**, 56–60.

Morris, J.S., Ohman, A., and Dolan, R.J. (1998). Conscious and unconscious emotional learning in the human amygdala. *Nature* **393**, 467–470.

Morris, J.S., Ohman, A., and Dolan, R.J. (1999). A subcortical pathway to the right amygdala mediating 'unseen' fear. *Proceedings of the National Academy of Sciences of the USA* **96**, 1680–1685.

Morris, J.S., Buchel, C., and Dolan, R.J. (2001a). Parallel neural responses in amygdala subregions and sensory cortex during implicit fear conditioning. *NeuroImage* **13**, 1044–1052.

Morris, J.S., DeGelder, B., Weiskrantz, L., and Dolan, R.J. (2001b). Differential extrageniculostriate and amygdala responses to presentation of emotional faces in a cortically blind field. *Brain* **124**, 1241–1252.

Morris, J.S., deBonis, M., and Dolan, R.J. (2002). Human amygdala responses to fearful eyes. *NeuroImage* **17**, 214–222.

Moutoussis, K., Keliris, G., Kourtzi, Z., and Logothetis N (2005). A binocular rivalry study of motion perception in the human brain. *Vision Research* **45**, 2231–2243.

Mowrer, O.H. (1939). A stimulus-response analysis of anxiety and its role as a reinforcing agent. *Psychological Review* **46**, 553–565.

Moyer, J.R., Deyo, R.A., and Disterhoft, J.F. (1990). Hippocampectomy disrupts trace eyeblink conditioning in rabbits. *Behavioral Neuroscience* **104**, 243–252.

Muller, N.G. and Knight, R.T. (2006). The functional neuroanatomy of working memory: contributions of human brain lesion studies. *Neuroscience* **139**, 51–58.

Nader, K. and LeDoux, J.E. (1997). Is it time to invoke multiple fear learning system? *Trends in Cognitive Science* **1**, 241–244.

Nauta, W.J.H. and Karten, H.J. (1970). A general profile of the vertebrate brain, with sidelights on the ancestry of cerebral cortex. In Schmitt, F.O. (ed.) *The Neurosciences: Second Study Program*, pp. 7–26. New York: Rockefeller University Press.

Nicotra, A., Critchley, H.D., Mathias, C.J., and Dolan, R.J. (2006). Emotional and autonomic consequences of spinal cord injury explored using functional brain imaging. *Brain* **129**, 718–728.

Nisbett, R.E. and Wilson, T.D. (1977). Telling more than we can know: verbal reports on mental processes. *Psychological Review* **84**, 231–259.

Norman, D.A. and Shallice T (1980). Attention to action: willed and automatic control of behavior. In Davidson, R.J., Schwartz, G.E., and Shapiro, D. (eds) *Consciousness and Self-regulation*, pp. 1–18. New York: Plenum.

North, N.T. and O'Carroll, R.E. (2001). Decision making in patients with spinal cord damage: afferent feedback and the somatic marker hypothesis. *Neuropsychologia* **39**, 521–524.

Nyberg, L., Ar, M., Houle, S., Nilsson, L.-G., and Tulving, E. (1996). Activation of medial temporal structures during episodic memory retrieval. *Nature* **380**, 715–717.

O'Keefe, J. (1985). Is consciousness the gateway to the hippocampal cognitive map? A speculative essay on the neural basis of mind. In Oakley, D.A. (ed.) *Brain and Mind*. New York: Methuen.

Oakley, D.A. (1999). Hypnosis and conversion hysteria: a unifying model. *Cognitive Neuropsychiatry* **4**, 243–265.

Ochsner, K. N. & Gross, J. J. (2005). The cognitive control of emotion. *Trends in Cognitive Sciences*, **9**(5), 242-249.

Ochsner, K.N., Ray, R.D., Cooper, J.C., Robertson, E.R., Chopra, S., Gabrieli, J.D., and Gross, J.J. (2004). For better or for worse: neural systems supporting the cognitive down- and up-regulation of negative emotion. *NeuroImage* **23**, 483–499.

Öhman, A. (2005). The role of the amygdala in human fear: automatic detection of threat. *Psychoneuroendocrinology* **30**, 953–958.

Öhman, A. and Mineka, S. (2001). Fears, phobias, and preparedness: toward an evolved module of fear and fear learning. *Psychological Reviews* **108**, 483–522.

Öhman, A. and Soares, J.J.F (1994). 'Unconscious anxiety': phobic responses to masked stimuli. *Journal of Abnormal Psychology* **103**, 231–240.

Öhman, A. and Soares, J.J. (1998). Emotional conditioning to masked stimuli: expectancies for aversive outcomes following nonrecognized fear-relevant stimuli. *Journal of Experimental Psychology. General* **127**, 69–82.

Olsson, A. and Phelps, E.A. (2004). Learned fear of 'unseen' faces after Pavlovian, observational, and instructed fear. *Psychological Science* **15**, 822–828.

Ortony, A., Clore, G.L., and Collins A (1988). *The Cognitive Structure of Emotions*. Cambridge University Press: Cambridge.

Panksepp, J. (1998). *Affective Neuroscience*. New York: Oxford University Press.

Papez, J.W. (1937). A proposed mechanism of emotion. *Archives of Neurology and Psychiatry* **79**, 217–224.

Pasternak, T. and Greenlee, M.W. (2005). Working memory in primate sensory systems. *Nature Reviews. Neuroscience* **6**, 97–107.

Pessoa, L., Kastner, S., and Ungerleider, L.G. (2002). Attentional control of the processing of neural and emotional stimuli. *Brain Research. Cognitive Brain Research* **15**, 31–45.

Petrovich, G.D., Risold, P.Y., and Swanson, L.W. (1996). Organization of projections from the basomedial nucleus of the amygdala: a PHAL study in the rat. *Journal of Comparative Neurology* **374**, 387–420.

Phelps, E.A. (2006). Emotion and cognition: insights from studies of the human amygdala. *Annual Review of Psychology* **57**, 27–53.

Phelps, E.A. and LeDoux, J.E. (2005). Contributions of the amygdala to emotion processing: from animal models to human behavior. *Neuron* **48**, 175–187.

Phelps, E.A., O'Connor, K.J., Cunningham, W.A., Funayama, E.S., Gatenby, J.C., Gore, J.C., and Banaji, M.R. (2000). Performance on indirect measures of race evaluation predicts amygdala activation. *Journal of Cognitive Neuroscience* **12**, 729–738.

Phelps, E.A., Delgado, M.R., Nearing, K.I., and LeDoux, J.E. (2004). Extinction learning in humans; role of the amygdala and vmPFC. *Neuron* **43**, 897–905.

Pinker, S. (1994). *The Language Instinct: How the Mind Creates Language*. New York: William Morrow and Co.

Pitkänen, A., Savander, V., and LeDoux, J.E. (1997). Organization of intra-amygdaloid circuitries in the rat: an emerging framework for understanding functions of the amygdala. *Trends in Neuroscience* **20**, 517–523.

Platt, M.L. and Glimcher, P.W. (1999). Neural correlates of decision variables in parietal cortex. *Nature* **400**, 233–238.

Plutchik, R. (1980). *Emotion: A Psychoevolutionary Synthesis*. New York: Harper & Row.

Posner, M.I. (1994). Attention: the mechanisms of consciousness. *Proceedings of the National Academy of Sciences of the USA* **91**, 7398–7403.

Posner, M.I. and Dehaene S (1994). Attentional networks. *Trends in Neuroscience* **1**, 75–79.

Posner, M.I. and Snyder, C.R.R (1975). Attention and cognitive control. In Solso, R.L. (ed.) *Information Processing and Cognition*, pp. 55–85. Hillsdale, NJ: Erlbaum.

Quirk, G.J., Russo, G.K., Barron, J.L., and Lebron, K. (2000). The role of ventromedial prefrontal cortex in the recovery of extinguished fear. *Journal of Neuroscience* **20**, 6225–6231.

Radley, J.J. and Morrison, J.H. (2005). Repeated stress and structural plasticity in the brain. *Ageing Research Reviews* **4**, 271–287.

Ranganath, C., Johnson, M.K., and D'Esposito M (2003). Prefrontal activity associated with working memory and episodic long-term memory. *Neuropsychologia* **41**, 378–389.

Reep, R. (1984). Relationship between prefrontal and limbic cortex: a comparative anatomical review. *Brain, Behavior and Evolution* **25**, 5–80.

Rees, G., Kreiman, G., and Koch, C. (2002). Neural correlates of consciousness in humans. *Nature Reviews. Neuroscience* **3**, 261–270.

Reinders, A.A., Nijenhuis, E.R., Paans, A.M., Korf, J., Willemsen, A.T., and den Boer, J.A. (2003). One brain, two selves. *NeuroImage* **20**, 2119–2125.

Repovs, G. and Baddeley, A. (2006). The multi-component model of working memory: explorations in experimental cognitive psychology. *Neuroscience* **139**, 5–21.

Robbins, T.W. (2005). Chemistry of the mind: neurochemical modulation of prefrontal cortical function. *Journal of Comparative Neurology* **493**, 140–146.

Robinson, M.D. (1998). Running from William James' bear: a review of preattentive mechanisms and their contributions to emotional experience. *Cognition and Emotion* **12**, 667–696.

Robinson, M.D. and Clore, G.L. (2002). Episodic and semantic knowledge in emotional self-report: evidence for two judgment processes. *Journal of Personal and Social Psychology* **83**, 198–215.

Rolls, E.T. (1999). *The Brain and Emotion*. Oxford: Oxford University Press.

Rolls, E.T. (2005). *Emotion Explained*. New York: Oxford University Press.

Romanski, L.M. (2004). Domain specificity in the primate prefrontal cortex. *Cognitive, Affective and Behavioral Neuroscience* **4**, 421–429.

Romanski, L.M., Tian, B., Fritz, J., Mishkin, M., Goldman-Rakic, P.S., and Rauschecker, J.P. (1999). Dual streams of auditory afferents target multiple domains in the primate prefrontal cortex. *Nature Neuroscience* **2**, 1131–1136.

Rosen, J.B. and Donley, M.P. (2006). Animal studies of amygdala function in fear and uncertainty: relevance to human research. *Biological Psychology* **73**, 49–60.

Rosenthal, D. (1990). A theory of consciousness. In *University of Bielefeld Mind and Brain Technical Report* 40. *Perspectives in Theoretical Psychology and Philosophy of Mind* (ZiF).

Rosenthal, D. (1993). Higher-order thoughts and the appendage theory of consciousness. *Philosophical Psychology* **6**, 155–166.

Ross, L., Green, D., and House P (1977). The 'false consensus effect': An ego-centric bias in social perception and attribution processes. *Journal of Experimental Social Psychology* **13**, 279–301.

Rowe, J.B., Toni, I., Josephs, O., Frackowiak, R.S., and Passingham, R.E. (2000). The prefrontal cortex: response selection or maintenance within working memory? *Science* **288**, 1656–1660.

Rozin, P. (1976). The evolution of intelligence and access to the cognitive unconscious. In Sprague, J.M. and Epstein, A.N. (eds) *Progress in Psychobiology and Physiological Psychology*. New York: Academic Press.

Russell, J.A. (1991). Culture and the categorization of emotions. *Psychological Bulletin* **110**, 426–450.

Sander, D., Grandjean, D., and Scherer, K.R. (2005a). A systems approach to appraisal mechanisms in emotion. *Neural Networks* **18**, 317–352.

Sander, D., Grandjean, D., Pourtois, G., Schwartz, S., Seghier, M.L., Scherer, K.R., and Vuilleumier, P. (2005b). Emotion and attention interactions in social cognition: brain regions involved in processing anger prosody. *NeuroImage* **28**, 848–858.

Sapir, E. (1921). *Language: An Introduction to the Study of Speech*. New York: Harcourt Brace.

Sarter, M.F. and Markowitsch, H.J. (1985). Involvement of the amygdala in learning and memory: a critical review, with emphasis on anatomical relations. *Behavioral Neuroscience* **99**, 342–380.

Sarter, M., Hasselmo, M.E., Bruno, J.P., and Givens, B. (2005). Unraveling the attentional functions of cortical cholinergic inputs: interactions between signal-driven and cognitive modulation of signal detection. *Brain Research. Brain Research Review* **48**, 98–111.

Schachter, S. (1975). Cognition and centralist-peripheralist controversies in motivation and emotion. In Gazzaniga, M.S. and Blakemore, C.B. (eds) *Handbook of psychobiology*, pp. 529–564. New York: Academic Press.

Schachter, S. and Singer, J.E. (1962). Cognitive, social, and physiological determinants of emotional state. *Psychological Review* **69**, 379–399.

Scherer, K.R. (1984). On the nature and function of emotion: a component process approach. In Scherer, K.R. and Ekman, P. (eds) *Approaches to emotion*, pp. 293–317. Hillsdale, NJ: Lawrence Erlbaum Associates.

Scherer, K.R. and Wallbott, H.G. (1994). Evidence for universality and cultural variation of differential emotion response patterning. *Journal of Personality and Social Psychology* **66**, 310–328.

Scherer, K.R., Walbott, H.G., and Summerfield, A.B. (1986). *Experiencing Emotion: A Crosscultural Study*. New York: Cambridge University Press.

Schmitz, T.W., Kawahara-Baccus, T.N., and Johnson, S.C. (2004). Metacognitive evaluation, self-relevance, and the right prefrontal cortex. *NeuroImage* **22**, 941–947.

Schneider, W. and Shiffrin, R.M. (1977). Controlled and automatic human information processing: I. Detection, search, and attention. *Psychological Review* **84**, 1–66.

Schneider, W., Dumais, S.T., and Shiffrin, R.M. (1984). Automatic and controlled processing and attention. In Parasuraman, P. and Davies, D.R. (eds) *Varieties of attention*, pp. 1–27. Orlando, FL: Academic Press.

Scoville, W.B. and Milner, B. (1957). Loss of recent memory after bilateral hippocampal lesions. *Journal of Neurology and Psychiatry* **20**, 11–21.

Shanks, D.R. and Lovibond, P.F. (2002). Autonomic and eyeblink conditioning are closely related to contingency awareness: reply to Wiens and Ohman (2002) and Manns *et al.* (2002). *Journal of Experimental Psychology. Animal Behavior Processes* **28**, 38–42.

Shweder, R.A. (1993). The cultural psychology of the emotions. In Lewis, M. and Haviland, J.M. (eds) *Handbook of Emotions*, pp. 417–431. New York: Guilford Press.

Simon, H.A. (1967). Motivational and emotional controls of cognition. *Psychological Review* **74**, 29–39.

Smith, C.A. and Ellsworth, P.C. (1985). Patterns of cognitive appraisal in emotion. *Journal of Personality and Social Psychology* **56**, 339–353.

Smith, E.E. and Jonides, J. (1999). Storage and executive processes in the frontal lobes. *Science* **283**, 1657–1661.

Smith, Y. and Paré, D. (1994). Intra-amygdaloid projections of the lateral nucleus in the cat: PHA-L anterograde labeling combined with postembedding GABA and glutamate immunocytochemistry. *Journal of Comparative Neurology* **342**, 232–248.

Sotres-Bayon, F., Cain, C.K., and LeDoux, J.E. (2006). Brain mechanisms of fear extinction: historical perspectives on the contribution of prefrontal cortex. *Biological Psychiatry* **60**(4), 329–336.

Squire, L.R. (1987). Memory: neural organization and behavior. In Plum, F. (ed.) *Handbook of Physiology, Section 1: The Nervous System. Vol. V. Higher Functions of the Brain*, pp. 295–371. Bethesda, MD: American Physiological Society.

Squire, L.R. and Zola, S.M. (1998). Episodic memory, semantic memory, and amnesia. *Hippocampus* **8**, 205–211.

Squire, L.R., Stark, C.E., and Clark, R.E. (2004). The medial temporal lobe. *Annual Review of Neuroscience* **27**, 279–306.

Srinivasan, R. and Petrovic, S. (2006). MEG phase follows conscious perception during binocular rivalry induced by visual stream segregation. *Cerebral Cortex* **16**, 597–608.

Stuss, D.T., Picton, W.T., and Alexander, M.P. (1998). Consciousness, self-awareness, and the frontal lobes. In Salloway, S., Malloy, P., and Duffy, J. (eds) *The Frontal Lobes and Neuropsychiatric Illness*. Washington, DC: American Psychiatric Association.

Sullivan, G.M., Apergis, J., Bush, D.E.A., Johnson, L.R., Hou, M., and LeDoux, J.E. (2004). Lesions in the bed nucleus of the stria terminalis disrupt corticosterone and freezing responses elicited by a contextual but not a specific cue-conditioned fear stimulus. *Neuroscience* **128**, 7–14.

Suzuki, W.A. (2006). Encoding new episodes and making them stick. *Neuron* **50**, 19–21.

Swanson, L.W. (1983). The hippocampus and the concept of the limbic system. In Seifert, W. (ed.) *Neurobiology of the Hippocampus*, pp. 3–19. London: Academic Press.

Tanaka, S. (2001). Computational approaches to the architecture and operations of the prefrontal cortical circuit for working memory. *Progress in Neuro-Psychopharmacology & Biological Psychiatry* **25**, 259–281.

Thomas, K.M., Drevets, W.C., Whalen, P.J., Eccard, C.H., Dahl, R.E., Ryan, N.D., and Casey, B.J. (2001). Amygdala response to facial expressions in children and adults. *Biological Psychiatry* **49**, 309–316.

Tomita, H., Ohbayashi, M., Nakahara, K., Hasegawa, I., and Miyashita, Y. (1999). Top-down signal from prefrontal cortex in executive control of memory retrieval. *Nature* **401**, 699–703.

Tomkins, S.S. (1962). *Affect, Imagery, Consciousness*. New York: Springer.

Tseng, W., Guan, R., Disterhoft, J.F., and Weiss, C. (2004). Trace eyeblink conditioning is hippocampally dependent in mice. *Hippocampus* **14**, 58–65.

Tulving, E. (1972). Episodic and semantic memory. In Tulving, E. and Donaldson, W. (eds) *Organization of Memory*, pp. 382–403. New York: Academic Press.

Tulving, E. (1983). *Memory and Consciousness*. Oxford: Clarendon Press.

Tulving, E. (2002). Episodic memory: from mind to brain. *Annual Review of Psychology* **53**, 1–25.

Tulving, E. and Markowitsch, H.J. (1998). Episodic and declarative memory: role of the hippocampus. *Hippocampus* **8**, 198–204.

Tulving, E., Kapur, S., Markowitsch, H.J., Craik, F.I., Habib, R., and Houle, S. (1994). Neuroanatomical correlates of retrieval in episodic memory: auditory sentence recognition. *Proceedings of the National Academy of Sciences of the USA* **91**, 2012–2015.

Ungerleider, L.G. and Mishkin M (1982). Two cortical visual systems. In Ingle, D.J., Goodale, M.A., and Mansfield, R.J.W. (eds) *Analysis of Visual Behavior*, pp. 549–586. Cambridge, MA: MIT Press.

Vargha-Khadem, F., Gadian, D.G., Watkins, K.E., Connelly, A., Van Paesschen, W., and Mishkin, M. (1997). Differential effects of early hippocampal pathology on episodic and semantic memory. *Science* **277**, 376–380.

Vargha-Khadem, F., Salmond, C.H., Watkins, K.E., Friston, K.J., Gadian, D.G., and Mishkin, M. (2003). Developmental amnesia: effect of age at injury. *Proceedings of the National Academy of Sciences of the USA* **100**, 10055–10060.

Velmans, M. (2000). *Understanding Consciousness*. Philadelphia: Routledge.

Vuilleumier, P., Armony, J.L., Driver, J., and Dolan, R.J. (2001). Effects of attention and emotion on face processing in the human brain: an event-related fMRI study. *Neuron* **30**, 829–841.

Vuilleumier, P., Armony, J.L., Clarke, K., Husain, M., Driver, J., and Dolan, R.J. (2002). Neural response to emotional faces with and without awareness: event-related fMRI in a parietal patient with visual extinction and spatial neglect. *Neuropsychologia* **40**, 2156–2166.

Vuilleumier, P., Richardson, M.P., Armony, J.L., Driver, J., and Dolan, R.J. (2004). Distant influences of amygdala lesion on visual cortical activation during emotional face processing. *Nature Neuroscience* **7**, 1271–1278.

Vygotsky, L. (1934). *Thought and Language*. Cambridge, MA: MIT Press.

Wegner, D. (2002). *The Illusion of Conscious Will*. Cambridge, MA: MIT Press.

Wegner, D.M. and Wheatley, T. (1999). Apparent mental causation. Sources of the experience of will. *American Psychologist* **54**, 480–492.

Weiskrantz, L. (1956). Behavioral changes associated with ablation of the amygdaloid complex in monkeys. *Journal of Comparative and Physiological Psychology* **49**, 381–391.

Weiskrantz, L. (1997). *Consciousness Lost And Found: A Neuropsychological Exploration*. New York: Oxford University Press.

Whalen, P.J., Rauch, S.L., Etcoff, N.L., McInerney, S.C., Lee, M.B., and Jenike, M.A. (1998). Masked presentations of emotional facial expressions modulate amygdala activity without explicit knowledge. *Journal of Neuroscience* **18**, 411–418.

Whalen, P.J., Shin, L.M., McInerney, S.C., Fischer, H., Wright, C.I., and Rauch, S.L. (2001). A functional MRI study of human amygdala responses to facial expressions of fear versus anger. *Emotion* **1**, 70–83.

Whalen, P.J., Kagan, J., Cook, R.G., Davis, F.C., Kim, H., Polis, S., McLaren, D.G., Somerville, L.H., McLean, A.A., Maxwell, J.S., and Johnstone T (2004). Human amygdala responsivity to masked fearful eye whites. *Science* **306**, 2061.

Wheeler, M.A., Stuss, D.T., and Tulving E (1997). Toward a theory of episodic memory: the frontal lobes and autonoetic consciousness. *Psychological Bulletin* **121**, 331–354.

Whorf, B.L. (1956). *Language, Thought, and Reality*. Cambridge, MA: MIT Press.

Wierzbicka, A. (1986). Does language reflect culture? Evidence from Australian English. *Language in Society* **5**, 349–374.

Wierzbicka, A. (1994). Emotion, language, and cultural scripts. In Kitayama, S. and Markus, H.R. (eds) *Emotion and Culture: Empirical Studies of Mutual Influence*, pp. 133–196. Washington, DC: American Psychological Association.

Williams, M.A., Morris, A.P., McGlone, F., Abbott, D.F., and Mattingley, J.B. (2004). Amygdala responses to fearful and happy facial expressions under conditions of binocular suppression. *Journal of Neuroscience* **24**, 2898–2904.

Wilson, T.D. (2002). *Strangers to Ourselves: Self-Insight and the Adaptive Unconscious*. Cambridge, MA: Harvard University Press.

Wolfe, J.M. and Bennett, S.C. (1997). Preattentive object files: shapeless bundles of basic features. *Vision Research* **37**, 25–43.

Yancey, S.W. and Phelps, E.A. (2001). Functional neuroimaging and episodic memory: a perspective. *Journal of Clinical and Experimental Neuropsychology* **23**, 32–48.

Zajonc, R. (1980). Feeling and thinking: preferences need no inferences. *American Psychologist* **35**, 151–175.

Zajonc, R.B. (1984). On the primacy of affect. *American Psychologist* **39**, 117–123.

Zeki, S. and Bartels, A. (1998). The autonomy of the visual systems and the modularity of conscious vision. *Philosophical Transactions of the Royal Society of London Series B Biological Sciences* **353**, 1911–1914.

Chapter 4

Emotion, higher-order syntactic thoughts, and consciousness

Edmund T. Rolls

4.1 Introduction

LeDoux (1996), in line with Johnson-Laird (1988) and Baars (1988) (see also Dehaene and Naccache 2001; Dehaene *et al*. 2006), emphasizes the role of working memory in consciousness, where he views working memory as a limited-capacity serial processor that creates and manipulates symbolic representations (p. 280). He thus holds that much emotional processing is unconscious, and that when it becomes conscious it is because emotional information is entered into a working memory system. However, LeDoux (1996) concedes that consciousness, especially its phenomenal or subjective nature, is not completely explained by the computational processes that underlie working memory (p. 281).

LeDoux (this volume) notes that the term working memory can refer to a number of different processes. In top-down attentional processing, a short-term memory is needed to hold online the subject of the attention, for example the position in space at which an object must be identified, or the object that must be found (Rolls and Deco 2002; Rolls 2008a). There is much evidence that this short-term memory is implemented in the prefrontal cortex by an attractor network implemented by associatively modifiable recurrent collateral connections between cortical pyramidal cells, which keep the population active during the attentional task (Rolls 2008a). This short-term memory then biases posterior perceptual and memory networks in the temporal and parietal lobes in a biased competition process (Miller and Cohen 2001; Rolls and Deco 2002; Deco and Rolls 2005a, 2005b; Rolls 2008a). The operation of this type of short-term memory acting using biased competition to implement top-down attention does not appear to be central to consciousness, for as LeDoux (this volume) agrees, prefrontal cortex lesions that have major effects on attention and short-term memory do not impair subjective feelings of consciousness (Rolls 2008a). Thus in the absence of any top-down modulation from a short-term memory to implement top-down attention by biased competition, consciousness in a landscape without top-down biasing can still occur. In this scenario, there is no top-down 'attentional spotlight' anywhere. The same evidence suggests that top-down attention itself is not a fundamental process that is necessary for consciousness, though of course if attention is directed towards particular perceptual events, this will increase the gain of the perceptual processing (Deco and Rolls 2005a, 2005b; Rolls 2008a), making the attended phenomena stronger.

In this chapter, I compare this approach with another approach to emotion and consciousness (Rolls 2005a). Emotion is considered first, and this then sets a framework for approaching the relation between affect and consciousness. I describe multiple routes to action, some of which involve implicit (unconscious) emotional processing, and one of which involves multiple-step planning and leads to a higher-order syntactic theory of consciousness. Then this theory of emotion and consciousness is compared with that of LeDoux (1996, this volume).

4.2 Emotions as states

Emotions can usefully be defined (operationally) as states elicited by rewards and punishers which have particular functions (Rolls 1999a, 2005a). The functions are defined below, and include working to obtain or avoid the rewards and punishers. A reward is anything for which an animal (which includes humans) will work. A punisher is anything that an animal will escape from or avoid. An example of an emotion might thus be happiness produced by being given a reward, such as a pleasant touch, praise, or winning a large sum of money. Another example of an emotion might be fear produced by the sound of a rapidly approaching bus, or the sight of an angry expression on someone's face. We will work to avoid such stimuli, which are punishing. Another example would be frustration, anger, or sadness produced by the omission of an expected reward such as a prize, or the termination of a reward such as the death of a loved one. Another example would be relief, produced by the omission or termination of a punishing stimulus such as the removal of a painful stimulus, or sailing out of danger. These examples indicate how emotions can be produced by the delivery, omission, or termination of rewarding or punishing stimuli, and go some way to indicate how different emotions could be produced and classified in terms of the rewards and punishments received, omitted, or terminated. A diagram summarizing some of the emotions associated with the delivery of reward or punishment or a stimulus associated with them, or with the omission of a reward or punishment, is shown in Fig 4.1.

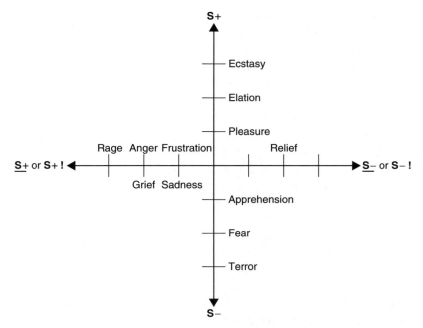

Fig. 4.1. Some of the emotions associated with different reinforcement contingencies are indicated. Intensity increases away from the centre of the diagram, on a continuous scale. The classification scheme created by the different reinforcement contingencies consists of (1) the presentation of a positive reinforcer (S+), (2) the presentation of a negative reinforcer (S−), (3) the omission of a positive reinforcer ($\underline{S+}$) or the termination of a positive reinforcer (S+!), and (4) the omission of a negative reinforcer ($\underline{S-}$) or the termination of a negative reinforcer (S−!).

Before accepting this approach, we should consider whether there are any exceptions to the proposed rule. Are any emotions caused by stimuli, events, or remembered events that are not rewarding or punishing? Do any rewarding or punishing stimuli not cause emotions? We will consider these questions in more detail below. The point is that if there are no major exceptions, or if any exceptions can be clearly encapsulated, then we may have a good working definition at least of what causes emotions. Moreover, it is worth pointing out that many approaches to or theories of emotion (Strongman 1996) have in common that part of the process involves 'appraisal' (Frijda 1986; Lazarus 1991; Oatley and Jenkins 1996). In all these theories the concept of appraisal presumably involves assessing whether something is rewarding or punishing. The description in terms of reward or punishment adopted here seems more tightly and operationally specified. I next consider a slightly more formal definition than rewards or punishments, in which the concept of reinforcers is introduced, and show how there has been a considerable history in the development of ideas along this line.

The proposal that emotions can be usefully seen as states produced by instrumental reinforcing stimuli follows earlier work by Millenson (1967), Weiskrantz (1968), Gray (1975, 1987), and Rolls (1986a, 1986b, 1990, 1999a, 2000a, 2005a). (Instrumental reinforcers are stimuli which, if their occurrence, termination, or omission is made contingent upon the making of a response, alter the probability of the future emission of that response.) Some stimuli are unlearned reinforcers (e.g. the taste of food if the animal is hungry, or pain); while others may become reinforcing by learning, because of their association with such primary reinforcers, thereby becoming 'secondary reinforcers'. This type of learning may thus be called 'stimulus–reinforcement association', and occurs via a process like classical conditioning. If a reinforcer increases the probability of emission of a response on which it is contingent, it is said to be a 'positive reinforcer' or 'reward'; if it decreases the probability of such a response it is a 'negative reinforcer' or 'punisher'. For example, fear is an emotional state which might be produced by a sound (the conditioned stimulus) that has previously been associated with an electric shock (the primary reinforcer).

The converse reinforcement contingencies produce the opposite effects on behaviour. The omission or termination of a positive reinforcer ('extinction' and 'time out' respectively, sometimes described as 'punishing') decreases the probability of responses. Responses followed by the omission or termination of a negative reinforcer increase in probability, this pair of negative reinforcement operations being termed 'active avoidance' and 'escape' respectively (Rolls 2005a).

This foundation has been developed (see Rolls 1986a, 1986b, 1990, 1999a, 2000a, 2005a) to show how a very wide range of emotions can be accounted for, as a result of the operation of a number of factors, including the following:

1 The *reinforcement contingency* (e.g. whether reward or punishment is given, or withheld) (see Fig. 4.1).
2 The *intensity* of the reinforcer (see Fig. 4.1).
3 Any environmental stimulus might have a *number of different reinforcement associations*. (For example, a stimulus might be associated both with the presentation of a reward and of a punisher, allowing states such as conflict and guilt to arise.)
4 Emotions elicited by stimuli associated with *different primary reinforcers* will be different.

5 Emotions elicited by *different secondary reinforcing stimuli* will be different from each other (even if the primary reinforcer is similar).

6 The emotion elicited can depend on whether an *active or passive behavioural response* is possible. (For example, if an active behavioural response can occur to the omission of a positive reinforcer, then anger might be produced, but if only passive behaviour is possible, then sadness, depression or grief might occur.)

By combining these six factors, it is possible to account for a very wide range of emotions (for elaboration see Rolls 2005a). It is also worth noting that emotions can be produced just as much by the recall of reinforcing events as by external reinforcing stimuli; that cognitive processing (whether conscious or not) is important in many emotions, for very complex cognitive processing may be required to determine whether or not environmental events are reinforcing. Indeed, emotions normally consist of cognitive processing which analyses the stimulus, and then determines its reinforcing valence; and then an elicited mood change if the valence is positive or negative. In that an emotion is produced by a stimulus, philosophers say that emotions have an object in the world, and that emotional states are intentional, in that they are about something. We note that a mood or affective state may occur in the absence of an external stimulus, as in some types of depression, but that normally the mood or affective state is produced by an external stimulus, with the whole process of stimulus representation, evaluation in terms of reward or punishment, and the resulting mood or affect being referred to as emotion.

It is worth raising the issue that some philosophers categorize fear in the example as an emotion, but not pain. The distinction they make may be that primary (unlearned or innate) reinforcers (for example pain) do not produce emotions, whereas secondary reinforcers (stimuli associated by stimulus–reinforcement learning with primary reinforcers) do. (An example is fear, which is a state produced by a secondary reinforcing stimulus such as the sight of an image associated by learning with a primary reinforcer such as pain.) They describe the pain as a sensation. But neutral stimuli (such as a table) can produce sensations when touched. Thus whether a stimulus produces a sensation or not does not seem to be a useful distinction that has anything to do with affective or emotional states. It accordingly seems to be much more useful to categorize stimuli according to whether they are reinforcing (in which case they produce emotions or affective states, produced by both primary and secondary reinforcers), or are not reinforcing (in which case they do not produce emotions or affective states such as pleasantness or unpleasantness). Clearly there is a difference between primary reinforcers and learned reinforcers; but this is most precisely caught by noting that this is the difference, and that it is whether a stimulus is reinforcing that determines whether it is related to affective states and emotion. These points are considered in more detail by Rolls (2005a), who provides many examples of primary versus secondary reinforcers, all of which elicit affective states.

4.3 The functions of emotion

The functions of emotion also provide insight into the nature of emotion. These functions, described more fully elsewhere (Rolls 1990, 1999a, 2005a), can be summarized as follows:

1. The *elicitation of autonomic responses* (e.g. a change in heart rate) and *endocrine responses* (e.g. the release of adrenaline/epinephrine). These prepare the body for action.

2. *Flexibility of behavioural responses to reinforcing stimuli*. Emotional (and motivational) states allow a simple interface between sensory inputs and action systems. The essence of this idea is that goals for behaviour are specified by reward and punishment evaluation. When an environmental stimulus has been decoded as a primary reward or punishment, or (after previous stimulus–reinforcer association learning) a secondary rewarding or punishing stimulus, then it becomes a goal for action. The animal can then perform any action (instrumental response) to obtain the reward, or to avoid the punisher. Thus there is flexibility of action, and this is in contrast with stimulus–response, or habit, learning in which a particular response to a particular stimulus is learned. The emotional route to action is flexible not only because any action can be performed to obtain the reward or avoid the punishment, but also because the animal can learn in as little as one trial that a reward or punishment is associated with a particular stimulus, in what is termed 'stimulus–reinforcer association learning'.

To summarize and formalize, two processes are involved in the actions being described. The first is stimulus–reinforcer association learning, and the second is instrumental learning of an operant response made to approach and obtain the reward or to avoid or escape from the punisher. Emotion is an integral part of this, for it is the state elicited in the first stage, by stimuli which are decoded as rewards or punishers, and this state has the property that it is motivating. The motivation is to obtain the reward or avoid the punisher, and animals must be built to obtain certain rewards and avoid certain punishers. Indeed, primary or unlearned rewards and punishers are specified by genes which effectively specify the goals for action. This is the solution which natural selection has found for how genes can influence behaviour to promote their fitness (as measured by reproductive success), and for how the brain could interface sensory systems to action systems, and is an important part of Rolls' theory of emotion (1990, 1999a, 2005a).

Selecting between available rewards with their associated costs, and avoiding punishers with their associated costs, is a process which can take place both implicitly (unconsciously), and explicitly using a language system to enable long-term plans to be made (Rolls 2005a, 2008a). These many different brain systems, some involving implicit evaluation of rewards, and others explicit, verbal, conscious, evaluation of rewards and planned long-term goals, must all enter into the selector of behaviour (see Fig. 4.2). This selector is poorly understood, but it might include a process of competition between all the

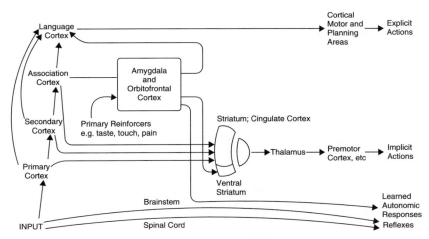

Fig. 4.2. Dual routes to the initiation of action in response to rewarding and punishing stimuli. The inputs from different sensory systems to brain structures such as the orbitofrontal cortex and amygdala allow these brain structures to evaluate the reward- or punishment-related value of incoming stimuli, or of remembered stimuli. The different sensory inputs enable evaluations within the orbitofrontal cortex and amygdala based mainly on the primary (unlearned) reinforcement value for taste, touch, and olfactory stimuli, and on the secondary (learned) reinforcement value for visual and auditory stimuli. In the case of vision, the 'association cortex' which outputs representations of objects to the amygdala and orbitofrontal cortex is the inferior temporal visual cortex. One route for the outputs from these evaluative brain structures is via projections directly to structures such as the basal ganglia (including the striatum and ventral striatum) to enable implicit, direct behavioural responses based on the reward- or punishment-related evaluation of the stimuli to be made. The second route is via the language systems of the brain, which allow explicit decisions involving multi-step syntactic planning to be implemented.

competing calls on output, and might involve the anterior cingulate cortex and basal ganglia in the brain (Rolls 2005a, 2008a) (see Fig. 4.2).

3   Emotion is *motivating*, as just described. For example, fear learned by stimulus–reinforcement association provides the motivation for actions performed to avoid noxious stimuli.

4   *Communication*. Monkeys, for example, may communicate their emotional state to others, by making an open-mouth threat to indicate the extent to which they are willing to compete for resources, and this may influence the behaviour of other animals. This aspect of emotion was emphasized by Darwin (1872), and has been studied more recently by Ekman (1982, 1993). He reviews evidence that humans can categorize facial expressions into the categories happy, sad, fearful, angry, surprised, and disgusted, and that this categorization may operate similarly in different cultures. As shown elsewhere, there are neural systems in the orbitofrontal cortex, amygdala and overlying temporal cortical visual areas which are specialized for the face-related aspects of this processing (Rolls 2005a, 2007b; Rolls *et al.* 2006).

5   *Social bonding*. Examples of this are the emotions associated with the attachment of the parents to their young, and the attachment of the young to their parents.

6   The current mood state can affect the *cognitive evaluation of events or memories* (see Oatley and Jenkins 1996). This may facilitate continuity in the interpretation of the reinforcing value of events in the environment. A hypothesis that back-projections from parts of the brain involved in emotion such as the orbitofrontal cortex and amygdala implement this is described in *Emotion Explained* (Rolls 2005a).

7   Emotion may facilitate the *storage of memories*. One way this occurs is that episodic memory (i.e. one's memory of particular episodes) is facilitated by emotional states (Rolls 2005a, 2008a). A second way in which emotion may affect the storage of memories is that the current emotional state may be stored with episodic memories, providing a mechanism for the current emotional state to affect which memories are recalled. A third way that emotion may affect the storage of memories is by guiding the cerebral cortex in the representations of the world which are set up (Rolls 2008a).

8   Another function of emotion is that by enduring for minutes or longer after a reinforcing stimulus has occurred, it may help to produce *persistent and continuing motivation and direction of behaviour*, to help achieve a goal or goals.

9   Emotion may trigger the *recall of memories* stored in neocortical representations. Amygdala back-projections to the cortex could perform this

for emotion in a way analogous to that in which the hippocampus could implement the retrieval in the neocortex of recent (episodic) memories (Rolls and Stringer 2001; Rolls 2008a).

4.4 Reward, punishment, and emotion in brain design: an evolutionary approach

The theory of the functions of emotion is further developed in *Emotion Explained* (Rolls 2005a). Some of the points made help to elaborate greatly on 4.3 above. Rolls (2005a) considers the fundamental question of why we and other animals are built to use rewards and punishments to guide or determine our behaviour. Why are we built to have emotions, as well as motivational states? Is there any reasonable alternative around which evolution could have built complex animals?

Rolls argues that a role of natural selection is to guide animals to build sensory systems that will respond to dimensions of stimuli in the natural environment along which actions can lead to better ability to pass genes on to the next generation, that is to increased fitness. The animals must be built by such natural selection to make actions that will enable them to obtain more rewards, that is to work to obtain stimuli that will increase their fitness. Correspondingly, animals must be built to make responses that will enable them to escape from, or learn to avoid, stimuli that will reduce their fitness. There are likely to be many dimensions of environmental stimuli along which responses can alter fitness. Each of these dimensions may be a separate reward–punishment dimension. An example of one of these dimensions might be food reward. It increases fitness to be able to sense nutrient need, to have sensors that respond to the taste of food, and to perform behavioural responses to obtain such reward stimuli when in that need or motivational state. Similarly, another dimension is water reward, in which the taste of water becomes rewarding when there is body fluid depletion (see Chapter 6 of *Emotion Explained*).

With many reward–punishment dimensions for which actions may be performed (see Table 2.1 of *Emotion Explained* for a non-exhaustive list!), a selection mechanism for actions performed is needed. In this sense, rewards and punishers provide a *common currency* for inputs to response selection mechanisms. Evolution must set the magnitudes of each of the different reward systems so that each will be chosen for action in such a way as to maximize overall fitness. Food reward must be chosen as the aim for action if a nutrient is depleted; but water reward as a target for action must be selected if current water depletion poses a greater threat to fitness than the current food depletion. This indicates that each reward must be carefully calibrated by evolution to have the right value in the common currency for the competitive selection process. Other types of behaviour, such as sexual behaviour, must be selected sometimes, but probably less frequently, in order to maximize fitness (as measured by gene transmission into the next generation). Many processes contribute to increasing the chances that a wide set of different environmental rewards will be chosen over a period of time, including not only need-related satiety mechanisms which decrease the rewards within a dimension, but also sensory-specific satiety mechanisms, which facilitate switching to another reward stimulus (sometimes within and sometimes outside the same main dimension), and attraction to novel stimuli. Finding novel stimuli rewarding is one way that organisms are encouraged to explore the multidimensional space in which their genes are operating.

The implication of this comparison is that operation by animals using reward–punishment systems tuned to dimensions of the environment that increase fitness provides a mode of operation that can work in organisms that evolve by natural selection. It is clearly a natural outcome of Darwinian evolution to operate using reward–punishment systems tuned to fitness-related dimensions of the environment, if arbitrary responses are to be made by the animals, rather than just preprogrammed movements such as tropisms and taxes. This view of brain design in terms of reward–punishment systems built by genes that gain their adaptive value by being tuned to a goal for action offers, I believe, a deep insight into how natural selection has shaped many brain systems, and is a fascinating outcome of Darwinian thought.

Part of the value of the approach to emotions described here, that they are states elicited by reinforcers that implement the functions described above, is that it provides a firm foundation for which systems to analyse in the brain, i.e. those brain systems involved in responding to reinforcers to implement these functions (*pace* a statement by LeDoux (this volume) who commented on just a part of this definition). This approach has been made use of extensively in *The Brain and Emotion* (Rolls 1999a), in *Emotion Explained* (Rolls 2005a), and elsewhere (Rolls 2000b, 2004b, 2006b, 2008b; O'Doherty *et al.* 2001b; Kringelbach and Rolls 2003;, 2004; de Araujo *et al.* 2005) where fuller details of the neuroscience are provided than can be included here.

4.5 To what extent is consciousness involved in the different types of processing initiated by emotional states?

It might be possible to build a computer which would perform the functions of emotions described above and in more detail by Rolls (2005a), and yet we might not want to ascribe emotional *feelings* to the computer. We might even build the computer with some of the main processing stages present in the brain, and implemented using neural networks which simulate the operation of the real neural networks in the brain (Rolls and Deco 2002; Rolls 2008a), yet we still might not wish to ascribe emotional feelings to this computer. In a sense, the functions of reward and punishment in emotional behaviour are described by the above types of process and their underlying brain mechanisms in structures such as the amygdala and orbitofrontal cortex as described by Rolls (2005a), but what about the subjective aspects of emotion, what about the feeling of pleasure? A similar point arises when we consider the parts of the taste, olfactory, and visual systems in which the reward value of the taste, smell, and sight of food are represented. One such brain region is the orbitofrontal cortex (Rolls 2004b, 2005a, 2006b). Although the neuronal representation in the orbitofrontal cortex is clearly related to the reward value of food, is this where the pleasantness (the subjective hedonic aspect) of the taste, smell, and sight of food is represented? Again, we could (in principle at least) build a computer with neural networks to simulate each of the processing stages for the taste, smell, and sight of food which are described by Rolls (2005a) (and more formally in terms of neural networks by Rolls 2008a and Rolls and Deco 2002), and yet would probably not wish to ascribe feelings of pleasantness to the system we have simulated on the computer.

What is it about neural processing that makes it feel like something when some types of information processing are taking place? It is clearly not a general property of processing in neural networks, for there is much processing, for example that concerned with the control of our blood pressure and heart rate, of which we are not aware. Is it, then, that awareness arises when a certain type of information processing is being performed? If so, what type of information processing? And how do emotional feelings, and sensory events, come to feel like anything? These feels are called qualia. These are great mysteries that have puzzled philosophers for centuries. They are at the heart of the problem of consciousness, for why it should feel like something at all is the great mystery. Other aspects of consciousness, such as the fact that often when we 'pay attention' to events in the world, we can process those events in some better way, that is process or access as opposed to phenomenal aspects of consciousness, may be easier to analyse (Allport 1988; Block 1995; Chalmers 1996). The puzzle of qualia, that is of the phenomenal aspect of consciousness, seems to be rather different from normal investigations in science, in that there is no agreement on criteria by which to assess whether we have made progress. So, although the aim of this chapter is to address the issue of consciousness, especially of qualia, in relation to emotional feelings and actions, what is written cannot be regarded as being establishable by the normal methods of scientific enquiry. Accordingly, I emphasize that the view on consciousness that I describe is only preliminary, and theories of consciousness are likely to develop considerably. Partly for these reasons, this theory of consciousness, at least, should not be taken to have practical implications.

4.6 A theory of consciousness

A starting point is that many actions can be performed relatively automatically, without apparent conscious intervention. An example sometimes given is driving a car. Such actions could involve control of behaviour by brain systems which are old in evolutionary terms such as the basal ganglia. It is of interest that the basal ganglia (and cerebellum) do not have back-projection systems to most of the parts of the cerebral cortex from which they receive inputs (Rolls and Treves 1998; Rolls 2005a). In contrast, parts of the brain such as the hippocampus and amygdala, involved in functions such as episodic memory and emotion respectively, about which we can make (verbal) declarations (hence declarative memory, Squire and Zola 1996) do have major back-projection systems to the high parts of the cerebral cortex from which they receive forward projections (Treves and Rolls 1994; Rolls 2008a). It may be that evolutionarily newer parts of the brain, such as the language areas and parts of the prefrontal cortex, are involved in an alternative type of control of behaviour, in which actions can be planned with the use of a (language) system which allows relatively arbitrary (syntactic) manipulation of semantic entities (symbols).

The general view that there are many routes to behavioural output is supported by the evidence that there are many input systems to the basal ganglia (from almost all areas of the cerebral cortex), and that neuronal activity in each part of the striatum reflects the activity in the overlying cortical area (Rolls 1994, 2005a). The evidence is consistent with the possibility that different cortical areas, each specialized for a different type of computation, have their outputs directed to the basal ganglia, which then select the strongest input, and map this into action (via outputs directed for example to the premotor cortex) (Rolls 2005a). Within this scheme, the language areas would offer one of many routes to action, but a route particularly suited to planning actions, because of the syntactic manipulation of semantic entities which may make long-term planning possible. A schematic diagram of this suggestion is provided in Fig. 4.2.

Consistent with the hypothesis of multiple routes to action, only some of which utilize language, is the evidence that split-brain patients may not be aware of actions being performed by the 'non-dominant' hemisphere (Gazzaniga and LeDoux 1978; Gazzaniga 1988, 1995; Cooney and Gazzaniga 2003).

Also consistent with multiple including non-verbal routes to action, patients with focal brain damage, for example to the prefrontal cortex, may perform actions, yet comment verbally that they should not be performing those actions (Rolls et al. 1994a, 1999b, 2005a; Hornak et al. 2003, 2004). The actions which appear to be performed implicitly, with surprise expressed later by the explicit system, include making behavioural responses to a no-longer-rewarded visual stimulus in a visual discrimination reversal (Rolls et al. 1994a; Hornak et al. 2004). In both these types of patient, confabulation may occur, in that a verbal account of why the action was performed may be given, and this may not be related at all to the environmental event which actually triggered the action (Gazzaniga and LeDoux 1978; Gazzaniga 1988, 1995; Rolls et al. 1994a, 2005a; LeDoux, this volume).

Also consistent with multiple (including non-verbal) routes to action is the evidence that in backward masking at short time delays between the stimulus and the mask, neurons in the inferior temporal visual cortex respond selectively to different faces, and humans guess which face was presented 50% better than chance, yet report having not seen the face consciously (Rolls 2003).

It is possible that sometimes in normal humans when actions are initiated as a result of processing in a specialized brain region, such as those involved in some types of rewarded behaviour, the language system may subsequently elaborate a coherent account of why that action was performed (i.e. confabulate). This would be consistent with a general view of brain evolution in which as areas of the cortex evolve, they are laid on top of existing circuitry connecting inputs to outputs, and in which each level in this hierarchy of separate input–output pathways may control behaviour according to the specialized function it can perform (see schematic in Fig. 4.2). (It is of interest that mathematicians may get a hunch that something is correct, yet not be able to verbalize why. They may then resort to formal, more serial and language-like, theorems to prove the case, and these seem to require conscious processing. This is a further indication of a close association between linguistic processing, and consciousness. The linguistic processing need not, as in reading, involve an inner articulatory loop.)

We may next examine some of the advantages and behavioural functions that language, present as the most recently added layer to the above system, would confer. One major advantage would be the ability to plan actions through many potential stages and to evaluate the consequences of those actions without having to perform the actions. For this, the ability to form propositional statements, and to perform syntactic operations on the semantic representations of states in the world, would be important. Also important in this system would be the ability to have second-order thoughts about the type of thought that I have just described (e.g. I think that he thinks that …), as this would allow much better modelling and prediction of others' behaviour, and therefore of planning, particularly planning when it involves others.[1] This capability for higher-order thoughts would also enable reflection on past events, which would also be useful in planning. In contrast, non-linguistic behaviour would be driven by learned reinforcement associations, learned rules, etc., but not by flexible planning for many steps ahead involving a model of the world including others' behaviour. (For an earlier view which is close to this part of the argument see Humphrey 1980.) The examples of behaviour by non-humans that may reflect planning may reflect much more limited and inflexible planning. For example, the dance of the honey-bee to signal to other bees the location of food may be said to reflect planning, but the symbol manipulation is not arbitrary. There are likely to be interesting examples of non-human primate behaviour, perhaps in the great apes, that reflect the evolution of an arbitrary symbol-manipulation system that could be useful for flexible planning, cf. Cheney and Seyfarth (1990). It is important to state that the language ability referred to here is not necessarily human verbal language (though this would be an example). What it is suggested is important to planning is the syntactic manipulation of symbols, and it is this syntactic manipulation of symbols which is the sense in which language is defined and used here.

It is next suggested that this arbitrary symbol-manipulation using important aspects of language processing and used for planning but not in initiating all types of behaviour is close to what consciousness is about. In particular, consciousness may *be* the state which arises in a system that can think about (or reflect on) its own (or other peoples') thoughts, that is in a system capable of second- or higher-order thoughts (Rosenthal 1986, 1990, 1993, 2004, 2005; Dennett 1991; Rolls 1995, 1997a, 1997b, 1999a, 2004a, 2005a, 2007a; Carruthers 1996; Gennaro 2004). On this account, a mental state is non-introspectively (i.e. non-reflectively) conscious if one has a roughly simultaneous thought that one is in that mental state. Following from this, introspective consciousness (or reflexive consciousness, or self-consciousness) is the attentive, deliberately focused consciousness of one's mental states. It is noted that not all of the higher-order thoughts need themselves be conscious (many mental states are not). However, according to the analysis, having a higher-order thought about a lower-order thought is necessary for the lower-order thought to be conscious. A slightly weaker position than Rosenthal's (and mine) on this is that a conscious

---

[1] Second-order thoughts are thoughts about thoughts. Higher-order thoughts refer to second-order, third-order, etc. thoughts about thoughts.

state corresponds to a first-order thought that has the *capacity* to cause a second-order thought or judgement about it (Carruthers 1996). (Another position which is close in some respects to that of Carruthers and the present position is that of Chalmers 1996, that awareness is something that has *direct availability for behavioural control*, which amounts effectively for him in humans to saying that consciousness is what we can report (verbally) about.) This analysis is consistent with the points made above that the brain systems that are required for consciousness and language are similar. In particular, a system which can have second- or higher-order thoughts about its own operation, including its planning and linguistic operation, must itself be a language processor, in that it must be able to bind correctly to the symbols and syntax in the first-order system. According to this explanation, the feeling of anything is the state which is present when linguistic processing that involves second- or higher-order thoughts is being performed.

It might be objected that this captures some of the process aspects of consciousness, what it is good for in an information processing system, but does not capture the phenomenal aspect of consciousness. I agree that there is an element of 'mystery' that is invoked at this step of the argument, when I say that it feels like something for a machine with higher-order thoughts to be thinking about its own first- or lower-order thoughts. But the return point (discussed further below) is the following: *if a human with second-order thoughts is thinking about its own first-order thoughts, surely it is very difficult for us to conceive that this would NOT feel like something?* (Perhaps the higher-order thoughts in thinking about the first-order thoughts would need to have in doing this some sense of continuity of self, so that the first-order thoughts would be related to the same system that had thought of something else a few minutes ago. But even this continuity aspect may not be a requirement for consciousness. Humans with anterograde amnesia cannot remember what they felt a few minutes ago; yet their current state does feel like something.)

It is suggested that part of the evolutionary adaptive significance of this type of higher-order thought is that is enables correction of errors made in first-order linguistic or in non-linguistic processing. Indeed, the ability to reflect on previous events is extremely important for learning from them, including setting up new long-term semantic structures. It was shown above that the hippocampus may be a system for such 'declarative' recall of recent memories. Its close relation to 'conscious' processing in humans (Squire and Zola 1996 have classified it as a declarative memory system) may be simply that it enables the recall of recent memories, which can then be reflected upon in conscious, higher-order, processing (Rolls and Kesner 2006; Rolls 2008a). Another part of the adaptive value of a higher-order thought system may be that by thinking about its own thoughts in a given situation, it may be able to better understand the thoughts of another individual in a similar situation, and therefore predict that individual's behaviour better (cf. Humphrey 1980, 1986; Barlow 1997).

As a point of clarification, I note that according to this theory, a language processing system (let alone a working memory; LeDoux this volume) is not *sufficient* for consciousness. What defines a conscious system according to this analysis is the ability to have higher-order thoughts, and a first-order language processor (that might be perfectly competent at language) would not be conscious, in that it could not think about its own or others' thoughts. One can perfectly well conceive of a system which obeyed the rules of language (which is the aim of much connectionist modelling), and implemented a first-order linguistic system, that would not be conscious. (Possible examples of language processing that might be performed non-consciously include computer programs implementing aspects of language, or ritualized human conversations, e.g. about the weather. These might require syntax and correctly grounded semantics, and yet be performed non-consciously. A more complex example, illustrating that syntax could be used, might be 'If A does X, then B will probably do Y, and then C would be able to do Z.' A first-order language system could process this statement. Moreover, the first-order language system could apply the rule usefully in the world, provided that the symbols in the language system—A, B, X, Y etc.—are grounded (have meaning) in the world.)

In line with the argument on the adaptive value of higher-order thoughts and thus consciousness given above, that they are useful for correcting lower-order thoughts, I now suggest that correction using higher-order thoughts of lower-order thoughts would have adaptive value primarily if the lower-order thoughts are sufficiently complex to benefit from correction in this way. The nature of the complexity is specific: that it should involve syntactic manipulation of symbols, probably with several steps in the chain, and that the chain of steps should be a one-off (or in American, 'one-time', meaning used once) set of steps, as in a sentence or in a particular plan used just once, rather than a set of well-learned rules. The first- or lower-order thoughts might involve a linked chain of 'if' ... 'then' statements that would be involved in planning, an example of which has been given above. It is partly because complex lower-order thoughts such as these which involve syntax and language would benefit from correction by higher-order thoughts, that I suggest that there is a close link between this reflective consciousness and language. The hypothesis is that by thinking about lower-order thoughts, the higher-order thoughts can discover what may be weak links in the chain of reasoning at the lower-order level, and having detected the weak link, might alter the plan, to see if this gives better success. In our example above, if it transpired that C could not do Z, how might the plan have failed? Instead of having to go through endless random changes to the plan to see if by trial and error some combination does happen to produce results, what I am suggesting is that by thinking about the previous plan, one might, for example, using knowledge of the situation and the probabilities that operate in it, guess that the step where the plan failed was that B did not in fact do Y. So by thinking about the plan (the first- or lower-order thought), one might correct the original plan, in such a way that the weak link in that chain, that 'B will probably do Y', is circumvented.

I draw a parallel with neural networks: there is a *'credit assignment'* problem in such multi-step syntactic plans, in that if the whole plan fails, how does the system assign credit or blame to particular steps of the plan. (In multilayer neural networks, the credit assignment problem is that if errors are being specified at the output layer, the problem arises about how to propagate back the error to earlier, hidden, layers of the network to assign credit or blame to individual synaptic connection; see Rolls and Deco 2002, Rumelhart *et al.* 1986 and Rolls 2008a.) The suggestion is that this is the function of higher-order thoughts and is why systems with higher-order thoughts evolved. The suggestion I then make is that if a system were doing this type of processing (thinking about its own thoughts), it would then be very plausible that it should feel like something to be doing this. I even suggest to the reader that it is not plausible to suggest that it would not feel like anything to a system if it were doing this.

Two other points in the argument should be emphasized for clarity. One is that the system that is having syntactic thoughts about its own syntactic thoughts (higher-order syntactic thoughts or HOSTs) would have to have its symbols grounded in the real world for it to feel like something to be having higher-order thoughts. The intention of this clarification is to exclude systems such as a computer running a program when there is in addition some sort of control or even overseeing program checking the operation of the first program. We would want to say that in such a situation it would feel like something to be running the higher-level control program only if the first-order program was symbolically performing operations on the world and receiving input about the results of those operations, and if the higher-order system understood what the first-order system was trying to do in the world. The issue of symbol grounding is considered further by Rolls (2005a). The symbols (or symbolic representations) are symbols in the sense that they can take part in syntactic processing. The symbolic representations are grounded in the world in that they refer to events in the world. The symbolic representations must have a great deal of information about what is referred to in the world, including the quality and intensity of sensory events, emotional states, etc.

The need for this is that the reasoning in the symbolic system must be about stimuli, events, and states, and remembered stimuli, events and states, and for the reasoning to be correct, all the information that can affect the reasoning must be represented in the symbolic system, including for example just how light or strong the touch was, etc. Indeed, it is pointed out in *Emotion Explained* that it is no accident that the shape of the multidimensional phenomenal (sensory, etc.) space does map so clearly onto the space defined by neuronal activity in sensory systems, for if this were not the case, reasoning about the state of affairs in the world would not map onto the world, and would not be useful. Good examples of this close correspondence are found in the taste system, in which subjective space maps simply onto the multidimensional space represented by neuronal firing in primate cortical taste areas. In particular, if a three-dimensional space reflecting the distances between the representations of different tastes provided by macaque neurons in the cortical taste areas is constructed, then the distances between the subjective ratings by humans of different tastes is very similar (Yaxley *et al.* 1990; Smith-Swintosky *et al.* 1991; Kadohisa *et al.* 2005). Similarly, the changes in human subjective ratings of the pleasantness of the taste, smell, and sight of food parallel very closely the responses of neurons in the macaque orbitofrontal cortex (see *Emotion Explained*).

The representations in the first-order linguistic processor that the HOSTs process include beliefs (for example 'Food is available', or at least representations of this), and the HOST system would then have available to it the concept of a thought (so that it could represent 'I believe [or there is a belief] that food is available'). However, as argued by Rolls (1999a, 2005a), representations of sensory processes and emotional states must be processed by the first-order linguistic system, and HOSTs may be about these representations of sensory processes and emotional states capable of taking part in the syntactic operations of the first-order linguistic processor. Such sensory and emotional information may reach the first-order linguistic system from many parts of the brain, including those such as the orbitofrontal cortex and amygdala implicated in emotional states (see Fig. 4.2 and *Emotion Explained*, Fig. 10.3). When the sensory information is about the identity of the taste, the inputs to the first-order linguistic system must come from the primary taste cortex, in that the identity of taste, independent of its pleasantness (in that the representation is independent of hunger) must come from the primary taste cortex. In contrast, when the information that reaches the first-order linguistic system is about the pleasantness of taste, it must come from the secondary taste cortex, in that there the representation of taste depends on hunger.

The second clarification is that the plan would have to be a unique string of steps, in much the same way as a sentence can be a unique and one-off (or one-time) string of words. The point here is that it is helpful to be able to think about particular one-off plans, and to correct them; and that this type of operation is very different from the slow learning of fixed rules by trial and error, or the application of fixed rules by a supervisory part of a computer program.

This analysis does not yet give an account for sensory qualia ('raw sensory feels', for example why 'red' feels red), for emotional qualia (e.g. why a rewarding touch produces an emotional feeling of pleasure), or for motivational qualia (e.g. why food deprivation makes us *feel* hungry). The view I suggest on such qualia is as follows. Information processing in and from our sensory systems (e.g. the sight of the colour red) may be relevant to planning actions using language and the conscious processing thereby implied. Given that these inputs must be represented in the system that plans, we may ask whether it is more likely that we would be conscious of them or that we would not. I suggest that it would be a very special-purpose system that would allow such sensory inputs, and emotional and motivational states, to be part of (linguistically based) planning, and yet remain unconscious. It seems to be much more parsimonious to hold that we would be conscious of such sensory, emotional and motivational qualia because they would be being used (or are available to be used) in this type of (linguistically based) higher-order thought processing, and this is what I propose.

The explanation for emotional and motivational subjective feelings or qualia that this discussion has led towards is thus that they should be felt as conscious because they enter into a specialized linguistic symbol-manipulation system that is part of a higher-order thought system that is capable of reflecting on and correcting its lower-order thoughts involved for example in the flexible planning of actions. It would require a very special machine to enable this higher-order linguistically based thought processing, which is conscious by its nature, to occur without the sensory, emotional, and motivational states (which must be taken into account by the higher-order thought system) becoming felt qualia. The qualia are thus accounted for by the evolution of the linguistic system that can reflect on and correct its own lower-order processes, and thus has adaptive value.

This account implies that it may be especially animals with a higher-order belief and thought system and with linguistic symbol manipulation that have qualia. It may be that much non-human animal behaviour, provided that it does not require flexible linguistic planning and correction by reflection, could take place according to reinforcement-guidance (using e.g. stimulus–reinforcement association learning in the amygdala and orbitofrontal cortex (Rolls 1999a, 2004b, 2005a), and rule-following (implemented e.g. using habit or stimulus–response learning in the basal ganglia (Rolls 2005a).

Such behaviours might appear very similar to human behaviour performed in similar circumstances, but would not imply qualia. It would be primarily by virtue of a system for reflecting on flexible, linguistic, planning behaviour that humans (and animals close to humans, with demonstrable syntactic manipulation of symbols, and the ability to think about these linguistic processes) would be different from other animals, and would have evolved qualia.

In order for processing in a part of our brain to be able to reach consciousness, appropriate pathways must be present. Certain constraints arise here. For example, in the sensory pathways, the nature of the representation may change as it passes through a hierarchy of processing levels, and in order to be conscious of the information in the form in which it is represented in early processing stages, the early processing stages must have access to the part of the brain necessary for consciousness (see Fig. 4.2). An example is provided by processing in the taste system. In the primate primary taste cortex, neurons respond to taste independently of hunger, yet in the secondary taste cortex, food-related taste neurons (e.g. responding to sweet taste) only respond to food if hunger is present, and gradually stop responding to that taste during feeding to satiety (Rolls 2005a, 2006a). Now the quality of the tastant (sweet, salt, etc.) and its intensity are not affected by hunger, but the pleasantness of its taste is decreased to zero (neutral) (or even becomes unpleasant) after we have eaten it to satiety (Rolls 2005a). The implication of this is that for quality and intensity information about taste, we must be conscious of what is represented in the primary taste cortex (or perhaps in another area connected to it which bypasses the secondary taste cortex), and not of what is represented in the secondary taste cortex. In contrast, for the pleasantness of a taste, consciousness of this could not reflect what is represented in the primary taste cortex, but instead what is represented in the secondary taste cortex (or in an area beyond it).

The same argument arises for reward in general, and therefore for emotion, which in primates is not represented early on in processing in the sensory pathways (nor in or before the inferior temporal cortex for vision), but in the areas to which these object analysis systems project, such as the orbitofrontal cortex, where the reward value of visual stimuli is reflected in the responses of neurons to visual stimuli (Rolls 2005a, 2006a). It is also of interest that reward signals (e.g. the taste of food when we are hungry) are associated with subjective feelings of pleasure (Rolls 2005a, 2006a). I suggest that this correspondence arises because pleasure is the subjective state that represents in the conscious system a signal that is positively reinforcing (rewarding), and that inconsistent behaviour would result if the representations did not correspond to a signal for positive reinforcement in both the conscious and the non-conscious processing systems.

Do these arguments mean that the conscious sensation of e.g. taste quality (i.e. identity and intensity) is represented or occurs in the primary taste cortex, and of the pleasantness of taste in the secondary taste cortex, and that activity in these areas is sufficient for conscious sensations (qualia) to occur? I do not suggest this at all. Instead, the arguments I have put forward above suggest that we are conscious of representations only when we have higher-order thoughts about them. The implication then is that pathways must connect from each of the brain areas in which information is represented about which we can be conscious, to the system which has the higher-order thoughts, which as I have argued above, requires language. Thus, in the example given, there must be connections to the language areas from the primary taste cortex, which need not be direct, but which must bypass the secondary taste cortex, in which the information is represented differently (Rolls 2005a). There must also be pathways from the secondary taste cortex, not necessarily direct, to the language areas so that we can have higher-order thoughts about the pleasantness of the representation in the secondary taste cortex. There would also need to be pathways from the hippocampus, implicated in the recall of declarative memories, back to the language areas of the cerebral cortex (at least via the cortical areas which receive backprojections from the amygdala, orbitofrontal cortex, and hippocampus, see Fig. 4.2, which would in turn need connections to the language areas).

One question that has been discussed is whether there is a causal role for consciousness (e.g. Armstrong and Malcolm 1984). The position to which the above arguments lead is that indeed conscious processing does have a causal role in the elicitation of behaviour, but only under the set of circumstances when higher-order thoughts play a role in correcting or influencing lower-order thoughts. The sense in which the consciousness is causal is then it is suggested, that the higher-order thought is causally involved in correcting the lower-order thought; and that it is a property of the higher-order thought system that it feels like something when it is operating. As we have seen, some behavioural responses can be elicited when there is not this type of reflective control of lower-order processing, nor indeed any contribution of language (see further Rolls (2003, 2005b) for relations between implicit and explicit processing). There are many brain processing routes to output regions, and only one of these involves conscious, verbally represented processing which can later be recalled (see Fig. 4.2).

It is of interest to comment on how the evolution of a system for flexible planning might affect emotions. Consider grief, which may occur when a reward is terminated and no immediate action is possible (see Rolls 1990, 2005a). It may be adaptive by leading to a cessation of the formerly rewarded behaviour and thus facilitating the possible identification of other positive reinforcers in the environment. In humans, grief may be particularly potent because it becomes represented in a system which can plan ahead, and understand the enduring implications of the loss. (Thinking about or verbally discussing emotional states may also in these circumstances help, because this can lead towards the identification of new or alternative reinforcers, and of the realization that, for example, negative consequences may not be as bad as feared.)

This account of consciousness also leads to a suggestion about the processing that underlies the feeling of free will. Free will would in this scheme involve the use of language to check many moves ahead on a number of possible series of actions and their outcomes, and then with this information to make a choice from the likely outcomes of different possible series of actions. (If in contrast choices were made only on the basis of the reinforcement value of immediately available stimuli, without the arbitrary syntactic symbol manipulation made possible by language, then the choice strategy would be much more limited, and we might not want to use the term free will, as all the consequences of those actions would not have been computed.) It is suggested that when this type of reflective, conscious, information processing is occurring and leading to action, the system performing this processing and producing the action would have to believe that it could cause the action, for otherwise inconsistencies would arise, and the system might no longer try to initiate action. This belief held by the system may partly underlie the feeling of free will. At other times, when other brain modules are initiating actions (in the implicit systems), the conscious processor (the explicit system) may confabulate and believe that it caused the action, or at least give an account (possibly wrong) of why the action was initiated. The fact that the conscious processor may have the belief even in these circumstances that it initiated the action may arise as a property of it being inconsistent for a system which can take overall control using conscious verbal processing to believe that it was overridden by another system. This may be the reason why confabulation occurs.

In the operation of such a free-will system, the uncertainties introduced by the limited information possible about the likely outcomes of series of actions, and the inability to use optimal algorithms when combining conditional probabilities, would be much more important factors than whether the brain operates deterministically or not. (The operation of brain machinery must be relatively deterministic, for it has evolved to provide reliable outputs for given inputs.)

I suggest that these concepts may help us to understand what is happening in experiments of the type described by Libet and many others in which consciousness appears to follow with a measurable latency the time when a decision was taken. This is what I predict, if the decision is being made by an implicit, perhaps reward–emotion or habit-related process, for then the conscious processor confabulates an account of or commentary on the decision, so that inevitably the conscious account follows the decision. On the other hand, I predict that if the rational (multistep, reasoning) route is involved in taking the decision, as it might be during planning, or a multistep task such as mental arithmetic, then the conscious report of when the decision was taken, and behavioural or other objective evidence on when the decision was taken, would correspond much more. Under those circumstances, the brain processing taking the decision would be closely related to consciousness, and it would not be a case of just confabulating or reporting on a decision taken by an implicit processor. It would be of interest to test this hypothesis in a version of Libet's task (Libet 2002) in which reasoning was required. The concept that the rational, conscious, processor is only in some tasks involved in taking decisions is extended further in the section on dual routes to action below.

I now consider some clarifications of the present proposal, and how it deals with some issues that arise when considering theories of the phenomenal aspects of consciousness. First, the present proposal has as its foundation the type of computation that is being performed, and suggests that it is a property of a HOST system used for correcting multistep plans with its representations grounded in the world that it would feel like something for a system to be doing this type of processing. To do this type of processing, the system would have to be able to recall previous multistep plans, and would require syntax to keep the symbols in each step of the plan separate. In a sense, the system would have to be able to recall and take into consideration its earlier multistep plans, and in this sense *report* to itself, on those earlier plans. Some approaches to consciousness take the ability to report on or make a *commentary* on events as being an important marker for consciousness (Weiskrantz 1997), and the computational approach I propose suggests why there should be a close relation between consciousness and the ability to report or provide a commentary, for the ability to report is involved in using higher-order syntactic thoughts to correct a multistep plan. Second, the implication of the present approach is that the type of linguistic processing or reporting need not be verbal, using natural language, for what is required to correct the plan is the ability to manipulate symbols syntactically, and this could be implemented in a much simpler type of mentalese or syntactic system (Fodor 1994; Jackendoff 2002; Rolls 2004a) than verbal language or natural language which implies a universal grammar. Third, this approach to consciousness suggests that the information must be being processed in a system capable of implementing

HOSTs for the information to be conscious, and in this sense is more specific than global workspace hypotheses (Baars 1988; Dehaene and Naccache 2001; Dehaene *et al.* 2006). Indeed, the present approach suggests that a workspace could be sufficiently global to enable even the complex processing involved in driving a car to be performed, and yet the processing might be performed unconsciously, unless HOST (supervisory, monitory, correcting) processing was involved. Fourth, the present approach suggests that it just is a property of HOST computational processing with the representations grounded in the world that it feels like something. There is to some extent an element of mystery about why it feels like something, why it is phenomenal, but the explanatory gap does not seem so large when one holds that the system is recalling, reporting on, reflecting on, and reorganizing information about itself in the world in order to prepare new or revised plans. In terms of the physicalist debate (see for a review Davies, this volume), an important aspect of my proposal is that it is a *necessary* property of this type of (HOST) processing that it feels like something (the philosophical description is that this is an absolute metaphysical necessity), and given this view, then it is up to one to decide whether this view is consistent with one's particular view of physicalism or not. Similarly, the possibility of a zombie is inconsistent with the present hypothesis, which proposes that it is by virtue of performing processing in a specialized system that can perform higher-order syntactic processing with the representations grounded in the world that phenomenal consciousness is necessarily present.

These are my initial thoughts on why we have consciousness, and are conscious of sensory, emotional, and motivational qualia, as well as qualia associated with first-order linguistic thoughts. However, as stated above, one does not feel that there are straightforward criteria in this philosophical field of enquiry for knowing whether the suggested theory is correct; so it is likely that theories of consciousness will continue to undergo rapid development; and current theories should not be taken to have practical implications.

4.7 Dual routes to action

According to the present formulation, there are two types of route to action performed in relation to reward or punishment in humans (see also Rolls 2003, 2005a). Examples of such actions include emotional and motivational behaviour.

The first route is via the brain systems that have been present in non-human primates such as monkeys, and to some extent in other mammals, for millions of years. These systems include the amygdala and, particularly well-developed in primates, the orbitofrontal cortex. These systems control behaviour in relation to previous associations of stimuli with reinforcement. The computation which controls the action thus involves assessment of the reinforcement-related value of a stimulus. This assessment may be based on a number of different factors. One is the previous reinforcement history, which involves stimulus–reinforcement association learning using the amygdala, and its rapid updating especially in primates using the orbitofrontal cortex. This stimulus–reinforcement association learning may involve quite specific information about a stimulus, for example of the energy associated with each type of food, by the process of conditioned appetite and satiety (Booth 1985). A second is the current motivational state, for example whether hunger is present, whether other needs are satisfied, etc. A third factor which affects the computed reward value of the stimulus is whether that reward has been received recently. If it has been received recently but in small quantity, this may increase the reward value of the stimulus. This is known as incentive motivation or the 'salted peanut' phenomenon. The adaptive value of such a process is that this positive feedback of reward value in the early stages of working for a particular reward tends to lock the organism into behaviour being performed for that reward. This means that animals that are, for example, almost equally hungry and thirsty will show hysteresis in their choice of action, rather than continually switching from eating to drinking and back with each mouthful of water or food. This introduction of hysteresis into the reward evaluation system makes action selection a much more efficient process in a natural environment, for constantly switching between different types of behaviour would be very costly if all the different rewards were not available in the same place at the same time. (For example, walking half a mile between a site where water was available and a site where food was available after every mouthful would be very inefficient.) The amygdala is one structure that may be involved in this increase in the reward value of stimuli early on in a series of presentations, in that lesions of the amygdala (in rats) abolish the expression of this reward-incrementing process which is normally evident in the increasing rate of working for a food reward early on in a meal (Rolls 2005a). A fourth factor is the computed absolute value of the reward or punishment expected or being obtained from a stimulus, e.g., the sweetness of the stimulus (set by evolution so that sweet stimuli will tend to be rewarding, because they are generally associated with energy sources), or the pleasantness of touch (set by evolution to be pleasant according to the extent to which it brings animals of the opposite sex together, and depending on the investment in time that the partner is willing to put into making the touch pleasurable, a sign which indicates the commitment and value for the partner of the relationship). After the reward value of the stimulus has been assessed in these ways, behaviour is then initiated based on approach towards or withdrawal from the stimulus. A critical aspect of the behaviour produced by this type of system is that it is aimed directly towards obtaining a sensed or expected reward, by virtue of connections to brain systems such as the basal ganglia and cingulate cortex (Rolls 2008b) which are concerned with the initiation of actions (see Fig. 4.2). The expectation may of course involve behaviour to obtain stimuli associated with reward, which might even be present in a chain.

Now part of the way in which the behaviour is controlled with this first route is according to the reward value of the outcome. At the same time, the animal may only work for the reward if the cost is not too high. Indeed, in the field of behavioural ecology, animals are often thought of as performing optimally on some cost-benefit curve (see e.g. Krebs and Kacelnik 1991). This does not at all mean that the animal thinks about the rewards, and performs a cost-benefit analysis using a lot of thoughts about the costs, other rewards available and their costs, etc. Instead, it should be taken to mean that in evolution, the system has evolved in such a way that the way in which the reward varies with the different energy densities or amounts of food and the delay before it is received can be used as part of the input to a mechanism which has also been built to track the costs of obtaining the food (e.g. energy loss in obtaining it, risk of predation, etc.), and to then select given many such types of reward and the associated cost, the current behaviour that provides the most 'net reward'. Part of the value of having the computation expressed in this reward-minus-cost form is that there is then a suitable 'currency', or net reward value, to enable the animal to select the behaviour with currently the most net reward gain (or minimal aversive outcome).

The second route in humans involves a computation with many 'if ... then' statements, to implement a plan to obtain a reward. In this case, the reward may actually be *deferred* as part of the plan, which might involve working first to obtain one reward, and only then to work for a second more highly valued reward, if this was thought to be overall an optimal strategy in terms of resource usage (e.g. time). In this case, syntax is required, because the many symbols (e.g. names of people) that are part of the plan must be correctly linked or bound. Such linking might be of the form: 'if A does this, then B is likely to do this, and this will cause C to do this ...'. The requirement of syntax for this type of planning implies that an output to language systems in the brain is required for this type of planning (see Fig. 4.2). Thus the explicit language system in humans may allow working for deferred rewards by enabling use of a one-off, individual, plan appropriate for each situation. Another building block for such planning operations in the brain may be the type of short-term memory in which the prefrontal cortex is involved. This short-term memory may be, for example in non-human primates, of where in space a response has just been made. A development of this type of short-term response memory system in humans to enable multiple short-term memories to be held in place correctly, preferably with the temporal order of the different items in the short-term memory coded correctly, may be another building block for the multiple step 'if ... then' type of computation in order to form a multiple-step plan. Such short-term memories are implemented in the (dorsolateral and inferior convexity) prefrontal cortex of non-human primates and humans (Goldman-Rakic 1996; Petrides 1996; Rolls 2008a), and may be part of the reason why prefrontal cortex damage impairs planning (Shallice and Burgess 1996).

Of these two routes (see Fig. 4.2), it is the second which I have suggested above is related to consciousness. The hypothesis is that consciousness is the state which arises by virtue of having the ability to think about one's own thoughts, which has the adaptive value of enabling one to correct long, multi-step syntactic plans. This latter system is thus the one in which explicit, declarative, processing occurs. Processing in this system is frequently associated with reason and rationality, in that many of the consequences of possible actions can be taken into account. The actual computation of how rewarding a particular stimulus or situation is or will be probably still depends on activity in the orbitofrontal cortex and amygdala, as the reward value of stimuli is computed and represented in these regions, and in that it is found that verbalized expressions of the reward (or punishment) value of stimuli are dampened by damage to these systems. (For example, damage to the orbitofrontal cortex renders painful input still identifiable as pain, but without the strong affective, 'unpleasant', reaction to it.) This language system which enables long-term planning may be contrasted with the first system in which behaviour is directed at obtaining the stimulus (including the remembered stimulus) which is currently most rewarding, as computed by brain structures that include the orbitofrontal cortex and amygdala. There are outputs from this system, perhaps those directed at the basal ganglia, which do not pass through the language system, and behaviour produced in this way is described as implicit, and verbal declarations cannot be made directly about the reasons for the choice made. When verbal declarations are made about decisions made in this first system, those verbal declarations may be confabulations, reasonable explanations, or fabrications, of reasons why the choice was made. These reasonable explanations would be generated to be consistent with the sense of continuity and self that is a characteristic of reasoning in the language system.

The question then arises of how decisions are made in animals such as humans that have both the implicit, direct reward-based, and the explicit, rational, planning systems (see Fig. 4.2) (Rolls 2008a). One particular situation in which the first, implicit, system may be especially important is when rapid reactions to stimuli with reward or punishment value must be made, for then the direct connections from structures such as the orbitofrontal cortex to the basal ganglia may allow rapid actions (Rolls 2005a). Another is when there may be too many factors to be taken into account easily by the explicit, rational, planning, system, when the implicit system may be used to guide action. In contrast, when the implicit system continually makes errors, it would then be beneficial for the organism to switch from automatic, direct, action based on obtaining what the orbitofrontal cortex system decodes as being the most positively reinforcing choice currently available, to the explicit, conscious control system which can evaluate with its long-term planning algorithms what action should be performed next. Indeed, it would be adaptive for the explicit system to regularly be assessing performance by the more automatic system, and to switch itself into control behaviour quite frequently, as otherwise the adaptive value of having the explicit system would be less than optimal.

There may also be a flow of influence from the explicit, verbal system to the implicit system, in that the explicit system may decide on a plan of action or strategy, and exert an influence on the implicit system which will alter the reinforcement evaluations made by and the signals produced by the implicit system (Rolls 2005a).

It may be expected that there is often a conflict between these systems, in that the first, implicit, system is able to guide behaviour particularly to obtain the greatest immediate reinforcement, whereas the explicit system can potentially enable immediate rewards to be deferred, and longer-term, multi-step, plans to be formed. This type of conflict will occur in animals with a syntactic planning ability, that is in humans and any other animals that have the ability to process a series of 'if ... then' stages of planning. This is a property of the human language system, and the extent to which it is a property of non-human primates is not yet fully clear. In any case, such conflict may be an important aspect of the operation of at least the human mind, because it is so essential for humans to correctly decide, at every moment, whether to invest in a relationship or a group that may offer long-term benefits, or whether to directly pursue immediate benefits (Rolls 2005a, 2008a).

The thrust of the argument (Rolls 2005a, 2008a) thus is that much complex animal including human behaviour can take place using the implicit, non-conscious, route to action. We should be very careful not to postulate intentional states (i.e. states with intentions, beliefs, and desires) unless the evidence for them is strong, and it seems to me that a flexible, one-off, linguistic processing system that can handle propositions is needed for intentional states. What the explicit, linguistic, system does allow is exactly this flexible, one-off, multi-step planning-ahead type of computation, which allows us to defer immediate rewards based on such a plan.

This discussion of dual routes to action has been with respect to the behaviour produced. There is of course in addition a third output of brain regions such as the orbitofrontal cortex and amygdala involved in emotion, that is directed to producing autonomic and endocrine responses (see Fig. 4.2). Although it has been argued by Rolls (2005a) that the autonomic system is not normally in a circuit through which behavioural responses are produced (i.e. against the James–Lange and related somatic theories), there may be some influence from effects produced through the endocrine system (and possibly the autonomic system, through which some endocrine responses are controlled) on behaviour, or on the dual systems just discussed which control behaviour.

4.8 Comparisons with other approaches to emotion and consciousness

The theory described here suggests that it feels like something to be an organism or machine that can think about its own (linguistic, and semantically based) thoughts. It is suggested that qualia, raw sensory and emotional subjective feelings, arise secondary to having evolved such a higher-order thought system, and that sensory and emotional processing feels like something because it would be unparsimonious for it to enter the planning, higher-order thought, system and *not* feel like something. The adaptive value of having sensory and emotional feelings, or qualia, is thus suggested to be that such inputs are important to the long-term planning, explicit, processing system. Raw sensory feels, and subjective states associated with emotional and motivational states, may not necessarily arise first in evolution. Some issues that arise in relation to this theory are discussed by Rolls (2000a, 2004a, 2005a); reasons why the ventral visual system is more closely related to explicit than implicit processing (because reasoning about objects may be important) are considered by Rolls (2003) and by Rolls and Deco (2002); and reasons why explicit, conscious, processing may have a higher threshold in sensory processing than implicit processing are considered by Rolls (2003, 2005a, 2005b).

I now compare this approach to emotion and consciousness with that of LeDoux which places some emphasis on working memory. A comparison with other approaches to emotion and consciousness is provided elsewhere (Rolls 2003, 2004a, 2005a, 2005b, 2007a).

A process ascribed to working memory is that items can be manipulated in working memory, for example placed into a different order. This process implies at the computational level some type of syntactic processing, for each item (or symbol) could occur in any position relative to the others, and each item might occur more than once. To keep the items separate yet manipulable into any relation to each other, just having each item represented by the firing of a different set of neurons is insufficient, for this provides no information about the order or more generally the relations between the items being manipulated (Rolls and Deco 2002; Rolls 2008a). In this sense, some form of syntax, that is a way to relate to each other the firing of the different populations of neurons each representing an item, is required. If we go this far (and LeDoux 1996, p. 280 does appear to), then we see that this aspect of working memory is very close to the concept I propose of syntactic thought in my HOST theory. My particular approach though makes it clear what the function is to be performed (syntactic operations), whereas the term working memory can be used to refer to many different types of processing (Repovs and Baddeley 2006), and is in this sense less well defined computationally. My approach of course argues that it is thoughts about the first-order thoughts that may be very closely linked to consciousness. In our simple case, the higher-order thought might be 'Do I have the items now in the correct reversed order? Should the X come before or after the Y?' To perform this syntactic manipulation, I argue that there is a special syntactic processor, perhaps in cortex near Broca's area, that performs the manipulations on the items, and that the dorsolateral prefrontal cortex itself provides the short-term store that holds the items on which the syntactic processor operates (Rolls 2008a). In this scenario, dorsolateral prefrontal cortex damage would affect the number of items that could be manipulated, but not consciousness or the ability to manipulate the items syntactically and to monitor and comment on the result to check that it is correct.

A property often attributed to consciousness is that it is *unitary*. LeDoux (this volume) might relate this to the limitations of a working memory system. The current theory would account for this by the limited syntactic capability of neuronal networks in the brain, which render it difficult to implement more than a few syntactic bindings of symbols simultaneously (McLeod *et al.* 1998; Rolls 2008a). This limitation makes it difficult to run several 'streams of consciousness' simultaneously. In addition, given that a linguistic system can control behavioural output, several parallel streams might produce maladaptive behaviour (apparent as e.g. indecision), and might be selected against. The close relation between, and the limited capacity of, both the stream of consciousness, and auditory–verbal short-term memory, may be that both implement the capacity for syntax in neural networks. My suggestion is that it is the difficulty the brain has in implementing the syntax required for manipulating items in working memory, and therefore for multi-step planning, and for then correcting these plans, that provides a close link between working memory concepts and my theory of higher-order syntactic processing. The theory I describe makes it clear what the underlying computational problem is (how syntactic operations are performed in the system, and how they are corrected), and argues that when there are thoughts about the system, i.e. HOSTs, and the system is reflecting on its first-order thoughts (cf. Weiskrantz 1997), then it is a property of the system that it feels conscious. As I argued above, first-order linguistic thoughts, which presumably involve working memory (which must be clearly defined for the purposes of this discussion), need not necessarily be conscious.

The theory of emotion described here is also different from LeDoux's approach to affect. LeDoux's (1996, this volume) approach to emotion is largely (to quote him) one of automaticity, with emphasis on brain mechanisms involved in the rapid, subcortical, mechanisms involved in fear. Much of the research described has been on the functions of the amygdala in fear conditioning. The importance of this system in humans may be less than in rats, in that human patients with bilateral amygdala damage do not present with overt emotional problems, although some impairments in fear conditioning and in the expression of fear in the face can be identified (Phelps 2004) (see Rolls 2005a). In contrast, the orbitofrontal cortex has developed greatly in primates, including humans, and major changes in emotion in humans follow damage to the orbitofrontal cortex, evident for example in reward reversal learning; in face expression identification; in behaviour, which can become disinhibited, uncooperative, and impulsive; and in altered subjective emotions after the brain damage (Rolls *et al.* 1994a; Hornak *et al.* 1996, 2003, 2004; Berlin *et al.* 2004; *et al.*; Berlin *et al.* 2005; Rolls 2005a). In addition, it is worth noting that the human amygdala is involved in reward-related processing, in that for example it is as much activated by rewarding sweet taste as by unpleasant salt taste (O'Doherty *et al.* 2001a). Also, although the direct, low road, inputs from subcortical structures to the amygdala have been emphasized, it should be noted that most emotions are not to stimuli that require no cortical processing (such as pure tones), but instead are to complex stimuli (such as the facial expression of a particular individual) which require cortical processing. Thus the cortical to amygdala connections are likely to be involved in most emotional processing that engages the amygdala (Rolls 2005a, 2000b). Consistent with this, amygdala face-selective neurons in primates have longer latencies (e.g. 130–180 ms) than those of neurons in the inferior temporal visual cortex (typically 90–110 ms) (Rolls 1984, 2005a; Leonard *et al.* 1985). Simple physical aspects of faces such as horizontal spatial frequencies reflecting the mouth might reach the amygdala by non-cortical routes, but inputs that reflect the identity of a face, as well as the expression, which are important in social behaviour, are likely to require cortical processing because of the complexity of the computations involved in invariant face and object identification (Rolls and Deco 2002; Rolls 2005a; Rolls and Stringer 2006). Although it is of interest that the amygdala can be activated by face stimuli that are not perceived in backward masking experiments (Phillips *et al.* 2004, LeDoux this volume), so too can the inferior temporal visual cortex (Rolls and Tovee 1994; Rolls *et al.* 1994b, 1999; Rolls 2003, 2005b), so the amygdala just follows the cortex in this respect. The implication here is that activation of the amygdala by stimuli that are not consciously perceived need not be due to subcortical routes to the amygdala. Temporal visual cortex activity also occurs when the stimulus is minimized by backward masking, and could provide a route for activity even when not consciously seen to reach the amygdala. It is suggested that the higher threshold for conscious awareness than for unconscious responses to stimuli, as shown by the larger neuronal responses in the inferior temporal cortex for conscious awareness to be reported, may be related to the fact that conscious processing is inherently serial because of the syntactic binding required, and may because it would be inefficient to interrupt this, have a relatively high threshold (Rolls 2003, 2005b).

Finally, I provide a short specification of what might have to be implemented in a neural network to implement conscious processing. First, a linguistic system, not necessarily verbal, but implementing syntax between symbols grounded in the environment would be needed (e.g. a mentalese language system). Then a higher-order thought system also implementing syntax and able to think about the representations in the first-order language system, and able to correct the reasoning in the first-order linguistic system in a flexible manner, would be needed. So my view is that consciousness can be implemented in neural networks (and that this is a topic worth discussing), but that the neural networks would have to implement the type of higher-order linguistic processing described in this chapter.

References

Allport, A. (1988). What concept of consciousness? In Marcel, A.J. and Bisiach, E. (eds) *Consciousness in Contemporary Science*, pp. 159–182. Oxford: Oxford University Press.

Armstrong, D.M. and Malcolm, N. (1984). *Consciousness and Causality*. Oxford: Blackwell.

Baars, B.J. (1988). *A Cognitive Theory of Consciousness*. New York: Cambridge University Press.

Barlow, H.B. (1997). Single neurons, communal goals, and consciousness. In Ito, M., Miyashita, Y., and Rolls, E.T. (eds) *Cognition, Computation, and Consciousness*, pp. 121–136. Oxford: Oxford University Press.

Berlin, H., Rolls, E.T., and Kischka, U. (2004). Impulsivity, time perception, emotion, and reinforcement sensitivity in patients with orbitofrontal cortex lesions. *Brain* **127**, 1108–1126.

Berlin, H., Rolls, E.T., and Iversen, S.D. (2005). Borderline personality disorder, impulsivity and the orbitofrontal cortex. *American Journal of Psychiatry* **162**, 2360–2373.

Block, N. (1995). On a confusion about a function of consciousness. *Behavioral and Brain Sciences* **18**, 227–247.

Booth, D.A. (1985). Food-conditioned eating preferences and aversions with interoceptive elements: learned appetites and satieties. *Annals of the New York Academy of Sciences* **443**, 22–37.

Carruthers, P. (1996). *Language, Thought and Consciousness*. Cambridge: Cambridge University Press.

Chalmers, D.J. (1996). *The Conscious Mind*. Oxford: Oxford University Press.

Cheney, D.L. and Seyfarth, R.M. (1990). *How Monkeys See the World*. Chicago: University of Chicago Press.

Cooney, J.W. and Gazzaniga, M.S. (2003). Neurological disorders and the structure of human consciousness. *Trends in Cognitive Science* **7**, 161–165.

Darwin, C. (1872) *The Expression of the Emotions in Man and Animals*, 3rd edn. Chicago: University of Chicago Press.

de Araujo, I.E.T., Rolls, E.T., Velazco, M.I., Margot, C., and Cayeux, I. (2005). Cognitive modulation of olfactory processing. *Neuron* **46**, 671–679.

Deco, G. and Rolls, E.T. (2005)a. Neurodynamics of biased competition and co-operation for attention: a model with spiking neurons. *Journal of Neurophysiology* **94**, 295–313.

Deco, G. and Rolls, E.T. (2005b). Attention, short-term memory, and action selection: a unifying theory. *Progress in Neurobiology* **76**, 236–256.

Dehaene, S. and Naccache, L. (2001). Towards a cognitive neuroscience of consciousness: basic evidence and a workspace framework. *Cognition* **79**, 1–37.

Dehaene, S., Changeux, J.P., Naccache, L., Sackur, J., and Sergent, C. (2006). Conscious, preconscious, and subliminal processing: a testable taxonomy. *Trends in Cognitive Science* **10**, 204–211.

Dennett, D.C. (1991). *Consciousness Explained*. London: Penguin.

Ekman, P. (1982). *Emotion in the Human Face*, 2nd edn. Cambridge: Cambridge University Press.

Ekman, P. (1993). Facial expression and emotion. *American Psychologist* **48**, 384–392.

Fodor, J.A. (1994). *The Elm and the Expert: Mentalese and its Semantics*. Cambridge, MA: MIT Press.

Frijda, N.H. (1986). *The Emotions*. Cambridge: Cambridge University Press.

Gazzaniga, M.S. (1988). Brain modularity: towards a philosophy of conscious experience. In Marcel, A.J. and Bisiach, E. (eds) *Consciousness in Contemporary Science*, pp. 218–238. Oxford: Oxford University Press.

Gazzaniga, M.S. (1995). Consciousness and the cerebral hemispheres. In Gazzaniga, M.S. (ed.) *The Cognitive Neurosciences*, pp. 1392–1400. Cambridge, MA: MIT Press.

Gazzaniga, M.S. and LeDoux, J. (1978). *The Integrated Mind*. New York: Plenum.

Gennaro, R.J. (ed.) (2004). *Higher Order Theories of Consciousness*. Amsterdam: John Benjamins.

Goldman-Rakic, P.S. (1996). The prefrontal landscape: implications of functional architecture for understanding human mentation and the central executive. *Philosophical Transactions of the Royal Society of London Series B Biological Sciences* **351**, 1445–1453.

Gray, J.A. (1975). *Elements of a Two-Process Theory of Learning.* London: Academic Press.

Gray, J.A. (1987). *The Psychology of Fear and Stress*, 2nd edn. Cambridge: Cambridge University Press.

Hornak, J., Rolls, E.T., and Wade, D. (1996). Face and voice expression identification in patients with emotional and behavioural changes following ventral frontal lobe damage. *Neuropsychologia* **34**, 247–261.

Hornak, J., Bramham, J., Rolls, E.T., Morris, R.G., O'Doherty, J., Bullock, P.R., and Polkey, C.E. (2003). Changes in emotion after circumscribed surgical lesions of the orbitofrontal and cingulate cortices. *Brain* **126**, 1691–1712.

Hornak, J., O'Doherty, J., Bramham, J., Rolls, E.T., Morris, R.G., Bullock, P.R., and Polkey, C.E. (2004). Reward-related reversal learning after surgical excisions in orbitofrontal and dorsolateral prefrontal cortex in humans. *Journal of Cognitive Neuroscience* **16**, 463–478.

Humphrey, N.K. (1980). Nature's psychologists. In Josephson, B.D. and Ramachandran, V.S. (eds) *Consciousness and the Physical World*, pp. 57–80. Oxford: Pergamon.

Humphrey, N.K. (1986). *The Inner Eye.* London: Faber.

Jackendoff, R. (2002). *Foundations of Language.* Oxford: Oxford University Press.

Johnson-Laird, P.N. (1988). *The Computer and the Mind: An Introduction to Cognitive Science.* Cambridge, MA: Harvard University Press.

Kadohisa, M., Rolls, E.T., and Verhagen, J.V. (2005). Neuronal representations of stimuli in the mouth: the primate insular taste cortex, orbitofrontal cortex, and amygdala. *Chemical Senses* **30**, 401–419.

Krebs, J.R. and Kacelnik A (1991). Decision making. In Krebs, J.R. and Davies, N.B. (eds) *Behavioural Ecology*, 3rd edn, pp. 105–136. Oxford: Blackwell.

Kringelbach, M.L. and Rolls, E.T. (2003). Neural correlates of rapid reversal learning in a simple model of human social interaction. *NeuroImage* **20**, 1371–1383.

Kringelbach, M.L. and Rolls, E.T. (2004). The functional neuroanatomy of the human orbitofrontal cortex: evidence from neuroimaging and neuropsychology. *Progress in Neurobiology* **72**, 341–372.

Lazarus, R.S. (1991). *Emotion and Adaptation.* New York: Oxford University Press.

LeDoux, J.E. (1996). *The Emotional Brain.* New York: Simon and Schuster.

Leonard, C.M., Rolls, E.T., Wilson, F.A.W., and Baylis, G.C. (1985). Neurons in the amygdala of the monkey with responses selective for faces. *Behavioural Brain Research* **15**, 159–176.

Libet, B. (2002). The timing of mental events: Libet's experimental findings and their implications. *Consciousness and Cognition* **11**, 291–299; discussion 304–233.

McLeod, P., Plunkett, K., and Rolls, E.T. (1998). *Introduction to Connectionist Modelling of Cognitive Processes.* Oxford: Oxford University Press.

Millenson, J.R. (1967). *Principles of Behavioral Analysis.* New York: Macmillan.

Miller, E.K., Cohen, J.D. (2001). An integrative theory of prefrontal cortex function. *Annual Review of Neuroscience* **24**, 167–202.

O'Doherty, J., Rolls, E.T., Francis, S., Bowtell, R., and McGlone, F. (2001a). The representation of pleasant and aversive taste in the human brain. *Journal of Neurophysiology* **85**, 1315–1321.

O'Doherty, J., Kringelbach, M.L., Rolls, E.T., Hornak, J., and Andrews, C. (2001b). Abstract reward and punishment representations in the human orbitofrontal cortex. *Nature Neuroscience* **4**, 95–102.

Oatley, K. and Jenkins, J.M. (1996). *Understanding Emotions*. Oxford: Backwell.

Petrides, M. (1996). Specialized systems for the processing of mnemonic information within the primate frontal cortex. *Philosophical Transactions of the Royal Society of London Series B Bilogical Sciences* **351**, 1455–1462.

Phelps, E.A. (2004). Human emotion and memory: interactions of the amygdala and hippocampal complex. *Current Opinions in Neurobiology* **14**, 198–202.

Phillips, M.L., Williams, L.M., Heining, M., Herba, C.M., Russell, T., Andrew, C., Bullmore, E.T., Brammer, M.J., Williams, S.C., Morgan, M., Young, A.W., and Gray, J.A. (2004). Differential neural responses to overt and covert presentations of facial expressions of fear and disgust. *NeuroImage* **21**, 1484–1496.

Repovs, G. and Baddeley, A. (2006). The multi-component model of working memory: explorations in experimental cognitive psychology. *Neuroscience* **139**, 5–21.

Rolls, E.T. (1984). Neurons in the cortex of the temporal lobe and in the amygdala of the monkey with responses selective for faces. *Human Neurobiology* **3**, 209–222.

Rolls, E.T. (1986a). A theory of emotion, and its application to understanding the neural basis of emotion. In Oomura, Y. (ed.) *Emotions. Neural and Chemical Control*, pp. 325–344. Basel: Karger.

Rolls, E.T. (1986b). Neural systems involved in emotion in primates. In Plutchik, R. and Kellerman, H. (eds) *Emotion: Theory, Research, and Experience. Vol. 3. Biological Foundations of Emotion*, pp. 125–143. New York: Academic Press.

Rolls, E.T. (1990). A theory of emotion, and its application to understanding the neural basis of emotion. *Cognition and Emotion* **4**, 161–190.

Rolls, E.T. (1994). Neurophysiology and cognitive functions of the striatum. *Revue Neurologique (Paris)* **150**, 648–660.

Rolls, E.T. (1995). A theory of emotion and consciousness, and its application to understanding the neural basis of emotion. In Gazzaniga, M.S. (ed.) *The Cognitive Neurosciences*, pp. 1091–1106. Cambridge, MA: MIT Press.

Rolls, E.T. (1997a). Consciousness in neural networks? *Neural Networks* **10**, 1227–1240.

Rolls, E.T. (1997b). Brain mechanisms of vision, memory, and consciousness. In Ito, M., Miyashita, Y., and Rolls, E.T. (eds) *Cognition, Computation, and Consciousness*, pp. 81–120. Oxford: Oxford University Press.

Rolls, E.T. (1999a). *The Brain and Emotion*. Oxford: Oxford University Press.

Rolls, E.T. (1999b). The functions of the orbitofrontal cortex. *Neurocase* **5**, 301–312.

Rolls, E.T. (2000a). Prècis of *The Brain and Emotion*. *Behavioral and Brain Sciences* **23**, 177–233.

Rolls, E.T. (2000b). Neurophysiology and functions of the primate amygdala, and the neural basis of emotion. In Aggleton, J.P. (ed.) *The Amygdala: A Functional Analysis*, 2nd edn, pp. 447–478. Oxford: Oxford University Press.

Rolls, E.T. (2003). Consciousness absent and present: a neurophysiological exploration. *Progress in Brain Research* **144**, 95–106.

Rolls, E.T. (2004a). A higher order syntactic thought (HOST) theory of consciousness. In Gennaro, R.J. (ed.) *Higher-Order Theories of Consciousness: An Anthology*, pp. 137–172. Amsterdam: John Benjamins.

Rolls, E.T. (2004b). The functions of the orbitofrontal cortex. *Brain and Cognition* **55**, 11–29.

Rolls, E.T. (2005a). *Emotion Explained*. Oxford: Oxford University Press.

Rolls, E.T. (2005b). Consciousness absent or present: a neurophysiological exploration of masking. In Ogmen, H. and Breitmeyer, B.G. (eds) *The First Half Second: The Microgenesis and Temporal Dynamics of Unconscious and Conscious Visual Processes*, pp. 89–108, Chapter 106. Cambridge, MA: MIT Press.

Rolls, E.T. (2006a). Brain mechanisms underlying flavour and appetite. *Philosophical Transactions of the Royal Society London Series B Biological Sciences* **361**, 1123–1136.

Rolls, E.T. (2006b). The neurophysiology and functions of the orbitofrontal cortex. In Zald, D.H. and Rauch, S.L. (eds) *The Orbitofrontal Cortex*, pp. 95–124. Oxford: Oxford University Press.

Rolls, E.T. (2007a). The affective neuroscience of consciousness: higher order linguistic thoughts, dual routes to emotion and action, and consciousness. In Zelazo, P., Moscovitch, M. and Thompson, E. (eds) *Cambridge Handbook of Consciousness*, pp. 831–859. Cambridge: Cambridge University Press.

Rolls, E.T. (2007b). The representation of information about faces in the temporal and frontal lobes. *Neuropsychologia* **45**, 125–143.

Rolls, E.T. (2008a). *Memory, Attention, and Decision-Making: A Unifying Computational Neuroscience Approach*. Oxford: Oxford University Press.

Rolls, E.T. (2008b). The anterior and midcingulate cortices and reward. In Vogt, B.A. (ed.) *Cingulate Neurobiology and Disease*. Oxford: Oxford University Press.

Rolls, E.T. and Deco, G. (2002). *Computational Neuroscience of Vision*. Oxford: Oxford University Press.

Rolls, E.T. and Kesner, R.P. (2006). A computational theory of hippocampal function, and empirical tests of the theory. *Progress in Neurobiology* **79**, 1–48.

Rolls, E.T. and Stringer, S.M. (2001). A model of the interaction between mood and memory. *Network: Computation in Neural Systems* **12**, 111–129.

Rolls, E.T. and Stringer, S.M. (2006). Invariant visual object recognition: a model, with lighting invariance. *Journal of Physiology (Paris)* **100**, 43–62.

Rolls, E.T. and Tovee, M.J. (1994). Processing speed in the cerebral cortex and the neurophysiology of visual masking. *Proceedings of the Royal Society of London Series B Biological Sciences* **257**, 9–15.

Rolls, E.T. and Treves, A. (1998). *Neural Networks and Brain Function*. Oxford: Oxford University Press.

Rolls, E.T., Hornak, J., Wade, D., and McGrath, J. (1994a). Emotion-related learning in patients with social and emotional changes associated with frontal lobe damage. *Journal of Neurology, Neurosurgery and Psychiatry* **57**, 1518–1524.

Rolls, E.T., Tovee, M.J., Purcell, D.G., Stewart, A.L., and Azzopardi, P. (1994b). The responses of neurons in the temporal cortex of primates, and face identification and detection. *Experimental Brain Research* **101**, 473–484.

Rolls, E.T., Tovee, M.J., and Panzeri, S. (1999). The neurophysiology of backward visual masking: information analysis. *Journal of Cognitive Neuroscience* **11**, 335–346.

Rolls, E.T., Critchley, H.D., Browning, A.S., and Inoue, K. (2006). Face-selective and auditory neurons in the primate orbitofrontal cortex. *Experimental Brain Research* **170**, 74–87.

Rosenthal, D.M. (1986). Two concepts of consciousness. *Philosophical Studies* **49**, 329–359.

Rosenthal, D.M. (1990). *A Theory of Consciousness*. Bielefeld: Zentrum für Interdisziplinaire Forschung.

Rosenthal, D.M. (1993). Thinking that one thinks. In Davies, M. and Humphreys, G.W. (eds) *Consciousness*, pp. 197–223. Oxford: Blackwell.

Rosenthal, D.M. (2004). Varieties of higher-order theory. In Gennaro, R.J. (ed.) *Higher Order Theories of Consciousness*. Amsterdam: John Benjamins.

Rosenthal, D.M. (2005). *Consciousness and Mind*. Oxford: Oxford University Press.

Rumelhart, D.E., Hinton, G.E., and Williams, R.J. (1986). Learning internal representations by error propagation. In Rumelhart, D.E., McClelland, J.L., and Group, T.P.R. (eds) *Parallel Distributed Processing: Explorations in the Microstructure of Cognition*. Cambridge, MA: MIT Press.

Shallice, T. and Burgess, P. (1996). The domain of supervisory processes and temporal organization of behaviour. *Philosophical Transactions of the Royal Society of London Series B Biological Sciences* **351**, 1405–1411.

Smith-Swintosky, V.L., Plata-Salaman, C.R., and Scott, T.R. (1991). Gustatory neural encoding in the monkey cortex: stimulus quality. *Journal of Neurophysiology* **66**, 1156–1165.

Squire, L.R. and Zola, S.M. (1996). Structure and function of declarative and nondeclarative memory systems. *Proceedings of the National Academy of Sciences of the USA* **93**, 13515–13522.

Strongman, K.T. (1996). *The Psychology of Emotion*, 4th edn. London: Wiley.

Treves, A. and Rolls, E.T. (1994). A computational analysis of the role of the hippocampus in memory. *Hippocampus* **4**, 374–391.

Weiskrantz, L. (1968). Emotion. In Weiskrantz, L. (ed.) *Analysis of Behavioural Change*, pp. 50–90. New York: Harper and Row.

Weiskrantz, L. (1997). *Consciousness Lost and Found*. Oxford: Oxford University Press.

Yaxley, S., Rolls, E.T., and Sienkiewicz, Z.J. (1990). Gustatory responses of single neurons in the insula of the macaque monkey. *Journal of Neurophysiology* **63**, 689–700.

Chapter 5

Conscious and unconscious visual processing in the human brain

A. David Milner

5.1 Introduction

Primate vision did not evolve as an end in itself. Darwinian considerations tell us that the phylogenesis of any brain system must be driven by the utility of the system for optimizing an animal's overt behaviour. It follows that the nature of the visual processing systems that we have inherited from our mammalian ancestry must have been dictated by their benefits for improving behavioural efficacy. It is ultimately what an animal *does*, rather than how it sees *per se*, that is critical in ensuring its longevity and maximizing its reproductive potential.

In practice, of course, human vision is rarely thought about in these terms. Instead, it is generally considered in terms of one's own perceptual experience—that is from an 'inside' (first-person) rather than an 'outside' (third-person) perspective. This bias predominates in the psychological and neuroscience literatures, as well as in everyday thinking and in philosophical discourse. Indeed the study of *visuomotor control*—that is, of how vision serves to guide our movements—is a quite recent development in psychology and neuroscience (Jeannerod 1988, 1997; Milner and Goodale 1995). It must be conceded, of course, that if visual science *restricted* itself to studying bodily acts, then it would never be able to throw any light on the fact that some, but not all, of the visual information guiding one's movements is reportable by the individual. Nor indeed could investigators get much of an inkling as to the mental or neural processes intervening between vision and action when a person sees something one minute but does not act upon that information until the next. It is obvious therefore that both a first- and a third-person perspective are needed if we are to arrive at anything like a full description of what is going on in the visual brain.

Broadly speaking, there are two ways in which visual information can influence and guide behaviour. One is immediate and direct. For example, information about the location and dimensions of an object can provide, through quasi-automatic visuomotor transformations, the means to program and guide an act of prehension, such that the hand reaches the target accurately and anticipates the object's physical dimensions so as to enable a smoothly executed grasp to take place. This kind of visual guidance needs to be quick and accurate, and evolution has ensured that it is. The other way in which vision can influence behaviour is much less direct, and depends upon the construction and storage of visual representations in the brain that reflect the structure and semantics of the scene facing the observer. The nature and intentions of subsequent actions will to varying degrees depend on the retrieval, and mental manipulation of, these representations. There is now extensive evidence from neurophysiology, and from lesion studies in human and non-human primates, that these two forms of visual guidance depend critically on two distinct and quasi-separate brain systems, the dorsal and ventral visual streams, respectively (Goodale and Milner 1992, 2004; Jeannerod and Rossetti 1993; Milner and Goodale 1995, 2006; Jacob and Jeannerod 2003). Each of these systems (see Fig. 5.1) consists of a number of interconnected visual areas located in the posterior half of the primate cerebral cortex (Morel and Bullier 1990; Baizer *et al.* 1991; Young 1992; Van Essen 2005), and both systems derive their principal visual inputs from the primary visual cortex (area V1).

Of course even when one's actions are prompted by careful thought about their targets and consequences, they still need some degree of guidance from the currently available retinal information for them to be executed successfully.

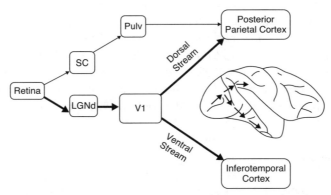

Fig. 5.1 The cortical visual streams. Schematic diagram showing major routes whereby retinal input reaches the dorsal and ventral streams. The inset shows the cortical projections on the right hemisphere of a macaque brain. LGNd, lateral geniculate nucleus, pars dorsalis; Pulv, pulvinar nucleus; SC, superior colliculus.

In fact, neither the ventral nor the dorsal visual system is ever likely to operate in isolation in guiding behaviour. It follows that the two systems need to work together in harmony; and one implication of this is that as far as possible they should sing from a similar (though not the same) visual 'hymn sheet'. That is to say, their visual contents at any time should be more or less mutually consistent. Inevitably, this makes it difficult for the scientist to separate their operation one from the other by using purely behavioural or psychological means. It is therefore no accident that the clearest evidence for the separate operation of the two systems comes from neuroscience, and from behavioural and cognitive neuroscience in particular. Broadly speaking, two complementary research strategies are available. (a) One can examine the operation of one system in relative isolation when the other is somehow damaged or temporarily disabled. Alternatively, (b) one can examine the neural activity of brain areas in one or the other system during visual perception or visually guided behaviour, either by directly recording from the constituent neurons or—increasingly— by using brain imaging techniques. The latter methodology, which nowadays chiefly involves the use of functional magnetic resonance imaging (fMRI), allows a glimpse inside the heads of human subjects in an entirely non-invasive fashion. Although until recently single-neuron recordings have been predominantly the preserve of animal research, and neuroimaging largely restricted to human research, both tools can in fact be used in both humans and animals, and the wider application of fMRI in particular is allowing increasingly close comparisons to be made between the visual brains of humans and their primate relatives (Tootell *et al.* 2003; Orban *et al.* 2004).

In this chapter I will consider evidence derived from both of these experimental strategies to inquire into the nature of the hymn sheets from which each visual processing stream is singing. Clearly the hymn sheets have to be different. In order to provide direct control of one's movements, the dorsal stream has to see the world egocentrically, that is to code visual information in direct relation to the observer's body coordinates; and the information thus coded has to be both up to date and immediately disposable. In contrast, the ventral stream specifically needs to encode the world in a way that will be useful not only in the short but also in the long term, for which purpose egocentric coding would be useless. Accordingly items of interest have to be encoded in relation to other items within the visual scene, with the result that accuracy is inevitably sacrificed to flexibility and durability. It is these different demands placed upon the two systems that no doubt caused them to evolve separately in the first place.

But these functional considerations derive from looking at the two systems from the outside, from a third-person perspective. How do they differ from a first-person perspective? If the visual coding strategies used by the two streams really are qualitatively different, then it is hard to imagine one's being conscious of both at the same time, since this would create severe problems of mutual interference—not a good recipe for biological fitness. Yet it certainly does happen on occasion that the products of the two streams will be downright contradictory, as when one stream is deceived by a visual illusion while the other is not (e.g. Aglioti *et al.* 1995; Dyde and Milner 2002; Króliczak *et al.* 2006). It is true that such contradictions will rarely arise in everyday life; but even when they can be demonstrated rather starkly in the laboratory, it is notable that no *awareness* of the contradiction is experienced by the observers.

We have argued (Milner and Goodale 1995, 2006; Goodale and Milner 2004) that the different functional properties of the two streams, driven by their disparate job descriptions, may have caused their visual products to differ *subjectively*. In other words, we believe that they differ from a first-person, as well as from a third-person perspective. On the one hand we argue that the ventral stream has the job of creating perceptual representations that can be stored and re-experienced, thereby providing informational content for planning, decision-making, and discourse. These representations are not based solely on the retinal input: they are also shaped by 'top-down' influences to varying extents from stored information within visual and semantic memory. Indeed there is increasing evidence that the borderline between perception and memory systems in the brain is a somewhat hazy one (Bussey and Saksida 2005; Buckley and Gaffan 2006). The dorsal stream, on the other hand, we consider to be a 'bottom-up' system depending from moment to moment on its current inputs from the eye. Having a memory would be counter-productive for a visuomotor system that has to deal with snapshots of visual stimulation that change constantly as the observer moves. By definition, we know that our perceptions (as normally understood) are, or usually are, consciously experienced by us. This places them in the same broad category as our recollections, thoughts, and dreams, with which indeed they interact (and sometimes blur into) on a regular basis. In contrast, dorsal-stream visual processing is encapsulated and somewhat inflexible: there are even separate subsystems dedicated to translating visual information into different movement domains such as reaching with a limb, moving the eyes, or opening the hand to grasp a target object (Milner and Goodale 1995; Goodale and Milner 2004). Our proposal is that *the visual products of dorsal stream processing are not available to conscious awareness*; that they exist only as evanescent raw materials to provide moment-to-moment sensory calibration of our movements, and then are gone.

The object of the present chapter is to discuss the current evidence for this claim. After all, despite the qualitative differences in the computations carried out in each visual stream, it remains conceivable *a priori* that both systems might generate products that are (or can be) conscious. This duality might remain absent from one's experience most of the time simply because the more dominant visual system regularly suppresses the phenomenal properties of the other.

Evidence from research on commissurotomized ('split-brain') patients has been cited in support of the possibility of such coexisting separate spheres of consciousness within a single brain (Puccetti 1981). A number of studies of epileptic patients who had undergone surgical section of the cerebral commissures during the 1960s and 1970s showed that each disconnected cerebral hemisphere could deliver responses that have all the hallmarks of intelligence (e.g. Levy *et al.* 1972; LeDoux *et al.* 1977). For example, in Levy *et al.*'s (1972) experiments where 'chimaeric faces' (a composite of two different half-face images) were presented briefly to the split-brain patients, normally one or the other hemisphere took over the job of reporting what was seen, depending on whether a verbal or a manual choice response was demanded in that particular set of trials. Each hemisphere was able to do this (and many other more complex cognitive tasks) quite successfully for its own visual hemifield, despite thereby generating contradictory responses from one set of trials to the next. Furthermore, when the instructions were switched immediately after stimulus presentation, the patient's other hemisphere could typically still respond successfully according to these *new* instructions. Still more impressively, the patient would often proceed to give a conflicting choice when switched *back* to the original form of response. These observations indicate that both the primed and the non-primed hemisphere in the original study had simultaneously processed their respective halves of the visual information. Indeed they suggest that both hemispheres could retain an active perceptual representation in the form of a visual working memory, which could then be 'attended to' on demand and translated into the currently required response.

Do results of this kind imply that there are two parallel streams of consciousness in the separated cerebral hemispheres? In considering this question, it may perhaps be useful to distinguish between 'access' and 'phenomenal' consciousness (Block 1995). To quote Block (1996): 'Phenomenal consciousness is just experience; access consciousness is a kind of direct control. More exactly, a representation is access-conscious if it is poised for direct control of reasoning, reporting and action.' Musing on this distinction in a commentary on Block's 1995 paper, Bogen (1997) suggested that the disconnected right hemisphere might provide an example of 'A-consciousness without P-consciousness'. It certainly does seem plausible to argue that two parallel streams of *access* consciousness must exist in the separated hemispheres, even if the extent of that access is somewhat less than in an intact whole brain. An example of this limitation would be that in most split-brain patients the right hemisphere has very limited access to language processing (see Bayne 2008; LeDoux, this volume). Clearly the very fact of surgical division necessarily means that neither the left nor right hemisphere is ever able to achieve *global* access to the brain's cognitive processing systems, and in this sense could never fully satisfy Baars' 'global workspace' conception of consciousness (Dehaene and Naccache 2001; Baars 2002; Baars *et al.* 2003).

With this proviso, my modified suggestion would be that each hemisphere is indeed access-conscious, and that each can also be phenomenally conscious—*but only one at a time.* The data suggest that phenomenal consciousness will usually be associated with the left hemisphere, which tends to take control, particularly when verbal cognition is required, even when it is less well-equipped to do so. But the right hemisphere can also be induced to take over by various means, and generally comes into its own when the patient has to deal with non-verbal stimuli and when making a motor response, such as pointing to a matching face in the Levy *et al.* (1972) experiment. This proposal, of bilateral access consciousness with unitary (but shifting) phenomenal consciousness, suitably hedged around, seems to me to fit the facts quite well. Nevertheless it is difficult to exclude two more radical views, each of which has had its adherents among the split-brain research fraternity.

First, several eminent split-brain researchers have expressed the belief that both of the patient's hemispheres have full phenomenal consciousness, and that the patient's brain can possess these two streams of consciousness simultaneously (Sperry 1974; LeDoux *et al.* 1977; Bogen 1997; Zaidel *et al.* 2003). As mentioned earlier, Puccetti (1981) took the argument one step further, and proposed that if there can be separate streams of consciousness in *separated* cerebral hemispheres, then perhaps there could be similar dual (or even multiple) states of consciousness in the intact brain too. Such an inference would be fully in the spirit of standard neuropsychological reasoning, summarized in Henry Head's dictum that 'through injury or disease, normal function is laid bare' (Head 1926). That is, the nature of normal function in one brain system can be uncovered by a loss or disconnection of other, more dominant, systems. But does it really make coherent sense to suppose that we might have two coexisting streams of phenomenal consciousness associated with a single intact brain (or a single intact person)? And even if it does make sense, how could we ever know whether such was or was not the case? I agree with Bayne's (2008) argument that there is nothing 'that it is like' (after Nagel 1974) to have dual simultaneous streams of consciousness. That is, I find it impossible to imagine experiencing two independent streams of phenomenal content simultaneously.

Intact Objects

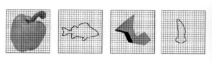

minus

Scrambled Objects

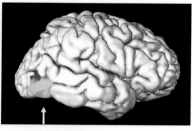

Lateral Occipital
area (area LO)
activation

Plate 5.1 The lateral occipital area (LO). Functional MRI has revealed the existence of a human cortical area that responds selectively to pictures of objects. This area shows a higher activation to intact pictures of objects than to pictures of non-objects such as faces or random shapes, or to fragmented versions of pictures of objects (as illustrated here). The diagram on the right illustrates the typical subtraction pattern resulting from responses to intact objects minus those to fragmented objects.

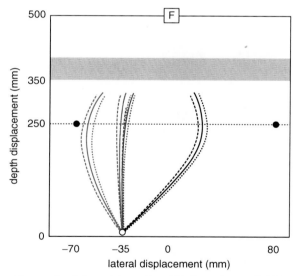

Plate 5.2 Reaching routes followed by patient V.E. (Experiment 1). V.E.'s spatially averaged trajectories in each condition (dotted lines indicate standard errors). The zero lateral coordinate is aligned with V.E.'s mid-sagittal axis. Green: reaches with only the right rod present; black: reaches with only the left rod present. Blue: reaches with both rods present, V.E. reports seeing both rods; red: reaches with both rods present, V.E. reports seeing only right rod. Reprinted from McIntosh et al. (2004a), *Proc. R Soc. B*, 172 15–20, with permission of the Royal Society.

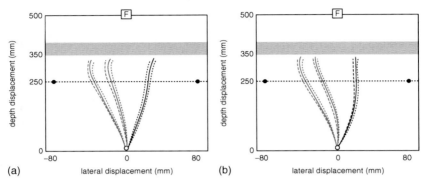

Plate 5.3 Reaching routes followed by patient V.E. (Experiment 2). V.E.'s spatially averaged trajectories in (a) the motor-verbal task and (b) the verbal-motor task (dotted lines indicate standard errors). The zero lateral coordinate is aligned with V.E.'s mid-sagittal axis. The colour coding is the same as in Plate 5.2. Reprinted from McIntosh et al. (2004a), Proc. R Soc. B 172, 15–20, with permission of the Royal Society.

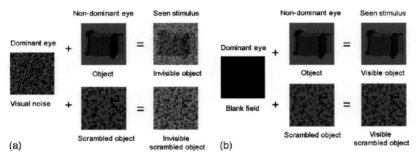

Plate 5.4 The stimuli in Fang and He's (2005) experiment. (a) The 'invisible' condition, in which awareness of intact or scrambled objects presented to the non-dominant eye was completely suppressed by dynamic, high-contrast, random textures presented to the dominant eye. The 'invisibility' of the suppressed images was validated in separate testing. (b) The 'visible' control condition, in which only intact or scrambled objects were presented to the non-dominant eye. Reprinted from Fang, F. and He, S., (2005) Nature Neuroscience 8, 1380–1385, by permission from Macmillan Publishers Ltd.

This argument, of course, does not present a problem for the more restricted 'access' consciousness view that I have proposed above: it seems quite possible to entertain the notion that while only one of the two hemispheres is able to achieve 'phenomenal' consciousness at any given moment, both hemispheres could be simultaneously enjoying separate *access* consciousness. None the less, I am not entirely convinced that the 'what it is like' argument provides a knock-out punch to the dual phenomenal consciousness view.

The second—less contentious—idea was put forward by Jerre Levy (1990) in discussing her data on the California split-brain patients (Levy *et al.* 1972; Levy and Trevarthen 1977), and has been rehabilitated by Bayne (2008). Levy argued (and still argues) that the switches between the disconnected hemispheres as to which controls behaviour at any moment are akin to normal switches in attention in everyday life. It is a common observation that an 'unheard' verbal comment, received while one is concentrating on something else, can attain consciousness when attention is retrospectively directed to it. In a similar way, according to Levy, the 'unattended' (i.e. temporarily deactivated) hemisphere is only able to engage in unconscious or 'preconscious' processing—until 'attention' is directed to it. In other words, although either hemisphere can be conscious, they are never both conscious at any given moment: there is an inbuilt reciprocal inhibition between them, such that only one is ever fully activated. Either the left or the right hemisphere seems to take control of the patient's (undivided) consciousness, on a fluctuating or time-sharing basis, depending on which is primed by the current context. This proposal avoids the awkward notion of two simultaneous streams of consciousness within a given person, and the corpus of evidence from split-brain research seems to be more or less consistent with it (Levy 1990; Bayne 2008).[1] On the other hand, however, I feel uneasy about the level of cognitive processing that one would have to assume was going on in a supposedly 'unconscious' hemisphere, in order to account for the results of Levy's own research on chimaeric faces described above. Would it really be possible, for example, for an *unconscious* hemisphere to entertain an active working memory representation

---

[1] Similar observations can be used to argue for an undivided *volition* in these patients, despite the apparent ability of either hemisphere to take control of action. To take a graphic example, in the early months after surgery, before the divided brain has adjusted to its new condition, conflicts between the two hemispheres may occur; but they always seem to occur *serially*. For example, one hand may zip up a pair of trousers, and then the other hand take control and unzip them. Yet the two hands never fight each other simultaneously. After fuller recovery, such conflicts do not normally occur at all, though they can be induced in the laboratory (J. Levy, personal communication 2006).

sufficient for a judgement on face identity to be made, even after several seconds? In contrast, an access-conscious (but not phenomenally conscious) hemisphere could presumably do this.

These competing accounts (and others: Bayne 2008) have been proposed with reference only to the two cerebral hemispheres; but in principle, variants of each of them could be applied to the dorsal and ventral visual streams *within* each hemisphere. That is, both streams could have the capacity to generate phenomenally conscious visual percepts, either simultaneously or successively; or only one of them may have this capacity. Of course if a dual-consciousness account were correct in the case of the two visual streams, then an obvious first expectation would be that severe damage to, or deactivation of, either stream would leave the other one still 'conscious'. In direct opposition to such a prediction, however, our research with patient D.F. indicates strongly that the dorsal stream processes a range of visual information quite unconsciously. D.F., who has a profound disorder of shape perception known as *visual form agnosia* as a result of bilateral damage to her ventral stream (Milner *et al.* 1991; James *et al.* 2003), can nonetheless use visual information quite normally to shape her hand and rotate her wrist when reaching out to pick up an object (Goodale *et al.* 1991; Milner *et al.* 1991; Carey *et al.* 1996). That is, she can efficiently *deploy* visual information to guide her actions, despite being unable to report that information, whether verbally or non-verbally. Moreover, a fMRI study has shown that when she does this her dorsal stream is activated in the usual way (James *et al.* 2003). [Conversely, patients with damage to their dorsal stream have difficulties in reaching for and grasping visual objects ('optic ataxia'), while remaining able to report their perceptions and to make discriminations based on them (Perenin and Vighetto 1983, 1988; Jeannerod *et al.* 1994; Milner *et al.* 2003).] In other words, D.F.'s dorsal stream is evidently performing its normal visuomotor role, yet without the processed visual information reaching her (access or phenomenal) consciousness.[2] As indicated earlier, the Milner and Goodale (1995) model goes beyond these observations (following the logic of Henry Head's dictum). We propose first that this separation between immediate visuomotor control and perception applies equally to the intact brain; and second that the dorsal stream, in

---

[2] This absence of visual awareness does not imply that D.F. has no awareness of her own *actions*, both visually to some degree (for example, she can perceive the motion of her own hand) and non-visually (see sections 5.3 and 5.4 below).

(James *et al*. 2003). Consequently D.F. has retained at least some crude perception of space and motion across the visual array (Steeves *et al*. 2004; Carey *et al*. 2006). Instead, we have carried out extensive testing of a patient with *visual extinction*. Extinction, which is a symptom often associated with spatial neglect, refers to a failure to detect a sensory stimulus presented on the side of space opposite to a brain lesion, but only when there is simultaneously a stimulus present on the 'good' (ipsilesional) side as well. The definition requires that a unilateral stimulus on either side can be detected alone, thus ruling out a simple sensory failure. Extinction is believed to be caused by a pathological imbalance of attentional resources, which reveals itself as a loss of awareness on the contralesional side of space under competitive conditions. As a rule, extinction occurs most reliably when the sensory stimulation is presented very briefly, presumably because longer presentations allow time for attention to be switched from one side to the other. This time-dependence of extinction makes it an ideal 'experiment of nature' for studying the causal role of sensory awareness in behaviour, since at intermediate stimulus durations the patient will sometimes detect, and sometimes not detect, the very same contralesional stimulus. By fixing the exposure duration judiciously, it thus becomes possible to collect and compare behavioural data on trials with identical physical stimulation but qualitatively different visual phenomenology.

Our patient, V.E., was aged in his early seventies at the time of testing, and a right parietotemporal stroke one year earlier had left him with a persistent left-side visual (and tactile) extinction, but with no other detectable neuropsychological impairment. In particular, we undertook careful screening for visual field loss, and found no indication of any areas of blindness.

5.2.2.1 Obstacle avoidance

In our first investigation with patient V.E., we examined his reaching behaviour in the presence of potential obstacles, using the arrangement illustrated in Fig. 5.3 (McIntosh *et al*. 2004a). On any given trial, one or two thin rods would be present on the testing board, and the task was simply to reach out and touch the grey strip at the back of the board, passing the hand between the rods whenever both were present. First, however, we needed to contrive the conditions such that V.E. would show extinction of the left rod. We therefore asked him to wear special 'shutter' spectacles with liquid-crystal lenses that could be switched between translucency and opacity instantaneously by means of an electrical pulse, so that his view of the board could be restricted reliably to a brief exposure. Figure 5.4 shows the results when he saw the scene either for 500 ms or 1000 ms. Our intention was to find an exposure time that would be brief enough to result in visual extinction on a substantial proportion of trials, but without being so brief that accurate reaching between the

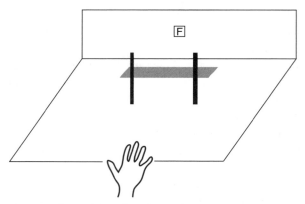

Fig. 5.3 Stimulus array for studying obstacle avoidance in extinction. The patient (V.E.) was simply asked to reach out from a fixed starting point on each trial and touch the grey strip at the back of the board with the right forefinger. Reprinted from McIntosh et al. (2004a), *Proc. R Soc. B* 172, 15–20, with permission of the Royal Society.

two rods would be difficult. As the figure shows, 1000 ms was always sufficient to allow V.E. to see the left-hand rod, even when both rods were present: that is, he never showed extinction. At 500 ms, however, he failed to detect the left-hand rod on a majority of occasions when both rods were present, despite his reporting its presence on 90% of occasions when it appeared alone. This exposure time was also sufficient to allow V.E., after a little practice, to reach between the two rods without colliding with either of them. In order to prevent such a collision ever happening in the experiment proper, however, we used an electromagnetic device that caused the rods to drop rapidly through holes in the board at the instant the spectacle lenses closed. Thus, without ever

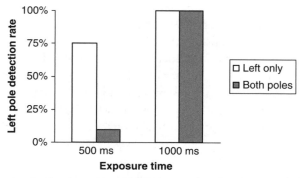

Fig. 5.4 Visual extinction in patient V.E. Histogram showing the rate of detection of the left rod, according to whether the right rod was also present, at two different exposure durations.

realizing it, V.E. always in fact made his reaches across a completely open board.

Unpredictably from trial to trial in the experiments proper, V.E. was presented with either the left rod alone, the right rod alone, both rods together, or neither rod present. On each trial, he made his reaches as quickly as possible, and also reported after doing so which rods (if any) he had seen on the board. As shown in Fig. 5.5 (see also Plate 5.2), his reaches veered strongly away from the left rod (average trajectory shown in black) or right rod (shown in green) whenever either was presented alone. The points at which these averaged reaches crossed the imaginary line joining the rods differed from one another at a highly reliable statistical level, and also did so from the reaches made when both rods were present (shown in blue and red). Crucially, when both rods were present, the subset of reaches made where V.E. reported seeing only the right one (red), were statistically indistinguishable from those where he reported seeing both rods (blue). They were grossly different from the reaches he made when only the right rod was *truly* present (shown in green), despite the fact that his reported

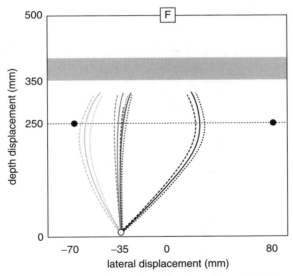

Fig. 5.5 Reaching routes followed by patient V.E. (Experiment 1). V.E.'s spatially averaged trajectories in each condition (dotted lines indicate standard errors). The zero lateral coordinate is aligned with V.E.'s mid-sagittal axis. Green: reaches with only the right rod present; black: reaches with only the left rod present. Blue: reaches with both rods present, V.E. reports seeing both rods; red: reaches with both rods present, V.E. reports seeing only right rod. Reprinted from McIntosh et al. (2004a), *Proc. R Soc. B,* 172 15–20, with permission of the Royal Society (see Plate 5.2).

visual experience on these trials was the same. Thus the reaches that V.E. made when he experienced visual extinction still took full account of the 'unseen' left rod, exactly like those he made when both rods were seen and reported.

But of course there was always a brief time lag between V.E.'s reach and his verbal report; so it could perhaps be argued that he 'forgot' seeing a rod on the left, even though he actually had been aware of it at the time of initiating his reach. In other words, his short-term memory of what he had seen might itself have suffered from a kind of extinction, whereby his memory of what he saw on the left was blotted out by his recall of what was on the right. We therefore carried out a further experiment in which we controlled for this possibility. In this experiment we asked V.E. to carry out the reaching task under two different conditions. In one condition, as before, he was asked to say after reaching what he had seen on that trial; and in the other condition, he was asked first to say what he saw, and then quickly to perform the reach. The results, shown in Fig. 5.6 (see also Plate 5.3), clearly allow us to rule out the 'memory failure' account of the previous experimental data. First, the reaching trajectories on the left side of the figure essentially replicate those shown in Fig. 5.5 (see also Plate 5.2); and the almost identical pattern shown on the right side of Fig. 5.6 (see also Plate 5.3) allows us to conclude that getting rid of the slight delay in reporting what he saw made absolutely no difference to this pattern. That is, his behaviour when he experienced extinction still closely resembled his behaviour when he reported seeing both rods; and it still differed completely from his behaviour when he *correctly* reported the presence of only the right rod.

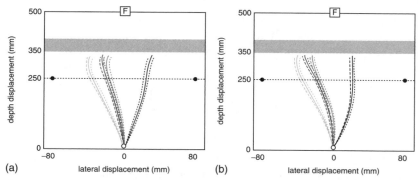

Fig. 5.6 Reaching routes followed by patient V.E. (Experiment 2). V.E.'s spatially averaged trajectories in (a) the motor-verbal task and (b) the verbal-motor task (dotted lines indicate standard errors). The zero lateral coordinate is aligned with V.E.'s mid-sagittal axis. The colour coding is the same as in Fig. 5.5. Reprinted from McIntosh et al. (2004a), *Proc. R Soc. B* 172, 15–20, with permission of the Royal Society (see Plate 5.3).

These results strongly support the idea that the visuomotor adjustments made to take account of potential obstacles during reaching are guided by visual information that is not, or does not have to be, conscious. In addition, we now have strong independent evidence that these adjustments to one's reaching behaviour depend crucially on the integrity of the dorsal stream. This evidence comes from a slightly different task, in which the potential obstacles are always present on each trial, but in varying locations (see Fig. 5.7). Since the left and right cylinder can appear in either of two locations, there are four possible configurations of the two, which are illustrated in the rows A, B, C, and D in Fig. 5.8. The connected open squares in Fig. 5.8 (left) show how healthy subjects behaved: each of them shifted their reaching responses lawfully and reliably in accordance with the configuration present on a given trial. The question we asked was whether two patients with bilateral damage to the dorsal stream (causing optic ataxia) would do the same. We found, as shown in Fig. 5.8 (left), that neither of our patients (I.G.: filled circles; A.T.: triangles) made any such systematic adjustments to their reaches at all. The reaches of both patients remained statistically invariant regardless of the configuration facing them when responding (Schindler *et al.* 2004). Thus we can conclude

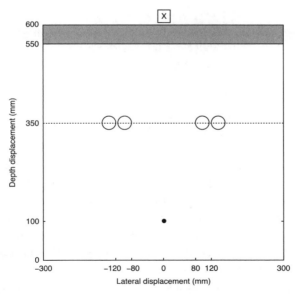

Fig. 5.7 Plan of the reaching board used in testing patients with optic ataxia. The open circles show the possible locations of the two cylinders, which were always presented one on the left and one on the right. The start position is shown by a black dot, and the fixation point is indicated by a cross. Reprinted from Schindler *et al.* (2004), with kind permission of Macmillan Publishers Ltd.

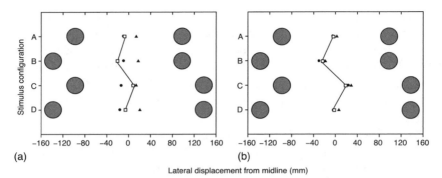

Fig. 5.8 Obstacle avoidance and gap bisection in patients with optic ataxia. The averaged responses in (a) the reaching task, and (b) a perceptual control task of spatial bisection are given for the two patients (A.T.: filled triangles) and (I.G.: filled circles), with the mean data for the eight control subjects shown as open squares. The responses plotted are the points where each response intersects the imaginary line joining the four possible cylinder locations. The dark grey circles depict the stimulus cylinder locations in the four configurations (A, B, C, D). Reprinted from Schindler et al. (2004), with kind permission of Macmillan Publishers Ltd.

that the integrity of the dorsal stream is necessary for this behaviour to survive after brain damage. In agreement with this, we have confirmed in other experiments that other neurological patients with severe visual difficulties, but whose dorsal stream is structurally spared, still perform normally on this task. As well as our patient with extinction (V.E.), the list includes patients with visuospatial neglect (McIntosh et al. 2004b), and also patients with visual form agnosia (D.F. and S.B.: Rice et al. 2006b).

In short, the combined data from all of these experiments provide decisive evidence that obstacle avoidance, of this unwitting variety at least, depends on the integrity of the dorsal stream, and that it proceeds without the need for conscious perception of the stimuli that guide it. Of course, there is no doubt that our healthy controls in these experiments *were* visually aware of both potential obstacles as they reached out between them. What the data from our extinction patient V.E. (McIntosh et al. 2004a) show is that they did not need to be. Indeed we argue that their awareness was generated in a quite separate brain system (either in the ventral stream itself or in 'higher' circuitry within the right inferior parietal lobule that receives its visual inputs from the ventral stream) from that which was guiding their movements through space (namely the dorsal stream and its associated premotor and subcortical areas).

5.2.2.2 Visual feedback during reaching

The second study we undertook with patient V.E. was designed to investigate the benefits of visual feedback received from the hand during reaching out towards an object. Reaching accuracy generally improves when one can see one's hand (Prablanc et al. 1979a, 1979b), particularly if the hand is visible during the deceleration phase of the movement, while it is 'zeroing in' on the target. In fact under normal viewing conditions, the brain continuously registers the visual locations of both the reaching hand and the target, incorporating these two visual elements within a single 'loop' that operates like a servomechanism to minimize their mutual separation in space (the 'error signal') as the movement unfolds. When the need to use such visual feedback is increased by the occasional introduction of unexpected perturbations in the location of the target during the course of a reach, a healthy subject will make the necessary adjustments to the parameters of his or her movement quite seamlessly (e.g. Goodale et al. 1986). In contrast, a patient with damage to the dorsal stream (patient I.G.—see above) was quite unable to take such new target information on board; she first had to complete the reach towards the original location, before then making a *post hoc* switch to the new target location (Pisella et al. 2000; Gréa et al. 2002). It should be noted, however, that I.G.'s deficit is not a failure to detect the changed location of the target (which she can certainly do), it is a failure to *use* information about the visual 'error signal' between hand and target. It thus seems very likely that the ability to exploit the error signal during reaching is dependent on the integrity of the dorsal stream.[4]

It should be noted here that even when the target is not perturbed, I.G. and other patients with optic ataxia still (by definition) show severe misreaching, particularly when the target is located peripherally in the visual field. This misreaching is partly, but certainly not only, due to their inability to use feedback from the hand: a major part of it is due to a motor programming deficit. This faulty programming is clearly evident, for example, in the large heading errors seen in optic ataxia even when visual feedback from the moving hand is not available, that is, in so-called open-loop reaching (Perenin and Vighetto 1983; Jeannerod 1988). The programming deficit is present during closed-loop pointing (Milner et al. 2003) as well, though it may be masked by the slow corrections made just before, or just after, the end of the reach. These offline

---

[4] There is evidence that although one may not be aware of the target shift in such experiments, one may nevertheless be aware of the *movement* one has made (Johnson and Haggard 2005). This certainly does not mean, however, that the dorsal stream itself mediates such action awareness (Tsakiris and Haggard, this volume). Indeed, Johnson and Haggard (2005) were unable to disrupt awareness of the movement by applying transcranial magnetic stimulation over the dorsal stream. Some independent mechanism for monitoring one's own movements, perhaps in premotor cortex or in *left* inferior parietal cortex, would be good alternative possibilities. (See also footnotes 9 and 11, and Frith, this volume.)

corrections seem unlikely to be implemented by the dorsal stream, and very probably depend on a conscious perception of the terminal error provided by the ventral stream. In contrast, the dorsal stream's job is to provide rapid online adjustments throughout the reach, thereby constantly updating the reach trajectory as it unfolds.

Are we aware of the visual information that the dorsal stream uses for these instantaneous online adjustments? First, there is good evidence that the 'target' side of the loop can be processed by the visual system quite unconsciously. Goodale *et al.* (1986) showed that when a shift in the object location is introduced without a healthy subject's knowledge (during the course of a quick 'saccadic' eye movement, when the observer is effectively blind), the reach is seamlessly adjusted in-flight, even when the hand itself is not visible during the reach.[5] But what about *self*-vision—does the dorsal stream need to register consciously the location of one's own hand as it moves towards the target? Schenk *et al.* (2005) addressed this question using a similar strategy as McIntosh *et al.* (2004a), by contriving a task in which patient V.E. would often, by virtue of his visual extinction, be unaware of the visual information coming from his hand while executing reaching movements. Since his extinction was for *left*-sided stimuli, we had to ask him to reach with his left hand, which despite his earlier stroke, he could by this stage do without difficulty. We attached a small LED to his index finger, and placed a similar LED on the table as the target for his reaches. The target always came on briefly at the beginning of each trial, but then during the reach itself one of the visual stimuli (target or finger), or both together, were switched on. Of course having both stimuli (target *and* finger) illuminated would 'close the loop', but would also tend to induce extinction of the hand LED. Simultaneous 500 ms presentations turned out again to be perfect for our purpose: this duration was long enough to provide significant feedback benefits for reaching accuracy, while at the same time being short enough to result in extinction of the hand LED on more than half of the test trials.

The results are shown in Fig. 5.9 (Schenk *et al.* 2005). Although the separation of the experiment into three phases is not important for present purposes, it is clear that throughout the experiment the benefits of visual feedback were strong and reliable. This was just as true in patient V.E. as in the three healthy control subjects we tested. Of particular relevance is that this benefit was equally apparent whether V.E. was aware of the visual feedback on a given trial or not: there were no statistical differences in reaching accuracy between the 'aware' trials and the 'unaware' ones. In other words, we can conclude that V.E.

---

[5] Of course large stimulus perturbations will normally be noticed consciously, and may induce a totally re-programmed movement, presumably through ventral-stream mediation.

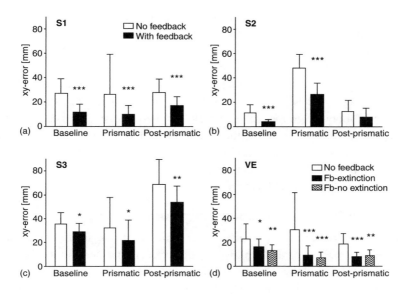

Fig. 5.9 The improvement in reaching accuracy resulting from visual feedback. Data are shown for three healthy subjects (**A–C**) and in patient V.E. (**D**). The bar graphs depict the mean x-y error and its standard deviation. The open bars show the results for the trials without visual feedback. The black bars show the results for the trials with visual feedback. In the case of patient V.E., the trials with visual feedback have been divided into those trials where the visual feedback information was extinguished (black bars) as compared to those where it was not extinguished (striped bars). The number of stars above the bars indicate the results from the t-tests which compared the pointing accuracy with and without visual feedback (no star: $p > 0.05$; *: $p < 0.05$; **: $p < 0.01$; ***: $p < 0.001$). In patient V.E., trials in which the visual feedback was extinguished, and trials in which it was not both differed significantly from trials where no visual feedback was provided. The two types of visual feedback trials (i.e. with and without extinction) did not differ significantly from each other. Reprinted from Schenk et al. (2005) with permission from Springer Science and Business Media.

had no need of conscious awareness of the light stimulus on his left hand as he reached out: it still benefited his visuomotor system to the same extent whether or not he was visually aware of its presence.

By extension, we can infer that even in our healthy controls, this same visuomotor control loop was operating on the basis of unconsciously processed information from their reaching hand, even though, unlike V.E., they were always able to report seeing the light on the hand whenever it was switched on. The logic behind this apparently paradoxical claim is similar to that in our interpretation above of V.E.'s behaviour in the obstacle avoidance experiment. There is no question of the controls being unaware of the light stimulus emanating from their reaching hand—clearly they were fully aware of it. Our claim is that the processing of that stimulus for visuomotor guidance took place in a *different brain system* from that which generated the conscious visual experiences reported by the controls. In keeping with this proposal, Sarlegna and colleagues (2004) have recently shown that healthy individuals will automatically modify their manual reach trajectory in response to an illusory shift in seen hand position, even when this shift is induced during a saccadic eye movement and is therefore not consciously detected (due to saccadic suppression). This supports our contention that the normal brain uses unconscious visual information about hand location during reaching.

5.2.3 Evidence from functional neuroimaging

All of the evidence I have presented so far comes from studying brain-damaged patients. Evidence of this kind is very powerful, particularly in its ability to decide between different causal accounts—in this instance regarding the role of unconscious visual processing in the guidance of behaviour. Explanations in cognitive neuroscience, however, are at their most convincing when they are supported by converging evidence from more than just one methodology. The directly converse approach to investigating how brain damage affects cognition and behaviour is to monitor the areas that are active during those same forms of cognition and behaviour in the *intact* brain. Currently the most effective way to do this in humans is to use fMRI (see Kanwisher and Duncan 2004), although this continues to have important technical limitations (notably in its temporal resolution). Of course, all techniques that record the brain activity that accompanies behaviour and cognition inevitably suffer from the critical limitation of being correlational—they can never, alone, provide conclusive evidence for particular causal accounts of the phenomena in question. But although fMRI cannot stand alone, it can provide powerful support for theories based on other methodologies.

In the case of several ventral-stream areas, there is now very strong evidence that neural activation levels correlate closely with visual awareness. First, it is a well-known observation that when observers are presented dichoptically with two conflicting images (for example, a face to one eye and a building to the other), they experience constantly changing percepts, of either the face alone or the house alone, depending from moment to moment on which eye's image is dominating perception. Yet despite these changing experiences, the two retinae (and the early parts of the visual system) are receiving unchanging visual stimulation throughout. Tong and his colleagues (1998) exploited this phenomenon of 'binocular rivalry' while measuring MRI activations in two eponymous areas of the human ventral stream known to be specialized for the processing of the stimuli used in their experiment: the fusiform face area (FFA) and parahippocampal place area (PPA), respectively. They found that activation in the FFA and PPA fluctuated in a reciprocal fashion, in close correspondence with a subject's current awareness of the face or the building (which the subject indicated by pressing a key whenever a change occurred). In other words, the level of activity in each area was directly correlated with the presence or absence of visual awareness for stimuli within its visual specialization. Presumably whatever brain process determined the switching between the two alternative percepts did so by modulating the relative activity levels in these two parts of the ventral stream.

It is notable, as a parenthetic aside, that this experiment was inspired by a striking earlier study by Sheinberg and Logothetis (1997), who trained monkeys to report on the content of their visual experience during binocular rivalry in a closely comparable experimental paradigm. They discovered that the activity of single neurons in the monkey's ventral stream that responded selectively (say to a face stimulus) exhibited correlated fluctuations in their activity according to what the monkey reported seeing. Each neuron would fire more when the monkey reported seeing the preferred stimulus for that cell, and less when the animal reported a switch to seeing the other stimulus. Evidently, not only did the monkey experience binocular rivalry much as we do, its fluctuating percepts were closely tied to individual stimulus-selective neurons. An equally remarkable piece of recent neurophysiological research goes a major step further, by providing a *causal* link between activity in ventral-stream neurons in the monkey's brain and the monkey's perceptual report. Afraz *et al.* (2006) trained their animals to categorize images as either a 'face' or a 'non-face'. They then artificially activated small clusters of neurons in the ventral stream by using electrical microstimulation, while the monkey viewed ambiguous visual images that varied in their 'noisiness'. Microstimulation of face-selective sites, but not other sites, strongly biased the monkeys' responses towards the 'face' category. This result seems to demonstrate that the level of activity of face-selective neurons directly determines the monkey's tendency to 'perceive' a face (even when there is no face present).[6]

---

[6] It is as though the monkey was experiencing a lower threshold for seeing a 'man in the moon'—the common human illusion of 'seeing' a face when looking at a totally inanimate object. The adaptive importance for primates of biological stimuli such as faces has evidently resulted in the evolution of specialized ventral-stream modules to process them. But this generally beneficial inheritance may bring with it a proneness to animistic 'false positives' in our perceptual experience and perhaps even to animistic beliefs more generally.

A second good example of how neuroimaging research has linked human ventral-stream activity with conscious perceptual content is provided by the moment-to-moment fluctuations in our perception of ambiguous stimuli such as the well-known Rubin face/vase figure (Fig. 5.10). Again, observers can be asked to record the occurrence of these changes by pressing a key, and it has been found that the FFA responds more strongly whenever they report seeing the face rather than the vase, despite the fact that an identical physical stimulus is present throughout (Hasson *et al.* 2001; Andrews *et al.* 2002). Such changes were not seen either in area PPA or in the 'object area' LO (see Fig. 5.2; see also Plate 5.1).

These and many other studies (e.g. Ferber *et al.* 2003; Andrews and Schluppeck 2004) provide strong evidence for a close association between different patterns of brain activity within the ventral stream and the contents of conscious perception. Notably, however, none of these studies tell us whether the differential fMRI activation associated with conscious versus unconscious processing is *restricted* to the ventral stream. Might not such differential effects also be present in the dorsal stream?

As mentioned earlier in this chapter, fMRI has uncovered specific parts of the human dorsal stream that are concerned with visuomotor control, including ones that are specialized for the visual processing of object shape, size, and orientation to guide our grasping actions when picking up objects. The best known part of this 'grasp' circuitry (area AIP) lies anteriorly within the intraparietal sulcus, close to the border between the dorsal stream and primary sensorimotor cortex. This area, first discovered in the monkey (Taira *et al.* 1990; Murata *et al.* 2000), shows selective activation in humans during visually guided grasping within the MRI scanner (Binkofski *et al.* 1999; Culham

Fig. 5.10 The Rubin face/vase figure.

et al. 2003). Significantly, the *causal* role of human AIP in determining effective grasp configurations under visually guidance has been documented through studies of patients with circumscribed parietal lobe damage (Binkofski *et al.* 1998), and of normal subjects using transcranial magnetic stimulation (Rice *et al.* 2006a). In both cases, a selective deficit in visually guided grasping has been demonstrated to follow disruption of area AIP. In addition, however, a more posterior dorsal-stream area, lying in the caudal intraparietal sulcus (cIPS), both in monkeys (Sakata *et al.* 1997) and in humans (James *et al.* 2002), is selectively activated by action-relevant information about the shape and orientation of objects, even when no overt action occurs (see review by Culham and Valyear 2006). In the monkey, this area is known to feed information forward to AIP, which in turn has reciprocal connections with a further grasp-related area (F5) in premotor cortex (Sakata *et al.* 1997). The crucial question for present purposes, therefore, is whether levels of activation in any of these visuomotor areas is ever associated with conscious visual experiences of shape, size or orientation.

Since they are concerned with processing object information for calibrating grasping, areas AIP and cIPS may be regarded as the dorsal-stream counterpart of the lateral occipital area (LO) in the ventral stream. As mentioned earlier, area LO is delineated by plotting the fMRI activation seen during the presentation of pictures of whole objects, and then subtracting the activation obtained from presenting scrambled versions of the same pictures (see Fig. 5.2; see Plate 5.1). In an interesting recent study, Fang and He (2005) have compared the activation of area LO with the object-related areas in the dorsal stream, while presenting images of objects to one eye that could not be consciously perceived, due to the presence of a simultaneous high-contrast dynamic noise pattern on the other eye (see Fig. 5.11; see Plate 5.4). They discovered, remarkably, that although observers were quite unaware of the object pictures as a result of this interocular suppression, the pictures still elicited substantial fMRI activation in the dorsal stream, and indeed this activation did not differ significantly from that recorded when the image was consciously perceived. In complete contrast, as would be expected from the previous data using binocular rivalry, Fang and He did find large differences in activation in the ventral stream (in and around area LO) between these 'aware' and 'unaware' conditions (Fig. 5.12). These results provide the first clear fMRI evidence that neural activity in the dorsal stream is not correlated with visual awareness. That is, visual shape information gets through and is fully processed in the dorsal stream, even when it is suppressed in the ventral stream, *and irrespective of whether or not it is consciously perceived.*

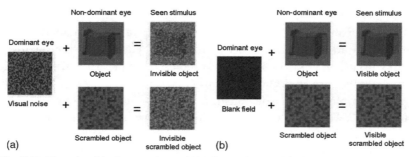

Fig. 5.11 The stimuli in Fang and He's (2005) experiment. (a) The 'invisible' condition, in which awareness of intact or scrambled objects presented to the non-dominant eye was completely suppressed by dynamic, high-contrast, random textures presented to the dominant eye. The 'invisibility' of the suppressed images was validated in separate testing. (b) The 'visible' control condition, in which only intact or scrambled objects were presented to the non-dominant eye. Reprinted from Fang, F. and He, S., (2005) *Nature Neuroscience* 8, 1380–1385, by permission from Macmillan Publishers Ltd (see Plate 5.4).

Fang and He's data nicely support the conclusions that emerge from studying brain-damaged patients, namely that visual processing in the dorsal stream proceeds quite independently of the concurrent conscious perception of the observer. In fact, we can reasonably assume that the very same object-processing systems in the dorsal stream that were activated in Fang and He's experiment must be functional in our agnosic patient D.F. too, thereby allowing

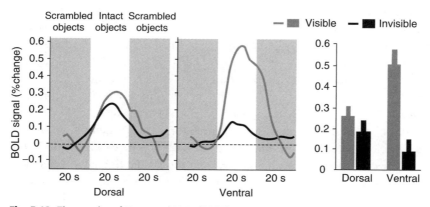

Fig. 5.12 The results of Fang and He's (2005) experiment. The time courses and the average fMRI response (percentage change BOLD signals) from dorsal and ventral object sensitive areas in 'visible' (grey curves and bars) and 'invisible' (black curves and bars) conditions. Data (mean ±SEm) were averaged across eight subjects. Reprinted from Fang, F. and He, S., (2005) *Nature Neuroscience* 8, 1380–1385, by permission from Macmillan Publishers Ltd.

her to perform perfectly well-formed grasping actions for objects of different shapes and sizes. Neither in D.F. nor in the healthy subjects of Fang and He's experiment do these systems provide any conscious visual percepts of object form. That conscious perception, when present, is evidently generated elsewhere—that is, within (or beyond) the ventral stream.

5.3 The ventral stream and conscious perception

Although I wish to maintain that the visual information that is processed in the dorsal stream does not reach conscious awareness, I would not wish to imply by this that all of the information that is processed in the ventral stream does reach conscious awareness; nor indeed that ventral-stream activity is in any direct way *identifiable* with conscious visual perception. If we were conscious of all the products of ventral-stream processing, we would be overwhelmed with information, and consciousness would be unable to serve any useful function: that is, it could not have the 'access' properties outlined by Block (1995). But irrespective of such theoretical arguments, empirical evidence from both human and non-human primate studies refute such a simple-minded notion. For example, in the monkey study of Sheinberg and Logothetis (1997) described earlier, even when one eye's stimulus is suppressed during rivalry, it still results in *some* activation of inferior temporal neurons selective for that stimulus, albeit much less than when the stimulus is dominant. Similarly, the fMRI changes observed in the experiments by Tong and colleagues (1998) and by Andrews and colleagues (2002) were not all-or-nothing: even when the areas under study were suppressed during perceptual fluctuations, they still responded significantly above base levels. It seems plausible to suppose that when one neuronal ensemble within the ventral stream is boosted, it exerts an active inhibitory influence to damp down, but not abolish, the local activity elicited by other stimuli. Presumably this damping-down happens only to the extent necessary to ensure that the favoured ensemble wins the competition for awareness. Most probably, these see-sawing fluctuations in neural activation are simply special cases of what happens all the time as one's visual attention jumps from one place or object in the environment to another (cf. Desimone and Duncan 1995). In other words, understanding how the brain controls our selective attention in general should help us to understand how it controls the contents of our visual consciousness in the particular case of stimulus ambiguity.

There is some empirical evidence in support of this idea. Functional MRI studies have revealed the activation of a network of brain areas, extending from superior parietal regions (in and around the intraparietal sulcus) to premotor regions in the frontal lobes, in synchrony with the perceptual fluctuations experienced in both binocular rivalry (Lumer et al. 1998), and when ambiguous figures are viewed (Kleinschmidt et al. 1998). To a first approximation at least, the areas that are active in this way correspond to the 'dorsal frontoparietal network' that is believed to control the switching of visuospatial attention between locations in the visual field in both humans (Corbetta et al. 2002) and monkeys (Goldberg et al. 2006; Wardak et al. 2006). And it is almost certainly no accident that in turn these same regions are centred around the cortical visuomotor areas dedicated for controlling saccadic eye movements, particularly the mutually interconnected areas LIP (lateral intraparietal area) and FEF (frontal eye field). For example, ingenious recent work with monkeys shows that mild microstimulation of the FEF can have facilitatory influences on visual neurons in the ventral stream (Moore et al. 2003). These and other recent findings (e.g. Cavanaugh and Wurtz 2004) show that subthreshold levels of activity within certain oculomotor systems (strong enough perhaps for 'preparing' an eye movement to a particular location, but not for executing it) can feed back to enhance local processing in perceptual systems representing the corresponding part of the visual field. This recent research provides striking new support for the 'premotor' theory of attention proposed some years ago by Rizzolatti and his colleagues (1985, 1994).

These findings also serve to make the important point that the systems that control the *switching* of attention are different from those that provide the experiential content of what we are currently attending to. It seems that the neural ensembles that code our perceptual representations in the ventral stream are subject to fluctuating levels of activation which are determined by a modulatory system that (paradoxically) is located within the *dorsal* stream and associated visuomotor structures in the frontal lobe and midbrain. Given this interaction between the streams it clearly makes no sense to say that 'visual attention' *per se* is localized in either stream, or even perhaps that it is localized at all.

These ideas about normal attention help us to understand what happens in brain-damaged patients who experience extinction (see section 5.2.2 above), although of course the visual awareness of such patients is affected more severely than that of healthy subjects during perceptual suppression, or during normal inattention (Mack and Rock 1998). Remarkably, despite their severe attentional impairments, many extinction patients retain a structurally intact ventral visual pathway. Rees et al. (2000) used fMRI to study this apparently paradoxical sparing of the ventral stream. They tested their patient with bilaterally presented pairs of images of faces or houses, and as expected, the image present in the patient's contralesional visual field was extinguished on the majority of trials. Yet although the face or house was absent from the patient's visual experience, it still resulted in significant (though reduced) activation in primary visual cortex and in the appropriate ventral stream area (FFA or PPA). Similar results were found in a study of a patient with combined hemispatial neglect and extinction (Vuilleumier *et al.* 2001). We can conclude, therefore, that in the case of extinction, as in normal inattention, ventral-stream activation *per se* does not provide a sufficient condition for visual awareness of the eliciting stimulus. One may speculate that the necessary neural enhancement is lacking in extinction patients due to a deafferentation or disruption of the parietofrontal attention-switching mechanisms we have discussed. There are, however, also other likely factors at work, including the general loss of cerebral arousal level that typifies patients with neglect (e.g. Robertson 2001).

It is clear from the discussion thus far that one can only claim that ventral-stream activity plays a *necessary* role in generating visual awareness. The obvious next step is to propose that a second necessary condition is the relative neural enhancement that we know introspectively as selective attention (Milner 1995). As indicated above, this enhancement seems to be governed by a network of areas concerned primarily with oculomotor control, including area LIP in the dorsal stream, area FEF in premotor cortex, and the superior colliculus, a subcortical area that is heavily connected with both LIP and FEF (Cavanaugh and Wurtz 2004).

It should be noted that such selective neural enhancement cannot *by itself* provide an account of the different capacities of the dorsal and ventral streams for generating conscious percepts, because similar enhancement has been observed in dorsal-stream visual areas as well (indeed it was discovered there before it was found in the ventral stream: Bushnell *et al.* 1981). There has to be something else special about the ventral stream, and this something may ultimately lie in its specialized computational characteristics. These characteristics, which are integral for the ventral stream's role in recognition and identification, may *ipso facto* provide the reasons why it has privileged access to awareness (Milner 1995). Among these special properties may be the use of relative ('scene-based') rather then absolute coding of visual properties such as size, orientation and location; 'object-based' rather than 'view-based' coding for establishing visual identity; and the constant two-way exchange of information with adjacent memory and semantic systems in the medial temporal lobe, which impose a top-down regulation of our perceptions (Milner and Goodale 1995, 2006). These various processing characteristics may well help to explain how ventral-stream processing has the potential to serve access consciousness, though of course they give no direct clues as to why this consciousness should ever have the *phenomenal* properties that it does.

We suggested some years ago (Goodale and Milner 1992; Milner and Goodale 1993; Milner 1995) that the presence of selective processing in both streams (as well as, no doubt, in many other parts of the brain too) is consistent with the existence of 'visuomotor attention' as well as 'perceptual attention'—the former associated with dorsal-stream visual processing, the latter with ventral-stream processing. After all, targets for action have to be selected from the visual array just as do the objects of perception. Yet in practice it seems highly likely that the two kinds of attention will work hand in hand in all normal circumstances, as a necessary part of ensuring that the two streams work together in harmony. This *de facto* coupling has been confirmed in a series of elegant experiments by Deubel and Schneider (2005), who have found that in several different task paradigms, whether involving ocular or manual responses, the spatial location of the action target always enjoys enhanced perceptual processing. Clear dissociations between perceptual and visuomotor selection, however, can be seen in neurological patients. For example, the ability of our extinction patient V.E., described earlier, to avoid obstacles (McIntosh *et al.* 2004a) was quite independent of his ability to report the relevant stimulus: visuomotor performance was *identical* between aware and unaware trials. This observation indicates that, under some circumstances, selection for action (taking account of the left obstacle during reaching) can occur independently of selection for perception (which was evidently absent on the trials when extinction occurred). Attention can also be separated from perception in cases of hemianopia following a lesion to the primary visual cortex. Thus Kentridge *et al.* (1999, 2004) have shown that the patient G.Y. is able to switch his attention between different parts of his 'blind' hemifield, thereby improving his detection and discrimination (i.e. his 'blindsight'), without any visual awareness whatever of the target stimuli. This phenomenon satisfies two essential requirements for it to count as true attention switching, in that (a) it can occur under voluntary (as well as involuntary) control, and (b) it is selectively beneficial for visual processing at the cued location.

The studies of Kentridge and his colleagues show that selective attention is not sufficient in itself for visual awareness. In the case of a blindsight patient like G.Y., we can attribute 'attention without awareness' to a visual deafferentation of the ventral stream by his brain damage: ventral-stream areas depend almost completely on visual input from primary visual cortex (Bullier *et al.* 1994). In other words, there simply is no current visual representation in G.Y.'s ventral stream that could benefit from enhancement. On the other hand, the parietofrontal–collicular network for controlling attention is presumably still functionally intact, and so may enable the patient to attend selectively by modulating visual processing in the dorsal stream (which still receives extensive visual input in blindsight: Bullier *et al.* 1994). Thus, as suggested by Kentridge *et al.* (2004), visual selective attention in G.Y.'s blind field may come as a by-product of processing that would normally underlie selective *visuomotor* attention.

When they are available together in combination, are these two apparently necessary conditions (ventral-stream processing and attentional selection) then *sufficient* for visual awareness? Here we enter the realms of pure speculation: territory where theories abound and directly supportive facts are relatively few. For example, it could be that another necessary condition for perceptual awareness is the presence of synchronous oscillations in the neural activity of disparate visual areas (Engel *et al.* 2001). Or it could be that 'recurrent processing', via the well-documented reciprocal back-projections from 'higher' to 'lower' visual areas, is also (or instead) necessary (Lamme 2003).[7] Both of these additional candidates for necessary conditions of consciousness are backed up by credible, if circumstantial, empirical evidence. Yet even if either or both of them turn out to be important determinants of visual awareness, their contributions would not explain the clear difference between the roles of ventral and dorsal streams in conscious perception, any more than does the contribution of selective attention, because all of these processes seem to occur to a more or less similar extent in both visual streams.

The same is also true, I believe, of the implicit knowledge of dynamic sensorimotor contingencies that seems to be built into our visual processing apparatus to allow the brain to anticipate the likely sensory consequences of different movements (Hurley 1998; O'Regan and Noë 2001). It seems *a priori* very likely that both visual streams will need to have such implicit information accessible to them in order to perform their roles effectively, for example in maintaining the necessary spatial stability in our visual life (e.g. Duhamel *et al.* 1992). A recent study supports this intuition, showing that subjects who viewed objects through a left–right reversing prism could undergo differential visuomotor adaptation, according to the kind of visual feedback (location or orientation) provided (Marotta *et al.* 2005). Also, the observed adaptation was specific to the action performed (reaching vs grasping), rather than to the visual features responded to. This strongly suggests that the recalibration was not occurring through a global or perceptual visual representation, but rather with

---

[7] In a recent paper Lamme (2006) not only reaffirms his view that recurrent processing is 'the key neural ingredient of consciousness', but even recommends that we should *redefine* consciousness as recurrent processing.

respect to individual visuomotor subsystems within the dorsal stream, probably through selective links with cerebellar systems (Baizer et al. 1999). In other words, it seems likely that implicit dynamic sensorimotor knowledge contributes to visual coding in the dorsal stream, just as it influences perceptual processing in the ventral stream. If so, then this kind of implicit knowledge cannot in itself be the critical determinant of conscious versus unconscious visual processing.

Finally, it needs to be emphasized that the necessary *participation* of the ventral stream does not imply that the ventral stream is itself the seat of visual consciousness. After all, retinal stimulation is necessary, under natural circumstances,[8] for visual consciousness, yet we do not imagine that the retinal image correlates directly with the content of conscious visual percepts. One plausible hypothesis for a structural correlate of our visual experience is suggested by the studies mentioned above by Rees and colleagues (2000) and Vuilleumier and colleagues (2001). Both of their patients had right inferior parietal damage, which is typically associated with neglect and/or extinction, and this damage clearly resulted in a loss or degradation of both access and phenomenal visual consciousness in their contralesional visual field. It has been suggested that in the intact human brain, the right inferior parietal lobule provides a flexible high-level representational medium within which the contents of visual consciousness are being constantly constructed and reconstructed, for example on different spatial scales according to the attentional needs of the moment (Milner 1997a 1997b, 1998; Driver and Mattingley 1998).[9] According to the hypothesis, this system could combine the products of several specialized processing areas within the ventral stream (FFA, LO, PPA, etc.), along with top-down information from memory, and information from other sensory modalities. This dynamic spatial representation is hypothesized to correspond on a quasi-1:1 basis with one's current visual experience, including the contents of one's visual working memory. But of course even if this speculation turned out to be correct, the proposed representational system (something perhaps not too far removed from a 'Cartesian movie theatre'[10]—*pace* Dennett 1991)

---

[8] I exclude here visual experiences caused by artificially stimulating the cortex, e.g. electrically or magnetically.

[9] It is important to note that the *left* inferior parietal lobule plays a quite different role in our mental life. It has long been known that inferior parietal damage on the left side of the brain causes apraxia, rather than spatial neglect. It is most probably concerned with constructing plans for action, rather than with constructing spatial representations. If put into practice, these plans are presumably then executed under the control of the dorsal stream (Milner and Goodale 1995, 2006; Glover 2004: see section 5.4).

[10] Any appeal that this cinematic metaphor may have is limited, like the present chapter, to discussions of *visual* consciousness.

could only be one part of a wider network of areas, including in particular the prefrontal cortex, whose activities would almost certainly impose constant constraints on what happens on-screen (e.g.Dehaene and Naccache 2001; Rees *et al.* 2002; Baars *et al.* 2003).

5.4 Some brief afterthoughts

I have tried in this chapter to set out my arguments for the unconscious character of the products of dorsal stream processing as unambiguously as possible. However, to forestall possible misunderstandings, it may help to make some brief further comments and clarifications.

First, I should emphasize that I have been referring throughout the chapter only to the *visual* products of dorsal-stream processing. That is, I have not been discussing at all our (non-visual) awareness of the *actions* we may make under visual guidance, nor our awareness of our own *intentions* in making those actions. These issues are addressed specifically in the commentary by Tsakiris and Haggard (this volume).[11] Nor have I been addressing questions of possible conscious access to the *visuomotor transformations* which provide the dorsal stream's *raison d'être*. This is a quite different question from the *visual* content of dorsal-stream processing, and probably deserves more philosophical attention than I am competent to offer. I will simply acknowledge that I know 'what it is like' to have visual experiences, and indeed to have motor experiences; but I do not know what it would be like to have distinct *visuomotor* experiences, whereby I might introspect upon the visuomotor transformations that underlie my actions.

But of course these visuomotor transformations in the dorsal stream are not the only ways in which the brain uses visual information to govern our actions: they form only the bottom layer of a vision/action hierarchy. When thinking about the visual guidance of behaviour more broadly, there is a need to distinguish between different levels of description of people's actions. In brief, we have intentions (to achieve a certain end, such as eating an apple); we select and plan actions (like picking up the apple rather than knocking it off

---

[11] Tsakiris and Haggard propose that action awareness is supported by the dorsal stream. I believe that a far more likely candidate for this role is the inferior parietal region of the left hemisphere, and most of the evidence adduced by Tsakiris and Haggard would be consistent with this idea. This region (see footnote 9 above) is quite separate from the visual projection areas that constitute the dorsal stream in humans, which are clustered in *superior* parts of the human parietal lobes, in and around the intraparietal sulcus. In fact in humans the *left* inferior parietal region not only lies outside the dorsal and ventral visual streams, it appears not to be a visual region at all.

the table); and in implementing these actions the brain puts together a smoothly orchestrated programme of component movements of the eyes, limbs, and body. All of these levels are, or can be, in different ways, highly dependent on visual input. But in our conceptualization of the division of labour within the visual system (Milner and Goodale 1995, 2006), the dorsal stream only comes into play at the third (i.e. lowest) of these levels of motor organization (cf. Clark 1999). In other words the dorsal stream is little more than a slave system for putting our plans and intentions into practice in real time. The ventral stream, on the other hand, is quite likely to play an important role in setting our goals and in selecting the appropriate actions to achieve them. As an example of the latter, one may consciously select a particular hand orientation to pick up an elongated object in anticipation of the way one intends to use the object. Patient D.F. is impaired in such visually guided action planning (Carey *et al.* 1996; Dijkerman *et al.* 2008), as would be predicted if these processes are mediated by the ventral stream, not the dorsal stream.

A neat illustration of the independence of the dorsal stream from the level of our intentions is given by the work of Pisella and her colleagues (2000). They have shown that part of the dorsal stream's job specification is to bypass our intentions completely under certain circumstances, through the operation of what they call 'automatic pilot' (see section 5.2.2.2). That is, the dorsal stream seems to have a kind of built-in servomechanism that ensures the arrival of the eye and the hand at a visual target even when that target shifts unpredictably during the movement. We cannot resist this momentum, no matter how hard we try. This automatic self-correction of our actions, in general terms, presumably has a high adaptive value for a visuomotor control system—an animal needs to be able to cope quickly with sudden movements of a prey object, for example, otherwise it could not survive in a real world where other creatures refuse to stay still. But this does not mean, of course, that the dorsal stream's activities are restricted to serving 'unintentional' or 'automatic' actions. It would certainly be true to say that the clearest examples of behaviour that is *dominated* by dorsal-stream control are provided by quick, well-practised actions, like catching, grasping, jabbing, throwing, knocking, or hitting. But even the most complex and circumspect actions will still need to recruit the dorsal stream for them to be executed smoothly under visual control. (For a recent discussion of these issues, see Milner and Goodale 2008.)

A second point that may be worth emphasizing is that I do not claim that the dorsal stream plays no role at all in shaping our visual consciousness. For example, although I have argued above that the dorsal stream does not play any *direct* part in furnishing the contents of our visual consciousness, it certainly does have an important role to play in determining the structure and (indirectly) the content, of our perceptions. The clearest example of this lies in the constant shaping and reshaping of the distribution of ventral stream activation that seems to be governed by dorsal-stream areas like LIP. Indeed without the ability to move one's gaze and attention around a scene (with or without the use of active exploratory body and hand movements), visual perception would lose its characteristic 'active' qualities (e.g. Findlay and Gilchrist 2003; Tsakiris and Haggard, this volume). An organism needs to be able to 'look' as well as to 'see' if it is to survive. Furthermore our perceptions (e.g. of depth) can change dramatically when we are able to move around the environment (e.g. Dijkerman *et al*. 1999; Wexler and Boxtel 2005). Intriguingly, however, a loss of this active aspect of vision seems not to destroy our ability to extract the gist of the scene (Michel and Hénaff 2005): this rather global aspect of perception seems to survive without the need for active interrogation of the visual array. It is also worth noting that when we watch a film or a TV drama, the fact that the shifts of viewpoint, parallax, perspective, and attentional focus are largely provided by the movements, cuts, and zooming actions of a camera rather than by our own head, body, or eye movements, presents no problem for our coherent perception of the events being depicted on the screen. Our ability to do this provides a testimony to the autonomy of visual consciousness as well as to the flexibility[12] of the ventral-stream system. Current knowledge suggests that the ventral stream is characterized by purely *visual* processing, and therefore may not provide the vehicle for non-visual components within our stream of consciousness—including awareness of our movements—however inseparable those may seem.[13]

It is worth re-acknowledging here that when we reach out to pick up an object, we may have full visual awareness of our arm moving and our hand configuring in a certain way; of the target object as having a certain shape and lying in a certain location; of the presence and nature of neighbouring objects; and so on. In other words, we may be fully *visually* aware of our actions as we

---

[12] As Tsakiris and Haggard (this volume) point out, the kind of relative, scene-based coding that allows us to understand events portrayed on a TV screen would not work for the visuomotor systems of the dorsal stream, which require the external world to be coded in egocentric spatial frames of reference. Indeed, the main reason for citing the example of watching a film or TV programme in a recent book (Goodale and Milner 2004) was precisely to illustrate this fundamental difference in the frames of reference needed for perception and action.

[13] The left and right inferior parietal cortices could perhaps play a joint role in integrating some of the perceptual and motor aspects of consciousness. But of course there doesn't have to be a single unitary 'seat' of consciousness, as many philosophers, e.g. Dennett (1991), have pointed out.

make them, along with the detailed environmental context in which they are made. But the essence of Milner and Goodale's interpretation is that this visual awareness accrues not from dorsal-stream processing, but from concurrent *ventral*-stream processing. Crucially, according to the model, such ventral-stream processing plays *no causal role* in the real-time visual guidance of the action, despite our strong intuitive inclination to believe otherwise. That is to say, we incorrectly but routinely make what Clark (2001) calls 'the assumption of experienced-based control'. Instead, that real-time guidance is provided through continuous visual sampling carried out by the dorsal stream, which has to code, non-semantically, those very same visual inputs (i.e. from hand and arm, from both target and non-target objects, etc.).

One important aspect of the dorsal stream's visual processing during our reaching and grasping movements—one that has received too little emphasis in the past—is its constant use of information about the moving limb itself (see section 5.2.2.2). The dorsal stream's role is supported by the discovery of individual neurons there that are responsive to the hand's position in space during reaching (Battaglia-Mayer *et al.* 2001). Recent work suggests that this neural information is used to code the 'error signal' between hand and target, within a hand-centred frame of reference (Buneo and Andersen 2006). In one intriguing study, Graziano *et al.* (2000) reported that some neurons in the monkey's parietal lobe respond not only to the sight of the animal's real arm, but also to a realistic false arm introduced by the experimenter. This work suggests that the dorsal stream is astute enough to prefer visual stimuli that have an arm-like appearance, and does not simply process anything that moves across the visual field in the general direction of the target. There is behavioural evidence that the configuration of the hand, as well as the movement of the arm, is visually monitored during reaching to grasp (Connolly and Goodale 1999), again presumably by neurons in the dorsal stream. It is therefore of interest that a recent fMRI study by Shmuelof and Zohary (2006) shows that the human dorsal stream is responsive to movie clips of a left or right hand grasping objects *as seen from the actor's perspective*. Shmuelof and Zohary showed that while the ventral stream too responded to these clips, there was an important difference. The ventral stream in each cerebral hemisphere responded to either hand (mostly when shown in the contralateral visual hemifield), whereas area AIP in the dorsal stream responded only to the contralateral hand (shown in either hemifield). It seems reasonable to assume that area AIP in Shmuelof and Zohary's study is responding in just the same way as it would when processing feedback from the observer's *own* contralateral hand during grasping (see section 5.2.2.2 above). To put it another way, the dorsal stream probably made the default assumption that the hand it saw was its own (i.e. that of the observer). The ventral stream of course would not have been fooled, and therefore neither would the observer.

If we accept that there are mechanisms in the dorsal stream that process the sight of the observer's own hand, then it becomes tempting to speculate as to the origins of the well-known 'mirror neurons' discovered by Rizzolatti and his colleagues, initially in the frontal premotor cortex of the monkey (Di Pellegrino *et al.* 1992; Gallese *et al.* 1996), and subsequently in and adjacent to the dorsal stream (specifically area 7a: Gallese *et al.* 2002). These neurons respond just prior to execution of the monkey's own actions, as would be expected, but also when the animal observes similar actions in others. How might these hybrid neuronal properties come about? It is easy, first, to see how a subset of neurons could arise in the dorsal stream that participate in the visuomotor transformations that elicit an action, and which also respond to the visual *feedback* from that same action. After all, if there are separate neurons lying cheek by jowl in the dorsal stream, some of which are associated with transforming visual environmental metrics into motor coordinates, and others of which are sensitive to the sight of the animal's own moving hand, then it seems intuitively quite likely that associative links will develop between the two through Hebbian learning. There is physiological evidence that such dual-function visuomotor neurons do in fact exist in the monkey's dorsal stream (Battaglia-Mayer *et al.* 2001). But of course these are not, *ipso facto*, true mirror neurons, because they respond to the sight of the observing animal's *own* movements, rather than to the sight of a conspecific's movements.[14]

It would nevertheless be a small step from such *self*-responsive neurons to the development of a subset among them whose visual responses generalize to *other* animals' actions too. In fact the visual tuning of many dorsal stream neurons may well be insufficiently fine to make reliable distinctions between (e.g.) first-person left-hand grasping and third-person (facing the observer) right-hand grasping. A human fMRI study by Pelphrey *et al.* (2004), in which subjects observed another person grasping, elicited a pattern of activations in precise agreement with this possibility. That is, the left AIP responded better to the sight of left-hand grasping (executed by a model facing the observer), and the right AIP better to right-hand grasping. This way of thinking may help to explain why mirror neurons have been found within visuomotor

---

[14] By the same token, the fMRI responses seen in Shmuelof and Zohary's (2006) experiment do not provide evidence for the activation of a true human 'mirror system' as the authors suggest.

regions like the dorsal stream and the premotor cortex, rather than, say, area STP in the anterior superior temporal sulcus, an offshoot of the ventral stream where neurons also respond to the sight of specific actions performed by others (Perrett et al. 1985).

I would not claim, however, that these simple ideas can account for the full gamut of mirror-neuron properties. For one thing, any mirror neurons that code high-level aspects of a conspecific's actions or intentions (Fogassi et al. 2005) would require computations of a semantic complexity that only the ventral stream could provide. For example, Perrett and others have shown that there are neurons within area STP and adjacent parts of the anterior ventral stream that code exactly such high-level semantics of another animal's actions (e.g. Jellema et al. 2000; see also Saxe et al. 2004 for related evidence in humans). Furthermore, area STP has interconnections with area 7a in the parietal lobe. Keysers and Perrett (2004) have argued that Hebbian associative learning could account for the properties of *all* mirror neurons, including these higher-level ones, given the extensive three-way synaptic interconnections between STS, the dorsal stream, and the premotor cortex. My present hypothesis is much less ambitious, but would have a consequence that Keysers and Perrett's account does not have: namely that a subset of rather unsophisticated mirror neurons should exist that have visual properties *uncorrelated with conscious awareness*. These, in my terms, would be 'true' dorsal-stream neurons, and their mirror properties would be very much secondary to—a corollary of—their visuomotor role. Actually, we do not yet know for sure whether even fully-fledged mirror neurons ever contribute to (access-) conscious visual percepts, though it seems likely that they do. An answer to that question must no doubt await the development of ever more sophisticated experimentation of the kind pioneered by Sheinberg and Logothetis (1997).

At the beginning of this essay I used the metaphor of two people singing from the same hymn sheet, and argued that the two visual streams necessarily sing from different, though in large part mutually consistent ones. In reality of course, each of the several subsystems within the dorsal stream—for guiding reaching, grasping, saccadic eye movements, pursuit eye movements, and locomotion—seems to have its own specialized supply of visual information, suited to its own particular purposes. The dorsal stream thus sings with more than one voice, though the resulting chorus is notably harmonious. To ask what 'it sees' may therefore be an unanswerable question in more ways than one. Nevertheless, my provisional answer (always subject to new empirical data) is that none of the dorsal stream's constituent visual representations is consciously accessible.

Acknowledgements

I am grateful to Chris Frith and Tony Atkinson and for their comments on a draft of this chapter; to my several collaborators in the work described here, particularly Robert McIntosh and Thomas Schenk; and to Manos Tsakiris and Patrick Haggard for their written commentary. I also thank the Medical Research Council and Leverhulme Trust for their financial support of our research, and the Warden and Fellows of All Souls College for providing me with an environment conducive to a fuller reflection on the implications of that research.

References

Afraz, S.-R., Kiani, R., and Esteky, H. (2006). Microstimulation of inferotemporal cortex influences face categorization. *Nature* **442**, 692–695.

Aglioti, S., Goodale, M.A. and DeSouza, J.F.X. (1995). Size-contrast illusions deceive the eye but not the hand. *Current Biology* **5**, 679–685.

Andrews, T.J. and Schluppeck, D. (2004). Neural responses to Mooney images reveal a modular representation of faces in human visual cortex. *NeuroImage* **21**, 91–98.

Andrews, T.J., Schluppeck, D., Homfray, D., Matthews, P. and Blakemore, C. (2002). Activity in the fusiform gyrus predicts conscious perception of Rubin's vase-face illusion. *NeuroImage* **17**, 890–901.

Baars, B.J. (2002). The conscious access hypothesis: origins and recent evidence. *Trends in Cognitive Sciences* **6**, 47–52.

Baars, B.J., Ramsøy, T.Z., and Laureys, S. (2003). Brain, conscious experience and the observing self. *Trends in Neurosciences* **26**, 671–675.

Baizer, J.S., Ungerleider, L.G., and Desimone, R. (1991). Organization of visual inputs to the inferior temporal and posterior parietal cortex in macaques. *Journal of Neuroscience* **11**, 168–190.

Baizer, J.S., Kralj-Hans, I., and Glickstein, M. (1999). Cerebellar lesions and prism adaptation in macaque monkeys. *Journal of Neurophysiology* **81**, 1960–1965.

Battaglia-Mayer, A., Ferraina, S., Genovesio, A., Marconi, B., Squatrito, S., Molinari, M., Lacquaniti, F., and Caminiti, R. (2001). Eye-hand coordination during reaching. II. An analysis of the relationships between visuomanual signals in parietal cortex and parieto-frontal association projections. *Cerebral Cortex* **11**, 528–544.

Bayne, T. (2008). The unity of consciousness and the commissurotomy syndrome. *Journal of Philosophy* in press.

Binkofski, F., Buccino, G., Stephan, K.M., Rizzolatti, G., Seitz, R.J., and Freund, H.-J. (1999). A parieto-premotor network for object manipulation: evidence from neuroimaging. *Experimental Brain Research* **128**, 210–213.

Binkofski, F., Dohle, C., Posse, S., Stephan, K.M., Hefter, H., Seitz, R.J., and Freund, H.-J. (1998). Human anterior intraparietal area subserves prehension. A combined lesion and functional MRI activation study. *Neurology* **50**, 1253–1259.

Block, N. (1995). On a confusion about a function of consciousness. *Behavioral and Brain Sciences* **18**, 227–247.

Block, N. (1996). How can we find the neural correlate of consciousness? *Trends in Neurosciences* **19**, 456–459.

Bogen, J.E. (1997). An example of access-consciousness without phenomenal consciousness? *Behavioral and Brain Sciences* **20**, 144.

Buckley, M.J. and Gaffan, D. (2006). Perirhinal cortical contributions to object perception. *Trends in Cognitive Sciences* **10**, 100–107.

Bullier, J., Girard, P., and Salin, P.-A. (1994). The role of area 17 in the transfer of information to extrastriate visual cortex. In Peters, A. and Rockland, K.S. (eds) *Cerebral Cortex, Volume 10: Primary Visual Cortex in Primates*, pp. 301–330. New York: Plenum.

Buneo, C.A. and Andersen, R.A. (2006). The posterior parietal cortex: Sensorimotor interface for the planning and online control of visually guided movements. *Neuropsychologia* **44**, 2594–2606.

Bushnell, M.C., Goldberg, M.E., and Robinson, D.L. (1981). Behavioral enhancement of visual responses in monkey cerebral cortex. I. Modulation in posterior parietal cortex related to selective visual attention. *Journal of Neurophysiology* **46**, 755–772.

Bussey, T.J. and Saksida, L.M. (2005). Object memory and perception in the medial temporal lobe: an alternative approach. *Current Opinion in Neurobiology* **15**, 730–737.

Carey, D.P., Harvey, M. and Milner, A.D. (1996). Visuomotor sensitivity for shape and orientation in a patient with visual form agnosia. *Neuropsychologia* **34**, 329–338.

Carey, D.P., Dijkerman, H.C., Murphy, K.J., Goodale, M.A., and Milner, A.D. (2006). Pointing to places and spaces in the visual form agnosic D.F. *Neuropsychologia* **44**, 1584–1594.

Cavanaugh, J. and Wurtz, R.H. (2004). Subcortical modulation of attention counters change blindness. *Journal of Neuroscience* **24**, 11236–11243.

Clark, A. (1999). Visual awareness and visuomotor action. *Journal of Consciousness Studies* **6**, 1–18.

Clark, A. (2001). Visual experience and motor action: are the bonds too tight? *Philosophical Review* **110**, 495–519.

Connolly, J.D. and Goodale, M.A. (1999). The role of visual feedback of hand position in the control of manual prehension. *Experimental Brain Research* **125**, 281–286.

Corbetta, M., Kincade, M.J., and Shulman, G.L. (2002). Two neural systems for visual orienting and the pathophysiology of unilateral spatial neglect. In Karnath, H.-O., Milner, A.D., and Vallar, G. (eds) *The Cognitive and Neural Bases of Spatial Neglect*, pp. 259–273. Oxford: Oxford University Press.

Culham, J.C., Danckert, S.L., DeSouza, J.F.X., Gati, J.S., Menon, R.S., and Goodale, M.A. (2003). Visually-guided grasping produces fMRI activation in dorsal but not ventral stream brain areas. *Experimental Brain Research* **153**, 180–189.

Culham, J.C. and Valyear, K.F. (2006). Human parietal cortex in action. *Current Opinion in Neurobiology* **16**, 205–212.

Dehaene, S. and Naccache, L. (2001). Towards a cognitive neuroscience of consciousness: basic evidence and a workspace framework. *Cognition* **79**, 1–37.

Dennett, D.C. (1991). *Consciousness Explained*. Boston: Little, Brown.

Desimone, R. and Duncan, J. (1995). Neural mechanisms of selective visual attention. *Annual Review of Neuroscience* **18**, 193–222.

Deubel, H. and Schneider, W.X. (2005). Attentional selection in sequential manual movements, movements around an obstacle and in grasping. In Humphreys, G.W.

and Riddoch, M.J. (eds) *Attention in Action: Advances from Cognitive Neuroscience*, pp. 69–91. Hove: Psychology Press.

Di Pellegrino, G., Fadiga, L., Fogassi, L., Gallese, V., and Rizzolatti, G. (1992). Understanding motor events: a neurophysiological study. *Experimental Brain Research* **91**, 176–180.

Dijkerman, H.C., Milner, A.D., and Carey, D.P. (1999). Prehension of objects oriented in depth: motion parallax restores performance of a visual form agnosic when binocular vision is unavailable. *Neuropsychologia* **37**, 1505–1510.

Dijkerman, H.C., McIntosh, R.D., Schindler, I., Nijboer, T.C.W., and Milner, A.D. (2008). Choosing between alternative wrist postures: action planning needs perception. Submiteed for publication.

Driver, J. and Mattingley, J.B. (1998). Parietal neglect and visual awareness. *Nature Neuroscience* **1**, 17–22.

Duhamel, J.-R., Colby, C.L., and Goldberg, M.E. (1992). The updating of the representation of visual space in parietal cortex by intended eye movements. *Science* **255**, 90–92.

Dyde, R.T. and Milner, A.D. (2002). Two illusions of perceived orientation: one fools all of the people some of the time, but the other fools all of the people all of the time. *Experimental Brain Research* **144**, 518–527.

Engel, A.K., Fries, P., and Singer, W. (2001). Dynamic predictions: oscillations and synchrony in top-down processing. *Nature Reviews Neuroscience* **2**, 704–716.

Fang, F. and He, S. (2005). Cortical responses to invisible objects in the human dorsal and ventral pathways. *Nature Neuroscience* **8**, 1380–1385.

Ferber, S., Humphrey, G.K., and Vilis, T. (2003). The lateral occipital complex subserves the perceptual persistence of motion-defined groupings. *Cerebral Cortex* 13, 716–721.

Findlay, J.M. and Gilchrist, I.D. (2003). *Active Vision*. Oxford: Oxford University Press.

Fogassi, L., Ferrari, P.F., Gesierich, B., Rozzi, S., Chersi, F., and Rizzolatti, G. (2005). Parietal lobe: from action organization to intention understanding. *Science* **308**, 662–667.

Gallese, V., Fadiga, L., Fogassi, L., and Rizzolatti, G. (1996). Action recognition in the premotor cortex. *Brain* **119**, 593–609.

Gallese, V., Fadiga, L., Fogassi, L., and Rizzolatti, G. (2002). Action representation and the inferior parietal lobule. In Prinz, W. and Hommel, B. (eds) *Attention and Performance XIX: Common Mechanisms in Perception and Action*, pp. 334–355. Oxford: Oxford University Press.

Glover, S. (2004). Separate visual representations in the planning and control of action. *Behavioral and Brain Sciences* **27**, 3–78.

Goldberg, M.E., Bisley, J.W., Powell, K.D., and Gottlieb, J. (2006). Saccades, salience and attention: the role of the lateral intraparietal area in visual behavior. *Progress in Brain Research* **155**, 157–175.

Goodale, M.A. and Milner, A.D. (1992). Separate visual pathways for perception and action. *Trends in Neurosciences* **15**, 20–25.

Goodale, M.A. and Milner, A.D. (2004). *Sight Unseen: An Exploration of Conscious and Unconscious Vision*. Oxford: Oxford University Press.

Goodale, M.A., Pélisson, D., and Prablanc, C. (1986). Large adjustments in visually guided reaching do not depend on vision of the hand or perception of target displacement. *Nature* **320**, 748–750.

Goodale, M.A., Milner, A.D., Jakobson, L.S., and Carey, D.P. (1991). A neurological dissociation between perceiving objects and grasping them. *Nature* **349**, 154–156.

Goodale, M.A., Jakobson, L.S., and Keillor, J.M. (1994). Differences in the visual control of pantomimed and natural grasping movements. *Neuropsychologia* **32**, 1159–1178.

Goodale, M.A., Meenan, J.P., Büthoff, H.H., Nicolle, D.A., Murphy, K.J., and Racicot, C.I. (1994). Separate neural pathways for the visual analysis of object shape in perception and prehension. *Current Biology* **4**, 604–610.

Graziano, M.S., Cooke, D.F., and Taylor, C.S. (2000). Coding the location of the arm by sight. *Science* **290**, 1782–1786.

Gréa, H., Pisella, L., Rossetti, Y., Desmurget, M., Tilikete, C., Grafton, S., Prablanc, C., and Vighetto, A. (2002). A lesion of the posterior parietal cortex disrupts on-line adjustments during aiming movements. *Neuropsychologia* **40**, 2471–2480.

Hasson, U., Hendler, T., Ben Bashat, D., and Malach, R. (2001). Vase or face? A neural correlate of shape-selective grouping processes in the human brain. *Journal of Cognitive Neuroscience* **13**, 744–753.

Head, H. (1926). *Aphasia and Kindred Disorders of Speech*. Cambridge: Cambridge University Press.

Humphrey, G.K., Goodale, M.A., and Gurnsey, R. (1991). Orientation discrimination in a visual form agnosic: evidence from the McCollough effect. *Psychological Science* **2**, 331–335.

Hurley, S.L. (1998). *Consciousness in Action*. Cambridge, MA: Harvard University Press.

Jacob, P. and Jeannerod, M. (2003). *Ways of Seeing: The Scope and Limits of Visual Cognition*. Oxford: Oxford University Press.

James, T.W., Humphrey, G.K., Gati, J.S., Menon, R.S., and Goodale, M.A. (2002). Differential effects of viewpoint on object-driven activation in dorsal and ventral streams. *Neuron* **35**, 793–801.

James, T.W., Culham, J., Humphrey, G.K., Milner, A.D., and Goodale, M.A. (2003). Ventral occipital lesions impair object recognition but not object-directed grasping: a fMRI study. *Brain* **126**, 2463–2475.

Jeannerod, M. (1988). *The Neural and Behavioural Organization of Goal-Directed Movements*. Oxford: Oxford University Press.

Jeannerod, M. (1997). *The Cognitive Neuroscience of Action*. Oxford: Blackwell.

Jeannerod, M. and Rossetti, Y. (1993). Visuomotor coordination as a dissociable visual function: experimental and clinical evidence. In Kennard, C. (ed.) *Visual Perceptual Defects (Baillière's Clinical Neurology, Vol.2, No.2)*, pp. 439–460. London Baillière: Tindall.

Jeannerod, M., Decety, J., and Michel, F. (1994). Impairment of grasping movements following bilateral posterior parietal lesion. *Neuropsychologia* **32**, 369–380.

Jellema, T., Baker, C.I., Wicker, B., and Perrett, D.I. (2000). Neural representation for the perception of the intentionality of actions. *Brain and Cognition* **44**, 280–302.

Johnson, H. and Haggard, P. (2005). Motor awareness without perceptual awareness. *Neuropsychologia* **43**(2), 227–237.

Kanwisher, N. and Duncan, J. (2004). *Attention and Performance XX: Functional Neuroimaging of Visual Cognition*. Oxford: Oxford University Press.

Kanwisher, N., McDermott, J., and Chun, M.M. (1997). The fusiform face area: A module in human extrastriate cortex specialized for face perception. *Journal of Neuroscience* **17**, 4302–4311.

Kanwisher, N., Woods, R.P., Iacoboni, M., and Mazziotta, J.C. (1997). A locus in human extrastriate cortex for visual shape analysis. *Journal of Cognitive Neuroscience* **9**, 133–142.

Keysers, C. and Perrett, D.I. (2004). Demystifying social cognition: a Hebbian perspective. *Trends in Cognitive Sciences* **8**, 501–507.

Kentridge, R.W., Heywood, C.A. and Weiskrantz, L. (1999). Attention without awareness in blindsight. *Proceedings of the Royal Society of London Series B Biological Sciences* **266**, 1805–1811.

Kentridge, R.W., Heywood, C.A., and Weiskrantz, L. (2004). Spatial attention speeds discrimination without awareness in blindsight. *Neuropsychologia* **42**, 831–835.

Kleinschmidt, A., Buchel, C., Zeki, S., and Frackowiak, R.S. (1998). Human brain activity during spontaneously reversing perception of ambiguous figures. *Proceedings of the Royal Society of London Series B Biological Sciences* **265**, 2427–2433.

Króliczak, G., Heard, P., Goodale, M.A., and Gregory, R.L. (2006). Dissociation of perception and action unmasked by the hollow-face illusion. *Brain Research*, **1080**, 9–16.

Lamme, V.A.F. (2003). Why visual attention and awareness are different. *Trends in Cognitive Sciences* **7**, 12–18.

Lamme, V.A.F. (2006). Towards a true neural stance on consciousness. *Trends in Cognitive Sciences* **10**, 494–501.

LeDoux, J.E., Wilson, D.H., and Gazzaniga, M.S. (1977). A divided mind: observations on the conscious properties of the separated hemispheres. *Annals of Neurology* **2**, 417–421.

Levy, J. (1990). Regulation and generation of perception in the asymmetric brain. In Trevarthen, C. (ed.) *Brain Circuits and Functions of the Mind. Essays in Honor of Roger W. Sperry*, pp. 231–248. Cambridge: Cambridge University Press.

Levy, J. and Trevarthen, C. (1977). Perceptual, semantic and phonetic aspects of elementary language processes in split-brain patients. *Brain* **100**, 105–118.

Levy, J., Trevarthen, C., and Sperry, R.W. (1972). Perception of bilateral chimeric figures following hemispheric deconnection. *Brain* **95**, 61–78.

Lumer, E.D., Friston, K.J., and Rees, G. (1998). Neural correlates of perceptual rivalry in the human brain. *Science* **280**, 1930–1934.

Mack, A. and Rock, I. (1998). *Inattentional Blindness*. Cambridge, MA: MIT Press.

Malach, R., Reppas, J.B., Benson, R.B., Kwong, K.K., Jiang, H., Kennedy, W.A., Ledden, P.J., Brady, T.J., Rosen, B.R., and Tootell, R.B.H. (1995). Object-related activity revealed by functional magnetic resonance imaging in human occipital cortex. *Proceedings of the National Academy of Sciences of the USA* **92**, 8135–8138.

Marotta, J.J., Keith, G.P., and Crawford, J.D. (2005). Task-specific sensorimotor adaptation to reversing prisms. *Journal of Neurophysiology* **93**, 1104–1110.

McIntosh, R.D., McClements, K.I., Schindler, I., Cassidy, T.P., Birchall, D., and Milner, A.D. (2004a). Avoidance of obstacles in the absence of visual awareness. *Proceedings of the Royal Society of London Series B Biological Sciences* **271**, 15–20.

McIntosh, R.D., McClements, K.I., Dijkerman, H.C., Birchall, D., and Milner, A.D. (2004b). Preserved obstacle avoidance during reaching in patients with left visual neglect. *Neuropsychologia* **42**, 1107–1117.

Michel, F. and Hénaff, M.A. (2004). Seeing without the occipito-parietal cortex: Simultagnosia as a shrinkage of the attentional visual field. *Behavioral Neurology* **15**, 3–13.

Milner, A.D. (1995). Cerebral correlates of visual awareness. *Neuropsychologia* **33**, 1117–1130.

Milner, A.D. (1997a). Vision without knowledge. *Philosophical Transactions of the Royal Society of London Series B Biological Sciences* **352**, 1249–1256.

Milner, A.D. (1997b). Neglect, extinction, and the cortical streams of visual processing. In Thier, P. and Karnath, H.-O. (eds) *Parietal Lobe Contributions to Orientation in 3D Space*, pp. 3–22. Heidelberg: Springer-Verlag.

Milner, A.D. (1998). Streams and consciousness: visual awareness and the brain. *Trends in Cognitive Sciences* **2**, 25–30.

Milner, A.D. and Goodale, M.A. (1993). Visual pathways to perception and action. *Progress in Brain Research* **95**, 317–337.

Milner, A.D. and Goodale, M.A. (1995). *The Visual Brain in Action*. Oxford: Oxford University Press.

Milner, A.D. and Goodale, M.A. (2006). *The Visual Brain in Action*, 2nd edn. Oxford: Oxford University Press.

Milner, A.D. and Goodale, M.A. (2008). Two visual systems re-viewed. *Neuropsychologia* **46**, 774–785.

Milner, A.D., Perrett, D.I., Johnston, R.S., Benson, P.J., Jordan, T.R., Heeley, D.W., Bettucci, D., Mortara, F., Mutani, R., Terazzi, E., and Davidson, D.L.W. (1991). Perception and action in 'visual form agnosia'. *Brain* **114**, 405–428.

Milner, A.D., Dijkerman, H.C., and Carey, D.P. (1999). Visuospatial processing in a pure case of visual-form agnosia. In Burgess, N., Jeffery, K.J., and O'Keefe, J. (eds) *The Hippocampal and Parietal Foundations of Spatial Cognition*, pp. 443–466. Oxford: Oxford University Press.

Milner, A.D., Dijkerman, H.C., McIntosh, R.D., Rossetti, Y., and Pisella, L. (2003). Delayed reaching and grasping in patients with optic ataxia. *Progress in Brain Research* **142**, 225–242.

Moore, T., Armstrong, K.M., and Fallah, M. (2003). Visuomotor origins of covert spatial attention. *Neuron* **40**, 671–683.

Morel, A. and Bullier, J. (1990). Anatomical segregation of two cortical visual pathways in the macaque monkey. *Visual Neuroscience* **4**, 555–578.

Murata, A., Gallese, V., Luppino, G., Kaseda, M., and Sakata, H. (2000). Selectivity for the shape, size, and orientation of objects for grasping in neurons of monkey parietal area AIP. *Journal of Neurophysiology* **83**, 2580–2601.

Nagel, T. (1974). What is it like to be a bat? *Philosophical Review* **83**, 435–450.

O'Regan, J.K. and Noë, A. (2001). A sensorimotor account of vision and visual consciousness. *Behavioral and Brain Sciences* **24**, 939–973.

Orban, G.A., Van Essen, D., and Vanduffel, W. (2004). Comparative mapping of higher visual areas in monkeys and humans. *Trends in Cognitive Sciences* **8**, 315–324.

Pelphrey, K.A., Morris, J.P., and McCarthy, G. (2004). Grasping the intentions of others: the perceived intentionality of an action influences activity in the superior temporal sulcus during social perception. *Journal of Cognitive Neuroscience* **16**, 1706–1716.

Perenin, M.-T. and Vighetto, A. (1983). Optic ataxia: a specific disorder in visuomotor coordination. In Hein, A. and Jeannerod, M. (eds) *Spatially Oriented Behavior*, pp. 305–326. New York: Springer-Verlag.

Perenin, M.-T. and Vighetto, A. (1988). Optic ataxia: a specific disruption in visuomotor mechanisms. I. Different aspects of the deficit in reaching for objects. *Brain* **111**, 643–674.

Perrett, D.I., Smith, P.A.J., Mistlin, A.J., Chitty, A.J., Head, A.S., Potter, D.D., Brönnimann, R., Milner, A.D., and Jeeves, M.A. (1985). Visual analysis of body movements by neurones in the temporal cortex of the macaque monkey: a preliminary report. *Behavioural Brain Research* **16**, 153–170.

Pisella, L., Gréa, H., Tilikete, C., Vighetto, A., Desmurget, M., Rode, G., Boisson, D., and Rossetti, Y. (2000). An 'automatic pilot' for the hand in human posterior parietal cortex: toward reinterpreting optic ataxia. *Nature Neuroscience* **3**, 729–736.

Prablanc, C., Échallier, J.F., Komilis, E., and Jeannerod, M. (1979a). Optimal response of eye and hand motor systems in pointing to a visual target. I. Spatio-temporal characteristics of eye and hand movements and their relationships when varying the amount of visual information. *Biological Cybernetics* **35**, 113–124.

Prablanc, C., Échallier, J.F., Jeannerod, M., and Komilis, E. (1979b). Optimal response of eye and hand motor systems in pointing at a visual target. II. Static and dynamic visual cues in the control of hand movement. *Biological Cybernetics* **35**, 183–187.

Puccetti, R. (1981). The case for mental duality: evidence from split-brain data and other considerations. *Behavioral and Brain Sciences* **4**, 93–123.

Rees, G., Wojciulik, E., Clarke, K., Husain, M., Frith, C., and Driver, J. (2000). Unconscious activation of visual cortex in the damaged right hemisphere of a parietal patient with extinction. *Brain* **123**, 1624–1633.

Rees, G., Kreiman, G., and Koch, C. (2002). Neural correlates of consciousness in humans. *Nature Reviews Neuroscience* **3**, 261–270.

Rice, N.J., Tunik, E., and Grafton, S.T. (2006a). The anterior intraparietal sulcus mediates grasp execution, independent of requirement to update: new insights from transcranial magnetic stimulation. *Journal of Neuroscience* **26**, 8176–8182.

Rice, N.J., McIntosh, R.D., Schindler, I., Mon-Williams, M., Démonet, J.-F., and Milner, A.D. (2006b). Intact automatic avoidance of obstacles in patients with visual form agnosia. *Experimental Brain Research* **174**, 176–188.

Rizzolatti, G., Gentilucci, M., and Matelli, M. (1985). Selective spatial attention: one center, one circuit, or many circuits? In Posner, M.I. and Marin, O.S.M. (eds) *Attention and Performance XI*, pp. 251–265. Hillsdale, NJ: Erlbaum.

Rizzolatti, G., Riggio, L., and Sheliga, B.M. (1994). Space and selective attention. In Umiltà, C., Moscovitch, M. (eds) *Attention and Performance XV*, pp. 231–265. *Conscious and Nonconscious Information Processing*. Cambridge, MA: MIT Press.

Robertson, I.H. (2001). Do we need the 'lateral' in unilateral neglect? Spatially nonselective attention deficits in unilateral neglect and their implications for rehabilitation. *NeuroImage* **14**, S85–90.

Sakata, H., Taira, M., Kusunoki, M., Murata, A., and Tanaka, Y. (1997). The parietal association cortex in depth perception and visual control of hand action. *Trends in Neurosciences* **20**, 350–357.

Sarlegna, F., Blouin, J., Vercher, J.L., Bresciani, J.P., Bourdin, C., and Gauthier, G.M. (2004). Online control of the direction of rapid reaching movements. *Experimental Brain Research* **157**, 468–471.

Saxe, R., Xiao, D.K., Kovacs, G., Perrett, D.I., and Kanwisher, N. (2004). A region of right posterior superior temporal sulcus responds to observed intentional actions. *Neuropsychologia* **42**, 1435–1446.

Schenk, T., Schindler, I., McIntosh, R.D., and Milner, A.D. (2005). The use of visual feedback is independent of visual awareness: evidence from visual extinction. *Experimental Brain Research* **167**, 95–102.

Schindler, I., Rice, N.J., McIntosh, R.D., Rossetti, Y., Vighetto, A., and Milner, A.D. (2004). Automatic avoidance of obstacles is a dorsal stream function: evidence from optic ataxia. *Nature Neuroscience* **7**, 779–784.

Sheinberg, D.L. and Logothetis, N.K. (1997). The role of temporal cortical areas in perceptual organization. *Proceedings of the National Academy of Sciences of the USA* **94**, 3408–3413.

Shmuelof, L. and Zohary, E. (2006). A mirror representation of others' actions in the human anterior parietal cortex. *Journal of Neuroscience* **26**, 9736–9742.

Sperry, R.W. (1974). Lateral specialization in the surgically separated hemispheres. In Schmitt, F.O. and Worden, F.G. (eds) *Third Neurosciences Study Program*. Cambridge, MA: MIT Press.

Steeves, J.K.E., Humphrey, G.K., Culham, J.C., Menon, R.S., Milner, A.D., and Goodale, M.A. (2004). Behavioral and neuroimaging evidence for a contribution of color and texture information to scene classification in a patient with visual form agnosia. *Journal of Cognitive Neuroscience* **16**, 955–965.

Taira, M., Mine, S., Georgopoulos, A.P., Mutara, A., and Sakata, H. (1990). Parietal cortex neurons of the monkey related to the visual guidance of hand movements. *Experimental Brain Research* **83**, 29–36.

Tong, F., Nakayama, K., Vaughan, J.T., and Kanwisher, N. (1998). Binocular rivalry and visual awareness in human extrastriate cortex. *Neuron* **21**, 753–759.

Tootell, R.B.H., Tsao, D., and Vanduffel, W. (2003). Neuroimaging weighs In humans meet macaques in 'primate' visual cortex. *Journal of Neuroscience* **23**, 3981–3989.

Van Essen, D.C. (2005). Corticocortical and thalamocortical information flow in the primate visual system. *Progress in Brain Research* **149**, 173–185.

Vuilleumier, P., Sagiv, N., Hazeltine, E., Poldrack, R.A., Swick, D., Rafal, R.D., and Gabrieli, J.D. (2001). Neural fate of seen and unseen faces in visuospatial neglect: combined event-related functional MRI and event-related potential study. *Proceedings of the National Academy of Science of the USA* **98**, 3495–3500.

Wardak, C., Ibos, G., Duhamel, J.R., and Olivier, E. (2006). Contribution of the monkey frontal eye field to covert visual attention. *Journal of Neuroscience* **26**, 4228–4235.

Weiskrantz, L. (1998). *Blindsight: A Case Study and Implications*. Oxford: Oxford University Press.

Wexler, M. and van Boxtel, J.J. (2005). Depth perception by the active observer. *Trends in Cognitive Sciences* **9**, 431–438.

Young, M.P. (1992). Objective analysis of the topological organization of the primate cortical visual system. *Nature* **358**, 152–155.

Zaidel, E., Iacoboni, M., Zaidel, D.W., and Bogen, J.E. (2003). The callosal syndromes. In Heilman, K.M. and Valenstein, E. (eds) *Clinical Neuropsychology*, pp. 347–403. Oxford: Oxford University Press.

Chapter 6

Vision, action, and awareness

Manos Tsakiris and Patrick Haggard

David Milner's argument hinges on a broad division between a ventral and a dorsal visual stream. We agree with him on this point. The weight of evidence for this division, much of it from David Milner's own elegant studies, now seems compelling. First, abundant neuropsychological and neurophysiological evidence supports the distinction between pathways for perception and pathways for action. Secondly, the two-pathways view has provided a valuable analytic tool for studying cognition, in keeping with the general finding in psychology that dissociation and individuation are the first steps in scientific understanding. Therefore, our main reaction to David Milner's chapter does not concern the division between the two visual streams *per se*, but rather the implications of this division for the study of consciousness.

6.1 Ventral-stream awareness: implications for consciousness

At first sight, the 'dual visual systems' view seems to favour a Cartesian, dualistic view of consciousness. Milner argues that conscious experience is associated with the ventral visual stream, and that the operation of the dorsal stream involves visual, or perhaps visuomotor, processing, but not conscious (visual) experience. His chapter in the present volume focuses more on the non-consciousness of the dorsal stream than on the consciousness of the ventral stream. However, the theory could also be judged by the implications that the dorsal–ventral dissociation has for the nature of consciousness in the ventral stream. Critically, the dorsal–ventral dissociation strongly separates conscious experience from interaction with the external world. The conscious visual perception generated along the ventral stream in fact recalls the perception of mental representations shown in a 'Cartesian theatre' (Milner, this volume). Within this 'theatre', the conscious subject is a relatively passive observer of the scenes represented. The subject is not himself on the stage, nor directing the action on the stage: those functions are specifically handled by the dorsal stream. This point has important consequences for the type of phenomenal experience the model is supposed to account for.

Representations in the ventral stream are internal mental states that represent the external world, corresponding, in the paradigm case, to discrete objects in the external world (see also Jeannerod and Jacob 2005) Accordingly, the phenomenology of the conscious ventral stream would necessarily have a strongly representational flavour. Goodale and Milner (2004) have used the metaphor of watching television as an example of perceptual experience (2004, p. 75): 'There is little doubt that the brain mechanisms that allow us to watch and understand TV are the very same mechanisms that allow us to perceive and understand the real world. In real life too, the brain uses stored knowledge of everyday objects, such as their size, to make inferences about the sizes of other objects and about their distance from us and from each other.' On closer inspection, this metaphor may not provide an accurate description of our conscious experience, or of the nature of representations used in vision *for* action. Perception involves the comparison and recognition of objects that are spatially represented in relation to each other, that is, in a coordinate system that is independent from the conscious agent. It is true that objects seen on a TV screen are represented in this way, and it may also be true that object recognition in everyday life may involve the same kind of spatial representation of the object that is 'perceiver-free'. However, action seems to require a representation of the external world in an egocentric spatial frame of reference. Objects perceived on a TV screen have little behavioural importance for the conscious perceiver. As such, they do not need to be represented in the egocentric frame of reference used for the planning and execution of our interactions with an object. An object moving on a screen does not move towards or away from me, as it only moves in relation to other objects on the screen. Conversely, an object in the world that moves towards my body will require a behavioural response on my behalf.

Possible interactions with the object would be excluded from the conscious content of percepts referring to the object, since such actions are handled by the non-conscious dorsal stream, not the ventral stream. This seems to be a logical consequence of Milner's view of consciousness, although he does not say so explicitly in this volume. However, that is not a view that we, or indeed many other psychologists, find easy to accept. In our view, the conscious experience that we have of external objects depends partly on the interactions that we have, or might have, with them. This strand of contemporary philosophy, going back to Merleau-Ponty (1962), links the phenomenology of perception to *intentionality*, whereby conscious perception is thought to be *about* things in the world and the possible interactions we may have with them, rather than about their mental representations. If interactions are relevant for conscious experience, and we believe they are, this seems to require a different kind of consciousness from the one involved in the vision-for-perception model of Milner and Goodale.

On this view, conscious visual experience for action would correspond to a practical, rather than a conceptual way of *seeing* (see Heidegger 1927); or to a pragmatic, rather than a semantic way of *seeing* (see Jacob and Jeannerod 2003). Phenomenology of perception in the ventral sense implies a way of seeing the world from the outside, as if you are looking at the television (Goodale and Milner 2004); an alternative view of conscious experience emphasizes that we see both for and in our own actions. If we follow Milner in agreeing that there is something 'it is like' to have purely visual experiences, or purely motor experiences, while at the same time denying the category of 'visuomotor experiences' (Milner, this volume), then we acknowledge that the dual visual system may not be able to account for the conscious experience of acting in the world. A theory or a model can only account for what it judges to fall within its explanatory scope, but it is also 'liable' for the positions that follow from it. The concept of conscious *agent*, as opposed to a conscious perceiver, who interacts with the (visual) environment, and for whom such interactions form an important part of conscious mental life, seems incompatible with the model. Given that the dual visual system is meant to be an account of purely visual consciousness, it may not admit the notion of a conscious agent, perceiving the world and acting in or upon it in parallel.

Unfortunately, very few experimental studies have directly compared purely perceptual aspects of consciousness and action-related consciousness. Here we review one of the few to do so, by Johnson and Haggard (2005). This study built on previous reports of rapid, unconscious reprogramming of reaching movements. Subjects may not perceive a jump in the position of a visual target that occurs during a saccade. They nevertheless adjust their reaching trajectory within a few hundred milliseconds (Castiello *et al*. 1991). Although previous studies drew attention to this contrast between lack of visual awareness and successful motor control, none, to our knowledge, had investigated whether subjects had a conscious experience associated with the rapid adjustment of their *action*. We therefore asked people to perform a visuomotor reaching task, which might randomly involve a target shift. They then *both* reported whether they perceived the target to shift or not, *and* reproduced the spatial path of the reaching movement they had just made. Our results revealed a surprising dissociation between perceptual awareness and motor awareness. Subjects clearly reproduced the adjustment they made in response to the target shift occurring in the original movement, even on trials where they denied any perceptual awareness that the target had shifted. It would be interesting to confront subjects with the paradoxical nature of this dissociation in future work: why did the subject adjust their movement to one side when, in their own experience, the target remained at the straight-ahead position?

In the meantime, we draw two inferences relevant to the present chapter. First, the operations of the dorsal stream can indeed produce awareness, in this case, awareness of one's own actions. The action awareness has the same spatial dimensions, in this case of lateral deviation, as the corresponding perceptual information. We imagine that patients like D.F. also have this kind of action awareness. Second, the awareness of one's own actions can be logically distinguished from perceptual awareness of the external world, under appropriate conditions, both in the experimental laboratory and in neuropsychological cases. An important point for future research would be to investigate how perceptual and motor awarenesses become integrated in normal circumstances. We do not normally perceive our motor response to an object without also perceiving the relevant features of the object. Haptic exploration (Lederman and Klatzky 1993) perhaps offers the most compelling example of the normal integration of perceptual and motor awareness. Haptic exploration results in a single conscious content which depends both on the exploratory movements of the hand and the physical properties of the object. Below we describe some of the neural circuits that may be involved in the integration of perceptual and action information to produce this unified aspect of conscious experience.

6.2 Neural basis of action awareness

Patients with visual agnosia, such as D.F., seem unable to report properties of objects, yet they have no significant problems in making skilled movements of the body, such as grasping, that would appear to require precisely the same information. This finding lies at the heart of Milner's proposed dissociation between ventral and dorsal streams. D.F.'s grasping behaviour may be subserved by automatic non-conscious motor processes involving the dorsal stream, yet it is (presumably) accompanied by a conscious phenomenology of acting upon the world. Phenomenology of action here refers to the conscious experience of one's own actions, namely, the awareness that one has of intending and performing voluntary goal-directed behaviours.

For centuries, the phenomenology of vision has dominated scientific thinking about mental processes. In the history of philosophy, one often finds the idea that the mind is above all a sensory apparatus whose basic mode of operation is visual perception of the world. This emphasis is perhaps not surprising given the rich phenomenal content of even the simplest visual experience. Inevitably, the quest for the nature of consciousness and more recently for the neural correlates of consciousness have evolved around the visual perception that may in principle occur in disembodied minds or simply in brains in a vat. When Milner and Goodale (2004) suggest that the experience of watching television 'convincingly mimics our experience of the world' (p. 74), or that 'the ventral stream corresponds to the human operator, who identifies items of potential importance on TV pictures received remotely from a video camera attached to the robot', they suggest that (a) there is no other form of perceptual experience or consciousness, and that (b) there is no consciousness at the level of the robot. Here we want to develop a different line of argument, which indeed figures in Turing's seminal paper on the mind (Turing 1950). On this view, a critical feature of the human mind, and a key determinant of the conscious experience that it generates, is the mind's connection to the human body. A model based on disembodied minds (or brains) cannot account for all forms of awareness.

First, an important part of conscious perceptual experience relates to the awareness of our own actions. Having received relatively little interest until recently, action awareness has steadily grown in importance in discussions of consciousness, and may play a particularly important role in the neuroscience of self-consciousness (Haggard 2005). Most authors agree that the primary phenomenology of action is thin and elusive (Metzinger 2003), and far less rich in content than visual experiences. We can produce remarkably sophisticated actions 'automatically', without strong or specific conscious content (Broadbent 1982). In addition, numerous studies show that fine-grained motor adjustment and corrections may occur outside awareness (but see our discussion on this point above). What is then left to be aware of when we are acting? Are we really aware that we are acting at all? We would emphatically answer 'yes' to the last point. An internally generated action is accompanied by a conscious experience of having intended and initiated this particular action. Moreover, we experience a certain kind of control over our movements, and actions, and we also monitor the effects of our actions. This experience of action, integrating intention and the perception of an action's effect, corresponds to the sense of agency that we have over our actions. This 'sense of agency' that we experience is of seminal importance for our self-awareness, as neuropsychological and neuropsychiatric conditions such as anosognosia for hemiplegia, anarchic hand syndrome, and schizophrenia suggest. The sense of agency, perhaps unlike the sense of the phenomenal visual field, works in the background: we often have a thin, recessive experience of what we are doing. But when our actions go wrong, the phenomenal consequences can be vivid indeed. This fact gives important information about how the mechanisms that produce phenomenology of action may work, and indeed what they are for.

Under normal circumstances, we experience agency when we voluntarily perform goal-directed movements. This statement implies three critical components that seem to underpin our sense of agency: (1) the movement needs to be intended and self-initiated; (2) it needs to be performed in a way coherent with the intention; (3) it needs to achieve the intended goal. The specific contribution of each component for the awareness of action and the experience of agency remain unclear. However, based on recent findings (Knoblich and Flach 2001, 2003; Knoblich et al. 2002; Haggard and Clark 2003; Tsakiris and Haggard 2003; Sato and Yasuda 2005; Tsakiris et al. 2005), we suggest at least some of these elements have implications for consciousness. We are often aware, in advance, of our intention to move and we may consciously monitor the effects of our actions on the outside world. Theories differ in the extent to which they emphasize the former, internal aspects of action awareness (Berti et al. 2005), or the latter distal aspects (Prinz 1992). On the internal account, awareness of action involves processing of events internal to the organism itself, such as preparatory intentions (Fried et al. 1991), or efference copy, or reafferent signals. The precise description of these individual representations may be relatively unimportant. What matters for action awareness is the interactions between them. Sensorimotor integration is fundamental for action awareness. We normally have an atomic experience, which we might describe as 'I did that' (Haggard and Johnson 2003). This experience appears to synthesize and compress the intention to act, the intended movement of the body itself, and the external effects of the action as relayed by the senses (Haggard et al. 2002; Haggard 2005; Sato and Yasuda 2005; Tsakiris and Haggard 2005). Crucially, this kind of action consciousness seems very different from the Cartesian theatre. It is embodied rather than cerebral. The interaction between the 'operator', the world, and the transmission of signals through the body of the 'operator' must be considered, in addition to the operator herself.

Converging evidence suggests that the posterior parietal cortex may underpin our conscious experience of action through the processing of central signals. An anterior–posterior functional differentiation within the parietal cortex for the processing of peripheral and centrally generated signals has been suggested by various research findings (Burbaud et al. 1991; Graziano and Botvinik 2001; Schwoebel e tal. 2002). Posterior parietal cortex has been linked to the planning of movements for a range of body parts (Gemba et al. 2004; for a review see Cohen and Andersen 2002), and also in online control of actions (Grea et al. 2002). The impaired performance of parietal patients in a self-recognition task has been attributed to an impaired ability to compare online sensory feedback with an internally generated representation of the planned movement (Sirigu *et al.* 1999).

Human neuroimaging studies have consistently showed activation in parietal cortex linked to the sense of agency (Fink *et al.* 1999; Ruby and Decety 2001; Farrer and Frith 2002; Leube *et al.* 2003a, 2003b; Lau *et al.* 2004a, 2004b). The role that David Milner proposed for the inferior parietal cortex is that of a 'high-level representational medium within which the contents of *visual* consciousness are being constantly constructed and reconstructed, for example on different spatial scales according to the attentional needs of the moment' (Milner, this volume, p. 200; italics added). Apart from the contribution of the inferior parietal cortex to visual consciousness, areas in the inferior parietal cortex, BA39 and BA40 have consistently been implicated in awareness of action, as the studies cited above suggest.

An interesting case study was reported by Schwoebel *et al.* (2002). Patient C.W. who suffered bilateral posterior parietal lesions performed movements when he imagined them even though he was unaware of these actions. According to the authors, the posterior parietal cortex is strongly involved in conscious awareness of movements and the intention to move. This is because C.W. did not actually intend to move when he imagined the movements. In another study, Sirigu and colleagues (2004) showed that parietal patients with lesions in BA39 and BA40 have an impaired time-awareness of intending an action, but not an impaired time-awareness of the actual movement.

Being aware of an intention and translating this intention into a motor plan seems to require consciousness, and moreover seems to engage areas in the posterior and inferior parietal cortices belonging to the dorsal stream. On this account, the claim that the posterior parietal cortex, as part of the dorsal stream is in fact a bottom-up system dependent on visual input seems unjustified. The dorsal stream also processes top-down information related to impending motor behaviour. These behaviours may involve object-oriented actions, but the object often appears to underdetermine the action. Thus, there is a clear role for top-down information within the dorsal stream. This information is processed to generate and elaborate complex motor plans, whose execution will bring about an intended state of body and object, such as lifting a glass to the lips. The processing of such information by the dorsal stream, and the control of the resulting action itself, have characteristic conscious correlates.

Thus, the dorsal stream is not restricted to automatic visuomotor transformations. It also underpins higher-level representations involved in the planning of complex actions, and the recognition and monitoring of one's own actions. These in turn may be used for understanding of other people's actions and possibly for distinguishing the self from the other. At this level, it is hard to make a clear-cut distinction between perception and action. For example, the 'visual' extrastriate body area, in the occipital cortex, has recently been implicated both in representation of one's own actions, and also in judging the agency of others (Astafiev et al. 2004; David et al. in press).

To conclude, Milner's chapter provides an elegant demonstration of the possibility of goal-directed action without visual awareness. However, it is important to remember that a characteristic form of conscious experience generally also accompanies goal-directed action. Though perhaps less phenomenally compelling than visual awareness, action awareness has a specific internal structure, identifiable neural substrates, and is amenable to scientific investigation. Interestingly, awareness of an action performed on an object can be dissociated from perceptual awareness of the object acted upon, at least in laboratory situations. Nevertheless, this dissociation merely highlights a more important but elusive question. How does our perceptual experience of objects in the world depend on, and become integrated with, our experience of interacting with them?

References

Astafiev, S.V., Stanley, C.M., Shulman, G.L., and Corbetta M (2004). Extrastriate body area in human occipital cortex responds to the performance of motor actions. *Nature Neuroscience* **7**, 542–548.

Berti, A., Bottini, G., Gandola, M., Pia, L., Smania, N., Stracciari, A., Castiglioni, I., Vallar, G., and Paulesu, E. (2005). Shared cortical anatomy for motor awareness and motor control. *Science* **309**, 488–491.

Broadbent, D.E. (1982). Task combination and selective intake of information. *Acta Psychologica* **50**, 253–290.

Burbaud, P., Doegle, C., Gross, C., and Bioulac, B. (1991). A quantitative study of neuronal discharge in areas 5, 2, and 4 of the monkey during fast arm movements. *Journal of Neurophsyiology* **66**, 429–443.

Castiello, U., Paulignan, Y., and Jeannerod, M. (1991). Temporal dissociation of motor responses and subjective awareness. A study in normal subjects. *Brain* **114**(6), 2639–2655.

Cohen, Y.E. and Andersen, R.A. (2002). A common reference frame for movement plans in the posterior parietal cortex. *Nature Reviews Neuroscience* **3**, 553–562.

David, N., Cohen, M.X., Newen, A., Bewernick, B.H., Shah, N.J., Fink, G.R., and Vogeeley K (in press). The extrastriate cortex distinguishes between the consequences of one's own and others' behaviour. *NeuroImage*, in press.

Farrer, C. and Frith, C.D. (2002). Experiencing oneself vs another person as being the cause of an action: the neural correlates of the experience of agency. *NeuroImage* **15**, 596–603.

Fried, I., Katz, A., McCarthy, G., Sass, K.J., Williamson, P., Spencer, S.S., and Spencer, D.D. (1991). Functional organization of human supplementary motor cortex studies by electrical stimulation. *Journal of Neuroscience* **11**, 3656–3666.

Fink, G.R., Marshall, J.C., Halligan, P.W., Frith, C.D., Driver, J., Frackowiak, R.S., and Dolan, R.J. (1999). The neural consequences of conflict between intention and the senses. *Brain* **122**(3), 497–512.

Gemba, H., Matsuura-Nakao, K., and Matsuzaki R (2004). Preparative activities in posterior parietal cortex for self-paced movement in monkeys. *Neuroscience Letters* **357**, 68–72.

Goodale, M.A. and Milner, A.D. (2004). *Sight Unseen: An Exploration of Conscious and Unconscious Vision*. Oxford: Oxford University Press.

Graziano, M.S.A and Botvinik, M.M. (2001). How the brain represents the body: insights from neurophysiology and psychology. In Prinz, W. and Hommel, B. (eds) *Common Mechanisms in Perception and Action, Attention and Performance XIX*. Oxford: Oxford University Press, 2002.

Grea, H., Pisella, L., Rossetti, Y., Desmurget, M., Tilikete, C., Grafton, S., *et al.* (2002). A lesion of the posterior parietal cortex disrupts on-line adjustments during aiming movements. *Neuropsychologia* **40**, 2471–24800.

Haggard, P. (2005). Conscious intention and motor cognition. *Trends in Cognitive Sciences* **9**(6), 290–295.

Haggard, P. and Clark, S. (2003). Intentional action: conscious experience and neural prediction. *Consciousness and Cognition* **12**, 695–707.

Haggard, P. and Johnson, H. (2003). Experiences of voluntary action. *Journal of Consciousness Studies* **10**, 72–84.

Haggard, P., Clark, S., and Kalogeras, J. (2002). Voluntary action and conscious awareness. *Nature Neuroscience* **5**, 382–385.

Heidegger, M. (1927). *Being and Time*, translated by Joan Stambaugh. Albany: State University of New York Press.

Jacob, P. and Jeannerod, M. (2003). *Ways of Seeing: The Scope and Limits of Visual Cognition*. Oxford: Oxford University Press.

Jeannerod, M. and Jacob P. (2005). Visual cognition: a new look at the two-visual systems model. *Neuropsychologia* **43**(2), 301–312.

Johnson, H. and Haggard, P. (2005). Motor awareness without perceptual awareness. *Neuropsychologia* **43**(2), 227–237.

Knoblich, G. and Flach, R. (2001). Predicting the effects of actions: interactions of perception and action. *Psychological Science* **12**, 467–472.

Knoblich, G. and Flach, R. (2003). Action identity: evidence from self-recognition, prediction, and coordination. *Consciousness and Cognition* **12**(4), 620–632.

Knoblich, G., Seigerschmidt, E., Flach, R., and Prinz, W. (2002). Authorship effects in the prediction of handwriting strokes: evidence for action simulation during action perception. *Quarterly Journal of Experimental Psychology A* **55**, 1027–1046.

Lau, H.C., Rogers, R.D., Haggard, P., and Passingham, R.E. (2004a). Attention to intention. *Science* **303**, 1208–1210.

Lau, H.C., Rogers, R.D., Ramnani, N., and Passingham, R.E. (2004b). Willed action and attention to the selection of action. *NeuroImage* **21**, 1407–1415.

Lederman, S.J. and Klatzky, R.L. (1993). Extracting object properties through haptic exploration. *Acta Psychologica* **84**(1), 29–40.

Leube, D.T., Knoblich, G., Erb, M., and Kircher, T.T.J (2003a). Observing one's hand become anarchic: an fMRI study of action identification. *Consciousness and Cognition* **12**, 597–608.

Leube, D.T., Knoblich, G., Erb, M., Grodd, W., Bartels, M., and Kircher, T.T.J (2003b). The neural correlates of perceiving one's own movements. *NeuroImage* **20**, 2084–2090.

Merleau-Ponty, M. (1962). *The Phenomenology of Perception*, translated by C. Smith. London: Routledge.

Metzinger, T. (2003). *Being No One. The Self-Model Theory of Subjectivity*. Cambridge, MA: MIT Press.

Prinz, W. (1992). Why don't we perceive our brain states? *European Journal of Cognitive Psychology* **4**, 1–20.

Ruby, P. and Decety, J. (2001). Effect of subjective perspective taking during simulation of action: a PET investigation of agency. *Nature Neuroscience* **4**, 546–550.

Sato, A. and Yasuda, A. (2005). Illusion of sense of self-agency: discrepancy between the predicted and actual sensory consequences of actions modulates the sense of self-agency, but not the sense of self-ownership. *Cognition* **94**(3), 241–255.

Schwoebel, J., Boronat, C.B., and Branch Coslett, H. (2002). The man who executed 'imagined' movements: evidence for dissociable components of the body schema. *Brain and Cognition* **50**, 1–16.

Sirigu, A., Daprati, E., Pradat-Diehl, P., Franck, N., and Jeannerod, M. (1999). Perception of self-generated movement following left parietal lesion. *Brain* **122**(10), 1867–1874.

Tsakiris, M. and Haggard, P. (2003). Awareness of somatic events associated with a voluntary action. *Experimental Brain Research* **149**, 439–446.

Tsakiris, M. and Haggard, P. (2005). Experimenting with the acting self. *Cognitive Neuropsychology* **22**, 387–407.

Tsakiris, M., Haggard, P., Franck, N., Mainy, N., and Sirigu, A. (2005). A specific role for efferent information in self-recognition. *Cognition* **96**, 215–231.

Tsakiris, M., Prabhu, G., and Haggard, P. (2006). Having a body versus moving your body: how agency structures body-ownership. *Consciousness and Cognition* **15**(2), 423–432.

Turing, A.M. (1950). Computing machinery and intelligence. *Mind* **59**, 433–460.

Chapter 7

The social functions of consciousness

Chris D. Frith

7.1 What have we learned from the experimental study of consciousness?

7.1.1 What is consciousness?

I might be expected to begin this essay by providing a precise definition of what I mean by consciousness. However, I believe that such a definition is premature. Currently the term consciousness embraces many different concepts. Eventually a definition will be narrowed down on the basis of the kinds of experiments I am about to discuss. In the very broadest terms there are two distinct aspects of consciousness. First, I can be conscious in the sense of being awake, rather than being asleep or in a coma. This dimension is referred to as the level of consciousness. Second, even when we are wide awake, there may be some things which we are conscious of (or aware of), while there are other things which we are not conscious of. The things we are conscious of when awake are referred to as the contents of consciousness. Here I shall concentrate on the contents of consciousness. In general, when we are awake our consciousness has contents, while when we are deeply asleep it does not. However, the relationship is not precise. When we are asleep, but dreaming, our consciousness has content. In contrast, patients in a persistent vegetative state go through a sleep–wake cycle, but appear to have no content of consciousness at any time.

7.1.2 The neural correlates of consciousness

The development of brain science over the last few decades has provided a new source of data important for our understanding of consciousness. This is the data generated by the search for the neural correlates of consciousness (Crick and Koch 1998; Frith *et al.* 1999). This search for the neural correlates of consciousness is guided by the following fundamental assumptions:

1. A change in the contents of consciousness is necessarily associated with a change in neural activity, but
2. a change in neural activity is not necessarily associated with a change in the contents of consciousness.

Point 2 reminds us that stimuli that do not reach consciousness can still elicit brain activity and influence our behaviour. There are many examples of this unconscious information processing. Indeed, some have claimed that we are never aware of the cognitive processes undertaken by the brain, only of their outcome (Velmans 1991). Particularly striking examples of unconscious processing can occur after brain damage. Patients with blindsight can locate stimuli that they cannot 'see'. Patient D.F., who is discussed in detail in David Milner's chapter, is unaware of the shape of an object, but can still shape her hand appropriately when required to pick the object up (Goodale *et al.* 1991). But unconscious processing also occurs in the normal brain. As discussed in Joe LeDoux's chapter, we flee from a frightening object, such as a snake, before we become aware that it is a snake (LeDoux 1993). Since Tony Marcel's demonstration of unconscious priming in reaction time tasks, many paradigms have been developed to show how stimuli of which we are unaware can nevertheless affect our behaviour (Marcel 1983; Merikle *et al.* 2001).

This distinction between conscious and unconscious processes raises a key question for neuroscience. This question was probably first asked by Donald MacKay in 1981, 'What kind of discontinuity might we expect to differentiate brain activity that mediates conscious experience from that which does not?'(MacKay 1981) The rapid development of brain imaging techniques since that time has generated much data that is directly relevant to this question. We now know that the brain contains 'essential nodes' associated with specific contents of consciousness (Zeki and Bartels 1999). For example, there are a number of circumscribed regions in the inferior temporal lobe, which are specialized for the perception of specific visual features or classes of objects. Area V4 in the posterior fusiform gyrus is specialized for the perception of colour. Activity in this region increases in response to coloured stimuli (Zeki *et al.* 1991) and damage to this region causes achromatopsia (Zeki 1990), in which the patient is no longer conscious of colour. Direct electrical stimulation of this area can cause experiences of colour (Lee *et al.* 2000) showing that input from the retina or primary visual cortex is not necessary for this experience. A similar story can be told for an adjacent region of the fusiform gyrus that is specialized for the perception of faces (the fusiform face area; Kanwisher and Yovel 2006). These brain regions are essential nodes for the perception of colour and faces respectively.

However, while these regions are necessary for particular contents of consciousness, they are not sufficient. One demonstration of this comes from the study of change blindness. When two pictures are presented in rapid alternation with a uniform grey screen displayed in between, quite large differences will frequently go unnoticed (Simons and Rensink 2005). In the study by Beck *et al.* (2001), the effect of changes in faces was investigated using this paradigm. On some trials there was no change in the identity of the face shown. In some trials the identity of the face was changed, but participants were not aware of this. In a third class of trials the face changed and participants were aware of the change.[1] A change in the face elicited increased activity in the fusiform face area, even when participants were unaware of the change. This observation suggests that increased activity in the fusiform face area is not sufficient to generate a conscious experience of a face. When participants were aware of the change in the face, then the increase in activity in the fusiform face area was greater, but in addition there was activity in the parietal and the frontal cortex. Similar results have been observed in a number of studies (Rees and Frith 2007). While activity in essential nodes determines the contents of the perceptual experience, concomitant activity in frontal and parietal cortex seems to be critical for awareness of this content.

Of course, increases in activity observed in brain imaging studies cannot prove that the associated brain region is necessary for the phenomenon being studied. However, evidence from studies of patients with lesions (Rees *et al.* 2002) and the application of TMS to normal volunteers (Beck *et al.* 2006) suggest that parietal cortex is necessary for awareness. Damage to right parietal cortex leads to neglect of the left side of space. The patient is unaware of events in that part of the visual field even though the relevant areas of visual cortex are intact (Nachev and Husain 2006). Imaging studies confirm that stimuli in the left visual field continue to elicit activity in visual cortex and in fusiform gyrus, even though the patient is unaware of the stimuli (Rees *et al.* 2002).

The regions of parietal and frontal cortex associated with consciousness are the same as those associated with selective attention (Rees and Lavie 2001).

---

[1] There is an implicit fourth condition in which subjects report a change when none actually occurred (false positives). In the study of Beck *et al.* such events were too rare to analyse. However, other studies suggest that false positives are associated with increased activity in the relevant essential node. The activity reflects the subjective experience rather than the physical stimulation (Ress and Heeger 2003).

This is consistent with the observation, going back at least to William James, that the contents of consciousness are dominated by whatever happens to be currently at the focus of our attention. This ability to focus attention depends upon regions in frontal and parietal cortex.

7.1.3 The problem of report

There is a fundamental problem with the studies I have described so far. When we compare conscious and unconscious processing of the same stimulus, such as the face in the change blindness experiment, we rely on our volunteer reporting whether or not she was aware of the stimulus. How can we separate the neural correlates of report from the neural correlates of consciousness? This problem is most easily seen in another widely used paradigm for the study of consciousness: binocular rivalry (Blake 1989).

If a face, for example, is presented to one eye while a grating is presented to the other eye, we do not see a mixture of face and grating. Instead we see the face for a few seconds and then our perception spontaneously switches to the grating. This alternation between face and grating carries on indefinitely and is largely outside the control of the observer. This phenomenon is important for the study of consciousness because the signal coming into the brain remains constant, while awareness changes. As with the change blindness experiment, we can use the rivalry paradigm to identify brain regions that respond to the stimulus of which we are currently unaware, and also brain regions that become more active when the stimulus enters awareness (Tong *et al.* 2006). Activity in parietal and frontal regions occurs at the point where there is switch from one percept to the other (Lumer *et al.* 1998). But, of course, we only know about these switches because our volunteer pressed a button to indicate when the switch occured. Some of the brain activity we see must be due to the need to report. Of course, the perceptual switches occur whether we report them or not, so is it possible to study binocular rivalry in the absence of report?

New developments in functional imaging have now made this possible. By looking at the details of the spatial pattern of activity in visual cortex, it is possible to determine which of two stimuli a volunteer is being presented with when no binocular rivalry is occurring. The patterns of activity so identified can then be used to predict which stimulus is being perceived in the rivalrous situation. This prediction agrees well with the volunteer's report (Haynes and Rees 2005).

This development is important for many reasons. We can use the technique to study brain changes associated with binocular rivalry in the absence of report. But more importantly, the technique could, in principle, be used to identify consciousness in cases where report is not available, such as human infants, animals, and patients with locked-in syndrome.

This brief review shows how rapidly the study of the neural correlates of conscious has developed in the last few years. From knowing almost nothing, we have now identified brain regions that are necessary for many different aspects of the contents of consciousness. And, as a result of these discoveries, we can use measures of brain activity to read minds (Haynes and Rees 2006). A number of answers have been given to MacKay's question about the characteristics of the neural activity associated with conscious experience. In addition to finding that activity in particular brain regions (an essential node) is associated with specific contents of consciousness, there is evidence that this activity has to be synchronized at a particular frequency (Schurger *et al.* 2006), that feedback from adjacent cortical areas is required (Lamme and Roelfsema 2000; Pascual-Leone and Walsh 2001) and that long-range cortical connectivity via the thalamus is involved (Tononi *et al.* 1998; Laureys *et al.* 2000).

And yet, in spite of all this progress, I remain unsatisfied about these answers. Proposals that the contents of consciousness could be equated with attention or the contents of working memory had been made before the advent of the programme for identifying the neural correlates of consciousness (NCC) (e.g. Shallice 1972). So what has all this new knowledge about the NCC added? These new data do not increase my understanding of the nature of consciousness or suggest how it might be better defined in terms of cognitive processes or forms of representation.

7.2 Consciousness and action

There is more to the contents of consciousness than perception of objects in the outside world. We are also aware of ourselves as agents who act upon that world. But how much are we really aware of these actions? David Milner's studies of patient D.F. show that complex reaching and grasping movements can be made without awareness of the stimuli that shape the movements (Goodale *et al.* 2004). There is further evidence that, in many situations, we are not aware of the movements either. If an adjustment has to be made, mid-flight, in a rapid hand or eye movement, then we are often not aware of the adjustment until after it has been made and are sometimes not even aware that an adjustment has been made (Prablanc and Martin 1992; Pisella *et al.* 2000).

Whenever we move, there are many sensory signals generated by receptors in skin and muscle that indicate how far our limbs have moved and their new positions relative to the rest of the body. Without such signals our ability to move is greatly impaired (Cole and Sedgwick 1992), but we have little awareness of these signals. For example, Fourneret and Jeannerod (1998) asked volunteers to move a cursor across a screen in a setting in which the volunteers' hands and arms were not visible. On some trials, the computer controlling the cursor introduced a distortion so that a straight movement on the screen required a deviant movement of the arm. Up to 15° of distortion, volunteers were not aware that their movements were deviant. This lack of awareness arises because sensory signals caused by our own movements are actively suppressed (Weiller et al. 1996). This is why we can't tickle ourselves. If someone else strokes the palm of my left hand much activity is elicited in brain regions concerned with tactile sensation. However, if I apply the same stroking movements to my left palm with my right hand, the activity elicited in these areas is considerable reduced (Blakemore et al. 1999).

I have very briefly reviewed the evidence that many aspects of motor control occur without awareness. Most of the time we are not aware of the sensory consequences of our movements or of the rapid movement corrections we are constantly making. This is perhaps a case where the experimental study of consciousness has thrown some light on the nature of consciousness (Hohwy and Frith 2004). This lack of awareness of certain aspects of motor control may explain why 'the phenomenology of volition is *thin* and it is *evasive*' (Metzinger 2006).

We have no problem in accepting the implication of these results; that consciousness has no causal role to play in making rapid corrections to movements. Indeed, there are good reasons for thinking this is an advantage since such corrections have to be made very rapidly. However, we have considerable problems with the idea that consciousness is not involved in deciding which movements to make and when to make them. Yet this is the implication of the much-discussed experiment by Libet and colleagues (1983). In this experiment EEG was measured while participants lifted their right index finger 'whenever they felt the urge to do so'. The occurrence of such voluntary movements is preceded by a gradual increase in negativity known as the readiness potential (Deecke 1990). When averaged over a number of trials the start of this increase in negativity can be detected for up to 1 second before the finger is actually moved. The novel feature of Libet's experiment was a method for measuring the subjective time at which the urge to move occurred. Participants observed a rapidly moving clock hand and reported the position of this hand at the moment they had the urge to respond. This was found to be about 200 ms *after* the first detectable change in brain activity (confirmed in more recent studies, e.g. Haggard et al. 1999). This result seems to imply that we are not aware of deciding to act until after this decision has already been made by some unconscious process in the brain.

As we shall see later on in this essay, there are other interpretations of the phenomenon discovered by Libet. However, there is further evidence that consciousness does not necessarily have a major role in choosing actions and making decisions. For example, Varraine and colleagues (2002) had participants walking on a treadmill. Once a stable pattern of walking had been achieved the system was perturbed by gradually increasing the resistance of the treadmill. Participants were instructed to change their walking pattern as soon as they detected this change in resistance. On some occasions they were asked to maintain their walking speed in spite of the increase of resistance. This required them to increase the horizontal force of their stride. On other occasions they were asked to keep the horizontal force constant. This required them to reduce their walking speed. Note that these responses to the change in resistance were not well practiced and automatic. They were arbitrary responses created by instructions. The results are reminiscent of the Libet experiment. Participants changed their walking pattern as instructed about 2 seconds after the change in resistance began. However, participants did not report being aware of the change in resistance for about another 4 seconds. An arbitrary motor response to a stimulus could be initiated without awareness.

But this is still a rather low-level action decision. Johansson and colleagues (2005) showed participants a series of pairs of female faces and asked them to choose which one they preferred. After each choice the participants were shown the chosen face again and asked to explain their choice. The special feature of this experiment is that, on 20% of trials, participants were not shown the face they had just chosen, but the one they had just rejected. Three-quarters of these deceptions were not detected. Most of the time participants seemed not to be aware of the choices they had just made.

Dijksterhuis and colleagues (2006) went one step further and showed that consciousness does not help us to make better choices. Participants in this experiment had to make a series of expensive decisions, such as which car to buy. The decisions were made simple or complex in terms of the number of aspects of the car that had to be taken into account. In addition, the authors had objective criteria for determining how good the decisions were. Participants were given 4 minutes to make their decision, but, on some occasions, they were prevented from thinking about their decision during this time by being given distracting tasks to do. The results showed that complex decisions were made better when participants did not get the chance to think about them.

7.3 **Consciousness and agency**

The evidence so far suggests that consciousness has surprisingly little role in the control of actions and decisions. Nevertheless, we have a strong sense of agency.

We are conscious of ourselves as making decisions and controlling our actions. Our sense of agency influences our experience of action. One way of manipulating our sense of agency is to use hypnosis. Under hypnosis susceptible participants can easily be made to believe that one of their arms is moving up and down 'by itself' (i.e. the movement is passive), rather than, as is actually the case, because they are actively moving it. In this situation activity in somatosensory regions of the brain goes with the belief rather than the reality. As we have already seen activity in this region is suppressed during active movements in comparison to passive movements. Under hypnosis this suppression no longer occurs during active movements if the participants believe these movements are passive (Blakemore et al. 2003).

Our sense of agency also affects the time at which events are experienced as occurring. Patrick Haggard and his colleagues (2002) performed a version of the Libet experiment in which participants pressed a button, causing a tone to occur 250 ms later. In terms of mental time these two events, the cause and its effect, were closer together. The button press was experienced as occurring slightly later and the tone as occurring slightly earlier than they occurred in physical time. Haggard has called this phenomenon, whereby actions and their effects are experienced together in time 'intentional binding'. One can speculate that a similar effect might be happening in Libet's experiment such that the urge to act and its consequence, the finger movement, are also pulled together in time. In support of his notion of intentional binding Haggard was able to show that, when movements are involuntary (caused by transcranial magnetic stimulation), the movement and the following tone are pulled apart in subjective time.

These intentional binding effects do not depend solely upon the experience of agency, but also on the interval between cause and effect. With delays of more than 500 ms binding seems to be much reduced. However, with short delays of ~100 ms the binding can create illusory reversals of temporal order. With a delay of 135 ms between key press and flash on 60% of trials, a flash occurring within 44 ms after the key press is perceived as occurring *before* the key press (Stetson et al. 2006).[2] From this perspective the subjective timing of events associated with action is altered by the context in which the events occur rather as the apparent length of lines is changed in visual illusions such as the Müller–Lyer effect. The Libet effect should be seen as an illusion of time rather than a demonstration of a lack of free will.

---

[2] This evidence for binding avoids some of the problems associated with the clock reading method used by Libet and Haggard.

7.3.1 Awareness of agency in self and others

When we perform any act, such as pressing a key, we have privileged information about the action, which is not available when we see someone else performing an act. These are the proprioceptive signals caused by the movement. If these signals have a role in our experience of agency then the experience of our own agency will be very different from our experience of the agency of others. However, as we have already seen, proprioceptive and other kinds of sensory input caused by our own actions have very limited access to consciousness. Wohlschläger and colleagues (2003) measured intentional binding to demonstrate that proprioception does not have a role in the perception of agency. They confirmed Haggard's result showing that binding was enhanced for voluntary actions and reduced for involuntary actions of the self. They then measured binding when participants watched actions being performed. When the actions were performed by another person enhanced binding occurred as for voluntary actions of the self, but, when the actions were performed by a machine, reduced binding occurred as with involuntary actions of the self. Thus, in terms of binding, we perceive agency in others just as we do in ourselves even though we have no access to the private sensations associated with the actions of others.[3]

The problem with this way of perceiving agency is that we can easily become confused about who is the agent: is it me or the other? In a series of experiments Daniel Wegner and colleagues have demonstrated these confusions and shown how they can go either way. We can think we are the agent when it is actually the other and vice versa. Wegner argues that the only data we have for linking intention to outcome is temporal contingency: The outcome follows the intention (Wegner 2002). But contingency need not mean causality. In the appropriate context, people can be fooled by contingency into making incorrect attributions of agency.

In one experiment, two participants rested their fingers on a computer mouse. They moved the mouse around continuously, with the instruction to stop the movement every 30 seconds or so. However, one of the participants was actually a stooge of the experimenter. Through headphones the real

---

[3] This phenomenon of intentional binding has interesting parallels with the properties of the mirror neurons identified in the inferior frontal cortex of the monkey. These neurons become active when the monkey performs an action (e.g. picking up a peanut) and also when the monkey sees someone else performing the same action. However, the action has to be goal directed. If the hand moves, but does not pick up the peanut, then the neuron does not fire. Likewise, if the peanut is picked up with a tool rather than a hand, the neuron does not fire (Gallese *et al.* 1996).

subject of the experiment was given the idea of stopping the mouse-controlled cursor at a particular position on the screen. Shortly afterwards the stooge was given the instruction to stop the cursor at that position. If the timing was right (i.e. the stop occurring between 1 and 5 seconds after the thought of stopping) then the subject believed that she had stopped the mouse even though it was actually the stooge (Wegner and Wheatley 1999).

In contrast, the technique of facilitated communication is a 'natural' experiment in which participants believe that their own actions are actually someone else's (Wegner et al. 2003). Facilitated communication was originally developed to help physically handicapped children to communicate with a keyboard even though they were not strong or dextrous enough to press the keys. The facilitator places her hand over those of the child in order to detect attempted presses and convert them into actual presses. This technique was apparently a great success, particularly for children with autism, in that many previously uncommunicative individuals were now apparently able to communicate. Unfortunately, controlled experiments showed that, in many cases, it was not the child who was communicating, but the facilitator. A situation was created in which, unbeknown to the facilitator, she was seeing different questions from the child. In these situations it was the questions seen by the facilitator that were answered.[4] The striking feature of these cases is that the facilitator firmly believed that the child was the agent.

The relative ease with which we can become confused about whether we, or some other person, is the agent behind some action confirms that the experience of agency has rather little dependency upon signals, such as efference copy and proprioception, to which the self has privileged access.

7.3.2 What is consciousness for? The importance of the awareness of agency

In this section of the chapter my aim is to demonstrate the advantage of detecting and representing agency in self and other in the same way.

If our experience of agency depended upon sensations to which we have privileged access, then this would always be a private experience. Our sense of our own agency would be qualitatively different from our sense of the agency of other people. But this is not the case. We experience agency in others in the

---

[4] Grave miscarriages of justice can arise when facilitated communication is used in court cases concerned with child abuse. As a result, the American Psychological Association has adopted the position 'that facilitated communication is a controversial and unproved communicative procedure with no scientifically demonstrated support for its efficacy'.

same way as we experience it in ourselves. This may sometimes cause confusion, but it also has enormous advantages.

Natalie Sebanz and colleagues used a standard choice reaction time task to study what happens when two people share the same task (joint action). The stimulus in this task was a pointing finger wearing either a red or a green ring. The task was to press the left button to the red ring or the right button to the green ring. The direction in which the finger pointed (left or right) was irrelevant. In the standard two-choice version of the task in which one participant presses both buttons, the irrelevant pointing direction interfered with performance. For example, participants were slower to make a left response when the finger pointed to the right. However, if the participant only had to respond to one button (i.e. press left to the red ring and ignore the green ring), then the response was faster and the interfering effect of the finger pointing direction disappeared. In the critical joint-action condition a second participant took part and pressed right for the green ring. In this condition, even though the task being performed by the original participant was identical (pressing left to the red ring), the interference effect of the finger-pointing direction returned although responses were still faster than in the two-choice version (Sebanz *et al.* 2003). This result suggests that, in the joint-action condition, participants automatically represent their partner's actions in the task as well as their own. This has the disadvantage of increasing interference, but the much greater advantage of aligning representations of the task between the two participants. Given such common representations, we know, not only what our partners will do, but also what they should do. In this way we can anticipate their actions and achieve a better degree of coordination (Sebanz *et al.* 2006).

Common representations of action and agency between self and others allows for greater coordination in joint task performance, but it also allows for greater cooperation in the sense of striving for common goals rather than competing for individual goals. The perception of agency has a critical effect on our behaviour in tasks that require cooperation.

In the 'public good' game a group of players are given equal sums of money, which they can choose to invest in the group. For example, if a player invests £10 and there are 4 players in the group, each player will be given £4. So the investing player loses £6, but the group as a whole gains £6. If everyone invests, then every one will gain. Inevitably in this game, however, 'defectors' or 'free riders' will emerge who choose to benefit from the generosity of others without putting in any money themselves. Once free riders have appeared in the group then willingness to invest decreases and every one loses out. Remarkably, this problem can be overcome by introducing 'altruistic punishment'. Players are allowed to punish free riders by having money (£3) taken away from them.

This is 'altruistic' punishment because it costs the punisher money (£1). Once altruistic punishment has been introduced free riding is reduced, more investment occurs, and everybody benefits (Fehr and Gachter 2002). Furthermore, in the long run, people will choose to be in groups where altruistic punishment occurs (Gürerk *et al.* 2006).

When we play trust games with people we have never met before we rapidly start showing emotional responses to those who persistently cooperate and those who persistently defect. We learn to like the cooperators and dislike the defectors (Singer *et al.* 2004, 2006). This liking and disliking may help to explain our willingness to apply altruistic punishment.[5] Reward systems in the brain are active when we punish defectors (de Quervain *et al.* 2004) or see defectors being punished (Singer *et al.* 2006).

What is the relevance of all this to the experience of agency? We only acquire our emotional response to cooperators and defectors if we perceive them to be agents. In the experiment by Tania Singer and her colleagues (2004) participants were told that some of the players in the trust game were forced to follow a response sequence determined by a computer (unintentional condition), while the other players decided for themselves whether to invest or not. Emotional responses to the faces of the players, in terms of behavioural ratings and brain activity, only emerged for the players who decided for themselves (the intentional condition).

This experiment shows the importance of the perception that we are interacting with agents who are acting deliberately. But, in this experiment, the participants were told whether or not the other players were agents. In real life we have to detect whether not the other players are acting deliberately. This is not always easy, especially when interactions occur at a distance. If someone fails to respond to my email I may conclude that they are lazy or uninterested when, in fact, their server has gone down. Such disruption to signals is called negative noise. Introducing such negative noise into a trust game has bad consequences. My partner might intend to send back a generous amount of money, but, through 'computer error', only a small amount is transferred to me. I will tend to attribute such events to deliberate meanness, and, thereafter, respond meanly myself (Tazelaar *et al.* 2004). These bad effects of negative noise can be overcome by direct communication or the adoption of a more generous attitude.

Our ability to detect agency in others is critical for successful group cooperation since it allows us to build up trust and apply altruistic punishment

---

[5] Altruistic punishment is not just costly in terms of money. Apply too much and you may get a reputation as a disagreeable person. Hence the appearance of 'second-order' free riders who let others do the punishing.

when necessary. There is no value in applying such sanctions to things that do not act deliberately. Given the importance of detecting agency in others it may well be that this is the primary mechanism to have evolved. The same mechanism was then used for representing agency in the self. This common form of representation (Prinz 1997) carries with it all the advantages for joint action that I have discussed above.

7.3.3 Is agency a critical feature of consciousness?

The study of consciousness and, in particular, the neural correlates of consciousness, is not just of theoretical interest. Such studies can have major practical relevance. In the first part of this chapter I raised the question as to how we could recognise consciousness in someone (or something) that is unable to make a report. Answering this question has enormous importance in the treatment of patients in a coma (Laureys *et al.* 2004). There are two states of apparent coma that are very difficult to distinguish; *persistent vegetative state* (PVS) and *locked-in syndrome*. Patients in a PVS go though a sleep–wake cycle. They may follow objects with their eyes, and may grimace, cry, or even laugh, but they are considered to be unconscious, in the sense of being unaware of their surroundings. Patients with locked-in syndrome are fully conscious but are unable to make any voluntary movements. Sometimes eye movements are spared. After recovering consciousness after a stroke, Parisian journalist Jean-Dominique Bauby found he could only control his left eyelid. He was able to create a channel of communication using eye blinks and in this way he wrote *The Diving Bell and the Butterfly*, a description of what it is like to be 'locked-in'.

The question is, how can we distinguish a person with PVS from someone with locked-in syndrome who has no voluntary control of any movement? This distinction is of immense practical importance, for it has major implications for how we interact with the patient. The locked-in patient may appear to be in a coma, but is actually conscious of everything that is going on around him. In such cases every effort should be put into creating a channel for communication.

Is the answer to use the 'mind-reading' abilities of brain imaging? I have already mentioned the new analysis techniques whereby it is possible to tell which percept is conscious in the binocular rivalry paradigm without the participant needing to make a report. In principle the same technique could be applied to a patient. If we could show the typical fluctuations in brain activity associated with rivalrous perception in a patient, would we conclude that the patient was conscious? Probably not. This pattern of activity may reflect the way the visual system of the brain deals with ambiguous stimuli, with or without consciousness.

A experiment of this kind has already been carried out by Adrian Owen and his colleagues (2005). It is already established that PVS patients respond to low-level sensory stimuli such as pain or simple sounds (Laureys 2005). But do some PVS patients respond to higher-level features such as meaning? Phrases were presented to the patients aurally, such as (a) '*dates and pears in the fruit bowl*' or (b) '*beer and cider on the shelf*'. In terms of their sounds, phrases as in example (a) contain semantically ambiguous words (*dates, pears*), while the phrases as in example (b) do not. Normal participants show greater activity in auditory association areas for the phrases with ambiguous words. This activity reflects recognition of a problem at the semantic level. Thus processing has occurred up to the semantic level. Owen *et al.* were able to show that PVS patients also show greater activity when confronted with the ambiguous phrases. This demonstrates that processing was occurring up the semantic level. But does it follow that they were conscious of the words? Unfortunately we cannot draw this conclusion. We already know that normal participants show evidence of semantic processing, both behaviourally and in terms of brain activity, for words they are not aware of seeing (e.g. Naccache and Dehaene 2001). Evidence that the brain is processing information, at however high a level, does not seem adequate to indicate this processing is accompanied by consciousness.

Owen and his colleagues have now reported a new experiment using a very different approach (Owen *et al.* 2006). Rather than have patients passively respond to stimulation they asked them to make an active response in the imagination. On some trials they were asked to imagine that they were playing tennis. On other trials they were asked to imagine that they were moving through all the rooms in their house. These tasks were chosen because they elicit very different patterns of brain activity in normal participants. One of the patients, when scanned, showed the same pattern of brain activity as the normal participants with a clear distinction between the two tasks being carried out in the imagination. Since this was not a highly practised task, unlike the word reading in the previous experiment, I believe that this was not a passive, automatic response to the words on the screen (e.g. 'Imagine playing tennis'). Rather, the brain activity reflected a deliberate and intentional response to the instructions. Even if this was not the case in this particular example, the implication is clear: for us to convince ourselves that the patient is conscious, but locked-in, rather than in a vegetative state, we need evidence of agency. Passive responding, however complex the processing involved, is not sufficient.

7.4 **Consciousness is for sharing**

I have argued that the key feature of our conscious experience of agency is that it can be applied to others as well as ourselves. This sharing of the experience of agency allows for sophisticated forms of joint action and enhances cooperation and altruism in social interactions. We recognize others as conscious if we perceive signs of agency. This account of the role and form of consciousness goes back to the original, pre-Cartesian meaning of the word as common knowledge.[6] To end this chapter I would like to speculate briefly about whether this idea of consciousness as sharing experience might apply to sensory perception as well as agency.

An important requirement for successful social cooperation, including both joint action and communication, is *common ground*. This common ground is partly created in the long term by learning and culture, but also needs to be rapidly established at the beginning of any cooperative endeavour (Barr 2004). Not only do we need a common goal in order to achieve the joint action, we also need to have a common view of the nature of the task we are carrying out (Roepstorff and Frith 2004). This applies not only at a concrete level; —it is better to be able to see what the other person is doing (Clark and Krych 2004)—but also at a conceptual level; we need, for example, to have an agreed vocabulary for describing the objects we are manipulating (Markman and Makin 1998). This representational alignment (Pickering and Garrod 2004) might even affect perception. For example, a particular shade of turquoise might be seen by one participant as green and by the other as blue. But, if a turquoise object had an important role in some task, the participants might agree to call it 'the green one'. There is clearly an advantage for our experience of the world to be aligned with that of other people.

Given this advantage, perhaps we should think of our conscious experience of the world as a joint endeavour. Our experience is constantly and subtly being modified through our interactions with others. If perception is a joint endeavour, then my sensations are just a part of the evidence that is being used to update my perception of the world. The perceptions of others in the group must also be taken into account. For optimal combination of all this information we need also to take account of how reliable our perceptions are (Knill and Pouget 2004). Note that this reliability is not determined solely by the noise in the signal. It is also depends upon the noise in my estimation of the cause of the signal.[7] If the reliability of my perceptions is low then I should

---

[6] The word first appears in Latin legal texts by writers such as Cicero. Here, *conscientia* can refer to the knowledge that a witness has of the action of someone else. But it can also refer to witnessing, and judging, our own actions.

[7] Here I am using the term 'sensation' to refer to the signals arising from my activated sense organs. Perception is my estimation of the cause the sensations, e.g. what the object in the outside world might be that is reflecting the light that is stimulating my retina.

give them much less weight and rely more on the reports of others. Lau (2007) has suggested that this knowledge of the reliability of our perceptions has a critical role in determining our criterion for saying that we are aware of a stimulus. He argues that it is this criterion that determines reports of awareness rather than the detectability of the stimulus.

I suggest that this criterion may be determined in part by social factors. Perhaps when the experimenter asks his subject, 'Can you see the stimulus?' she actually answers the question, 'Is the stimulus visible?' and takes into account what others might report. It is not just our experience of agency; all the contents of consciousness are the outcome of a social endeavour.

Acknowledgements

I am most grateful to the Warden and Fellows of All Souls College for giving me the opportunity and providing the ideal environment for developing these ideas. My research is supported by the Wellcome Trust, the Danish National Research Foundation, and the Arts and Humanities Research Council within the CNCC programme of the European Research Foundation. I am grateful to Uta Frith, Celia Hayes, and Rosalind Ridley for the comments on earlier versions of this essay.

References

Barr, D.J. (2004). Establishing conventional communication systems: Is common knowledge necessary? *Cognitive Science* **28**(6), 937–962.

Beck, D.M., Rees, G., Frith, C.D., and Lavie, N. (2001). Neural correlates of change detection and change blindness. *Nature Neuroscience* **4**(6), 645–650.

Beck, D.M., Muggleton, N., Walsh, V., and Lavie, N. (2006). Right parietal cortex plays a critical role in change blindness. *Cerebral Cortex* **16**(5), 712–717.

Blake, R. (1989). A neural theory of binocular rivalry. *Psychological Reviews* **96**(1), 145–167.

Blakemore, S.J., Frith, C.D., and Wolpert, D.M. (1999). Spatio-temporal prediction modulates the perception of self- produced stimuli. *Journal of Cognitive Neuroscience* **11**(5), 551–559.

Blakemore, S.J., Oakley, D.A., and Frith, C.D. (2003). Delusions of alien control in the normal brain. *Neuropsychologia* **41**(8), 1058–1067.

Clark, H.H. and Krych, M.A. (2004). Speaking while monitoring addressees for understanding. *Journal of Memory and Language* **50**(1), 62–81.

Cole, J.D. and Sedgwick, E.M. (1992). The perceptions of force and of movement in a man without large myelinated sensory afferents below the neck. *Journal of Physiology* **449**, 503–515.

Crick, F. and Koch, C. (1998). Consciousness and neuroscience. *Cerebral Cortex* **8**(2), 97–107.

de Quervain, D.J., Fischbacher, U., Treyer, V., Schellhammer, M., Schnyder, U., Buck, A., et al. (2004). The neural basis of altruistic punishment. *Science* **305**(5688), 1254–1258.

Deecke, L. (1990). Electrophysiological correlates of movement initiation. *Revue Neurologique (Paris)* **146**(10), 612–619.

Dijksterhuis, A., Bos, M.W., Nordgren, L.F., and van Baaren, R.B. (2006). On making the right choice: the deliberation-without-attention effect. *Science* **311**(5763), 1005–1007.

Fehr, E. and Gachter, S. (2002). Altruistic punishment in humans. *Nature* **415**(6868), 137–140.

Fourneret, P. and Jeannerod, M. (1998). Limited conscious monitoring of motor performance in normal subjects. *Neuropsychologia* **36**(11), 1133–1140.

Frith, C., Perry, R., and Lumer, E. (1999). The neural correlates of conscious experience: an experimental framework. *Trends in Cognitive Science* **3**(3), 105–114.

Gallese, V., Fadiga, L., Fogassi, L., and Rizzolatti, G. (1996). Action recognition in the premotor cortex. *Brain* **119** (2), 593–609.

Goodale, M.A., Milner, A.D., Jakobson, L.S., and Carey, D.P. (1991). A neurological dissociation between perceiving objects and grasping them. *Nature* **349**(6305), 154–156.

Goodale, M.A., Westwood, D.A., and Milner, A.D. (2004). Two distinct modes of control for object-directed action. *Progress in Brain Research* **144**, 131–144.

Gürerk, O., Irlenbusch, B., and Rockenbach, B. (2006). The competitive advantage of sanctioning institutions. *Science* **312**(5770), 108–111.

Haggard, P., Newman, C., and Magno, E. (1999). On the perceived time of voluntary actions. *British Journal of Psychology* **90**, 291–303.

Haggard, P., Clark, S., and Kalogeras, J. (2002). Voluntary action and conscious awareness. *Nature Neuroscience* **5**(4), 382–385.

Haynes, J.D. and Rees, G. (2005). Predicting the stream of consciousness from activity in human visual cortex. *Current Biology* **15**(14), 1301–1307.

Haynes, J.D. and Rees, G. (2006). Decoding mental states from brain activity in humans. *Nature Reviews. Neuroscience* **7**(7), 523–534.

Hohwy, J. and Frith, C. (2004). Can neuroscience explain consciousness? *Journal of Consciousness Studies* **11**(7–8), 180–198.

Johansson, P., Hall, L., Sikstrom, S., and Olsson, A. (2005). Failure to detect mismatches between intention and outcome in a simple decision task. *Science* **310**(5745), 116–119.

Kanwisher, N. and Yovel, G. (2006). The fusiform face area: a cortical region specialized for the perception of faces. *Philosophical Transactions Royal Society of London Series B Biological Sciences* **361**(1476), 2109–2128.

Knill, D.C. and Pouget, A. (2004). The Bayesian brain: the role of uncertainty in neural coding and computation. *Trends in Neuroscience* **27**(12), 712–719.

Lamme, V.A. and Roelfsema, P.R. (2000). The distinct modes of vision offered by feedforward and recurrent processing. *Trends in Neuroscience* **23**(11), 571–579.

Lau, H.C. (2007). A higher-order Bayesian decision theory of consciousness. *Progress in Brain Research*. **168**, 35–48.

Laureys, S. (2005). The neural correlate of (un)awareness: lessons from the vegetative state. *Trends in Cognitive Science* **9**(12), 556–559.

Laureys, S., Faymonville, M.E., Luxen, A., Lamy, M., Franck, G., and Maquet, P. (2000). Restoration of thalamocortical connectivity after recovery from persistent vegetative state. *Lancet* **355**(9217), 1790–1791.

Laureys, S., Owen, A.M., and Schiff, N.D. (2004). Brain function in coma, vegetative state, and related disorders. *Lancet Neurology* **3**(9), 537–546.

LeDoux, J.E. (1993). Emotional memory: in search of systems and synapses. *Annals of the New York Academy of Science* **702**, 149–157.

Lee, H.W., Hong, S.B., Seo, D.W., Tae, W.S., and Hong, S.C. (2000). Mapping of functional organization in human visual cortex—electrical cortical stimulation. *Neurology* **54**(4), 849–854.

Libet, B., Gleason, C.A., Wright, E.W., and Pearl, D.K. (1983). Time of conscious intention to act in relation to onset of cerebral-activity (readiness-potential)—the unconscious initiation of a freely voluntary act. *Brain* **106**, 623–642.

Lumer, E.D., Friston, K.J., and Rees, G. (1998). Neural correlates of perceptual rivalry in the human brain. *Science* **280**(5371), 1930–1934.

MacKay, D.M. (1981). Neural basis of cognitive experience. In Szekely, G., Labos, E., and Dammon, S. (eds), *Neural Communication and Control*, pp. 315–332. Budapest: Akademiai Kiado.

Marcel, A.J. (1983). Conscious and unconscious perception: Experiments on visual masking and word recognition. *Cognitive Psychology* **15**(2), 197–237.

Markman, A.B. and Makin, V.S. (1998). Referential communication and category acquisition. *Journal of Experimental Psychology: General* **127**(4), 331–354.

Merikle, P.M., Smilek, D., and Eastwood, J.D. (2001). Perception without awareness: perspectives from cognitive psychology. *Cognition* **79**(1–2), 115–134.

Metzinger, T. (2006). Conscious volition and mental representation. In Sebanz, N. and Prinz, W. (eds), *Disorders of volition*, pp. 19–48. Cambridge, MA: MIT Press.

Naccache, L. and Dehaene, S. (2001). The priming method: imaging unconscious repetition priming reveals an abstract representation of number in the parietal lobes. *Cerebral Cortex* **11**(10), 966–974.

Nachev, P. and Husain, M. (2006). Disorders of visual attention and the posterior parietal cortex. *Cortex* **42**(5), 766–773.

Owen, A.M., Coleman, M.R., Menon, D.K., Johnsrude, I.S., Rodd, J.M., Davis, M.H., et al. (2005). Residual auditory function in persistent vegetative state: a combined PET and fMRI study. *Neuropsychological Rehabilitation* **15**(3–4), 290–306.

Owen, A.M., Coleman, M.R., Boly, M., Davis, M.H., Laureys, S., and Pickard, J.D. (2006). Detecting awareness in the vegetative state. *Science* **313**(5792), 1402.

Pascual-Leone, A. and Walsh, V. (2001). Fast backprojections from the motion to the primary visual area necessary for visual awareness. *Science* **292**(5516), 510–512.

Pickering, M.J. and Garrod, S. (2004). Toward a mechanistic psychology of dialogue. *Behavioral and Brain Sciences* **27**(2), 169–225.

Pisella, L., Grea, H., Tilikete, C., Vighetto, A., Desmurget, M., Rode, G. et al. (2000). An 'automatic pilot' for the hand in human posterior parietal cortex: toward reinterpreting optic ataxia. *Nature Neuroscience* **3**(7), 729–736.

Prablanc, C. and Martin, O. (1992). Automatic-control during hand reaching at undetected 2-dimensional target displacements. *Journal of Neurophysiology* **67**(2), 455–469.

Prinz, W. (1997). Perception and action planning. *European Journal of Cognitive Psychology* **9**(2), 129–154.

Rees, G. and Frith, C. (2007). A systematic survey of methodologies for indetifying the neural correlates of consciousness. In Velmans, M. and Schneider, S. (eds), *A Companion to Consciousness*. Oxford: Blackwell.

Rees, G. and Lavie, N. (2001). What can functional imaging reveal about the role of attention in visual awareness? *Neuropsychologia* **39**(12), 1343–1353.

Rees, G., Wojciulik, E., Clarke, K., Husain, M., Frith, C. and Driver, J. (2002). Neural correlates of conscious and unconscious vision in parietal extinction. *Neurocase* **8**(5), 387–393.

Ress, D. and Heeger, D.J. (2003). Neuronal correlates of perception in early visual cortex. *Nature Neuroscience* **6**(4), 414–420.

Roepstorff, A. and Frith, C. (2004). What's at the top in the top-down control of action? Script-sharing and 'top-top' control of action in cognitive experiments. *Psychology Research* **68**(2–3), 189–198.

Schurger, A., Cowey, A., and Tallon-Baudry, C. (2006). Induced gamma-band oscillations correlate with awareness in hemianopic patient GY. *Neuropsychologia* **44**(10), 1796–1803.

Sebanz, N., Knoblich, G., and Prinz, W. (2003). Representing others' actions: just like one's own? *Cognition* **88**(3), B11–21.

Sebanz, N., Bekkering, H., and Knoblich, G. (2006). Joint action: bodies and minds moving together. *Trends in Cognitive Science* **10**(2), 70–76.

Shallice, T. (1972) Dual functions of consciousness. *Psychological Review* **79**(5), 383–393.

Simons, D.J. and Rensink, R.A. (2005). Change blindness: past, present, and future. *Trends in Cognitive Science* **9**(1), 16–20.

Singer, T., Kiebel, S.J., Winston, J.S., Dolan, R.J., and Frith, C.D. (2004). Brain responses to the acquired moral status of faces. *Neuron* **41**(4), 653–662.

Singer, T., Seymour, B., O'Doherty, J.P., Stephan, K.E., Dolan, R.J., and Frith, C.D. (2006). Empathic neural responses are modulated by the perceived fairness of others. *Nature* **439**(7075), 466–469.

Stetson, C., Cui, X., Montague, P.R., and Eagleman, D.M. (2006). Motor-sensory recalibration leads to an illusory reversal of action and sensation. *Neuron* **51**(5), 651–659.

Tazelaar, M.J.A., Van Lange, P.A.M., and Ouwerkerk, J.W. (2004). How to cope with 'noise', in social dilemmas: the benefits of communication. *Journal of Personality and Social Psychology* **87**(6), 845–859.

Tong, F., Meng, M., and Blake, R. (2006). Neural bases of binocular rivalry. *Trends in Cognitive Science* **10**(11), 502–511.

Tononi, G., Srinivasan, R., Russell, D.P., and Edelman, G.M. (1998). Investigating neural correlates of conscious perception by frequency-tagged neuromagnetic responses. *Proceedings of the National Academy of Sciences of the USA* **95**(6), 3198–3203.

Varraine, E., Bonnard, M., and Pailhous, J. (2002). The top down and bottom up mechanisms involved in the sudden awareness of low level sensorimotor behavior. *Brain Research: Cognitive Brain Research* **13**(3), 357–361.

Velmans, M. (1991). Is human information processing conscious? *Behavioural and Brain Sciences* **14**(4), 651–668.

Wegner, D.M. (2002). *The illusion of conscious will*. Cambridge, MA: MIT Press.

Wegner, D.M. and Wheatley, T. (1999). Apparent mental causation—sources of the experience of will. *American Psychologist* **54**(7), 480–492.

Wegner, D.M., Fuller, V.A., and Sparrow, B. (2003). Clever hands: uncontrolled intelligence in facilitated communication. *Journal of Personal and Social Psychology* **85**(1), 5–19.

Weiller, C., Juptner, M., Fellows, S., Rijntjes, M., Leonhardt, G., Kiebel, S., *et al.* (1996). Brain representation of active and passive movements. *NeuroImage* **4**(2), 105–110.

Wohlschlager, A., Haggard, P., Gesierich, B., and Prinz, W. (2003). The perceived onset time of self- and other-generated actions. *Psychological Science* **14**(6), 586–591.

Zeki, S. (1990). A century of cerebral achromatopsia. *Brain* **113**(6), 1721–1777.

Zeki, S. and Bartels, A. (1999). Toward a theory of visual consciousness. *Consciousness and Cognition* **8**(2), 225–259.

Zeki, S., Watson, J.D., Lueck, C.J., Friston, K.J., Kennard, C., and Frackowiak, R.S. (1991). A direct demonstration of functional specialization in human visual cortex. *Journal of Neuroscience* **11**(3), 641–649.

Chapter 8

Are we studying consciousness yet?

Hakwan C. Lau

8.1 Introduction

It is now over a decade and a half since Christof Koch and the late Francis Crick first advocated the now popular NCC project (Crick and Koch 1990), in which one tries to find the neural correlate of consciousness (NCC) for perceptual processes. Here we critically take stock of what has actually been learned from these studies. Many authors have questioned whether looking for the neural correlates would eventually lead to an *explanatory* theory of consciousness, while the proponents of NCC research maintain that focusing on correlates is a strategically sensible first step, given the complexity of the problem (Crick and Koch 1998; Crick and Koch 2003). My point here is not to argue whether studying the NCC is useful, but rather, to question whether we are really studying the NCC at all. I argue that in hoping to sidestep the difficult conceptual issues, we have sometimes also missed the phenomenon of perceptual consciousness itself.

8.2 Stimulus confound vs performance confound

In neuroimaging, one standard strategy for identifying the NCC is to compare a condition where the subjects can perceive a stimulus against a condition where the subjects cannot. One example is visual masking, in which a visual target such as a word is rendered invisible by presenting irrelevant visual patterns before and after. Masking is an interesting paradigm because, when the visual targets are masked and thus not seen, they are nonetheless presented to the retina. The retinal information is only rendered 'unconscious' because of the masks presented before and/or after it. Functional magnetic resonance imaging (fMRI) has been used to compare the neural activity when subjects were presented with masked (and thus invisible) words, against the neural activity when subjects were presented with unmasked (and thus visible) words (Dehaene *et al.* 2001). It was found that, even when the words were masked, there was still activity in the visual areas, although its intensity was reduced. This means that in both the masked and unmasked conditions, the stimuli (the words) were present and to a certain degree processed in the brain. This is a critical aspect of the rationale of the experiment: what we are looking at is a difference in consciousness, but not a difference in the absence or presence of the external stimuli.

Some other paradigms allow us to maintain the constancy of the stimuli in an even more elegant fashion. In fact, much of our effort in the NCC project is spent dealing with this 'stimulus confound'. In other words, we try to make sure that any difference in the neural activity is not merely due to a difference in the stimulus. To this end neuroimagers employ various psychophysical paradigms. For example, in change blindness (Simons and Rensink 2005) subjects frequently fail to detect a change in the visual scene. Researchers have compared the neural activity when the subjects detect the change with the activity when the subjects fail to detect the change (misses) (Beck *et al.* 2001).

Another elegant paradigm for maintaining stimulus constancy is binocular rivalry (Blake and Logothetis 2002), where two different images are presented, one to each eye. Under suitable conditions, the percept of the subject flips between the two images, making the subject see one image for a few seconds, and then the other for a few seconds, and so on back and forth. Researchers on the NCC have tried to look for neural activity that reflects this dynamic change of percept, in the confidence that such activity could not be due to a change in the retinal input (Blake and Logothetis 2002). If the retinal input is the same, what changes in the percept must takes place inside our mind, inside our consciousness—so the logic goes. By mapping these changes we hope to uncover the NCC.

Amid converging neuroimaging evidence, a consensus has now emerged that activity in a 'frontoparietal network' is an important component of the NCC (Rees *et al.* 2002), in addition to activity in modality specific areas (such as occipital and temporal areas for visual stimuli). It has been proposed that neurons in the prefrontal and parietal cortices form a global neuronal workspace for consciousness (Dehaene *et al.* 2003). When information from any modality enters this workspace, it becomes conscious (Baars 1988). The emergence of this consensus could be considered a major achievement of neuroimaging research, which allows us to monitor neural activity across the whole brain. Before this method became available, studies of neuroanatomy and cortical lesions led to the distinction between the dorsal (upper) and the ventral (lower) stream in visual information processing (Ungerleider and Haxby 1994). Based on studies of patients with lesions to either the parietal or temporal cortex, it has been suggested that the dorsal stream, which is routed mainly through the superior parietal cortex, is important for the online control of action, and that the processing is not necessarily conscious (Goodale and Milner 1992). According to this influential view, visual consciousness depends more on the ventral stream, which is routed mainly through the temporal cortex. Converging evidence from whole-brain neuroimaging has begun to undermine this view, highlighting the additional contributions of dorsal areas to visual consciousness.

However, it is important that we understand how specific these activations in the 'frontoparietal network' are. Tasks that are difficult, or simply require high attention, typically activate these regions (Cabeza and Nyberg 2000; Duncan and Owen 2000; Hon *et al.* 2006). Reaction times also correlate with activity in these regions (e.g. Göbel *et al.* 2004). It has also been reported that activity in this 'frontoparietal network' reflects the chewing of gum (Takada and Miyamoto 2004); there was a significant difference in activity when subjects actually chewed gum, as compared to pseudo-chewing (without gum), which was supposed to control for jaw movements. Given that such a wide variety of tasks could activate this 'network', which vaguely maps to a large part of the upper half of the brain, extra caution should be exercised in interpreting its supposed relationship with consciousness. In particular, although NCC researchers are extremely careful in controlling for the stimulus confound, it might still be possible that the results could be explained by other experimental confounds.

Given the role of the dorsal areas in action, the potential significance of a confound in performance should not be overlooked. In studying early visual areas which receive information from the eyes (via the thalamus), controlling for the visual stimulus confound is clearly important. However, in neuroimaging we often investigate neural activity from the whole brain, including higher cognitive areas. This means that merely controlling for stimulus constancy is not necessarily enough. Here I argue that performance, i.e. the ability to detect or discriminate visual targets, as measured by accuracy or reaction times in forced-choice tasks, is a critical confound in NCC research. Many findings in the NCC project could potentially reflect differences in performance, rather than perceptual consciousness itself.

For example, in using visual masking to render words invisible, we often give subjects a forced-choice discrimination task to make sure that the mask is working properly. To claim that the words in the masked condition are really invisible, we may try to demonstrate that subjects can only distinguish the words from meaningless letter strings at chance level, i.e. 50% correct. In the unmasked condition, on the other hand, we argue that the words are visible because subjects can perform the discrimination at nearly 100% correct.

When we compare the neural activity underlying the two conditions, how do we know that the results reflect a difference in consciousness of the words rather than the difference in the performance (50% vs 100% correct)? Similarly in a change blindness experiment, we compare trials where subjects see a change (hits) with the trials where subject do not see the change (misses). This means that we are essentially comparing 100% correct performance (hits) against 0% correct performance (misses).

One might think that binocular rivalry does not suffer from this problem, because subjects were not required to perform a task (apart from reporting their perceptual changes in some experiments; sometimes they are not even required to do this). However, it has been demonstrated that when subjects are asked to detect a small physical change in the presented images (such as an increase of contrast, or inception of a dot), they are more likely to be able to detect the change in the image they are seeing at the moment than the change in the image they do not see (Blake and Camisa 1979). In other words, visual consciousness correlates with performance; conscious perception of an image predicts superior performance in change detection. Therefore, when we look for a change in activity that reflects the change in percept, it could be argued that it reflects the changes in detection performance instead.

One could argue that this is only *potential* performance, as we usually do not give subjects a detection task in binocular rivalry. But this is the same as arguing that we can eliminate a confound by not measuring it. Performance here does not refer to the actual motor action. It refers to the *ability* to detect or discriminate visual targets, or in other words, the effectiveness of visual information processing. These are what performance indexes such as reaction times or accuracies are meant to measure. Even if we do not measure the difference, it is there. And it could be the explanation of why we find a difference in neural activity.

8.3 Task performance as a measure of consciousness

One may respond to the foregoing criticisms that the effectiveness of information processing is the same thing as consciousness. The argument would be that high performance in a psychophysical task is the hallmark of, or even one and the same as, perceptual consciousness. If one is conscious of the stimuli, one can perform well accordingly. If one is unconscious of the stimuli, one performs at the chance level. I shall argue that these claims are not true. In psychophysical experiments, performance is usually measured by the subjects' report of, or best guesses about, the stimulus (what is it?); this is different from their report about their conscious experience (what do you *see*?).

There is empirical evidence that the two can be dissociated, and this means that task performance cannot be one and the same as perceptual consciousness.

First, task performance can exceed the level of consciousness. This is dramatically demonstrated in blindsight patients (Weiskrantz 1986, 1999). After a lesion to the primary visual area, these patients report a lack of perceptual consciousness in the affected region of the visual field. However, when forced to guess the identity or presence of certain stimuli, they can perform well above chance level, sometimes to an impressive range of 80–90% correct. If we take performance as a measure of consciousness, we would have to say that blindsight subjects do not lack visual consciousness in their affected visual field. But they insist that they do. Even in very well controlled laboratory experiments, using painstaking psychophysical measures, blindsight subjects show clear signs of disturbed visual consciousness (Weiskrantz 1986, 1999; Azzopardi and Cowey 1997; Cowey 2004).

Second, task performance may underestimate the level of consciousness. In a classic experiment on iconic memory, Sperling (1960) briefly presented an array of three rows of four letters to subjects. When asked to report all three rows of letters, subjects could only report a fraction. In another condition, immediately after the disappearance of the letters, they were cued to report letters from a specific row. In this condition they could perform nearly perfectly, whichever row they were cued to report. This means that, somehow, at some point after the disappearing of the letters, the subjects had access to all letters, at least for a limited time. One could argue that the images of all letters had entered perceptual consciousness, which seems to be phenomenologically plausible (Block 2007). According to this view, when the letters were briefly presented, subjects saw all of them. When asked to report all 12 letters, the subjects only fail because of a failure of memory. One could generalize this to other tasks. One could always consciously see something, but fail to report it later due to a failure of memory. In general, if the reporting procedure in the task is not optimal, such as that it takes too long, as is probably the case in the Sperling (1960) experiment, or that attention is distracted, one might fail to report the full content of what has entered consciousness.

Finally, whereas consciousness is unified, task performance is not. We typically believe that we cannot be conscious and unconscious of the same stimulus at the same time. This is because consciousness gives one coherent, unified perspective. However, Marcel (1993) has asked subjects to perform a visual task using different reporting modalites, such as manual, verbal, and eye-blink responding. The different modalities gave different performance levels. For example, sometimes the subjects get the answer right using blinks of the eye, whereas they do not with a manual response. This finding is very counterintuitive, but it has recently been replicated by my colleagues (Oliver Hulme and Barrie Roulston, personal communications). One could argue that the results challenge the concept of the unity of consciousness, but a more plausible interpretation is that whichever the reporting modality, perceptual consciousness is the same. It is only that *unconscious* guessing may be more effective in a particular reporting modality than in others. Task performance, especially in forced-choice settings (which psychophysicists favour), can reflect an unknown and varying degree of influence by unconscious guessing, and is therefore sometimes an inconsistent measure of consciousness, as demonstrated in this example.

To sum up, when considered as a measure of consciousness, task performance sometimes captures too much, sometimes too little, and sometimes gives contradictory results. Given that task performance is such a poor measure of consciousness, why do we still use it in many neuroimaging experiments?

8.4 Objective vs subjective measures

Task performance, in particular performance in forced-choice tasks, is favoured by many researchers because it seems to be a relatively objective measure. After all, it is one main goal of science to try to characterize things in an objective fashion. If we ask subjects whether they consciously see something or not, their answer might be biased by their personal strategy or their interpretation of the phrase 'consciously see'. Liberal subjects may frequently say yes, even when they only see a hint of a shadow. Conservative subjects may frequently say no, until they see the target clearly and vividly. It is therefore not a bad idea to ask subjects the relatively objective question of what is presented instead. Even if they are not sure, we force them to make the best guess. This way we can hopefully exhaust all the information the subjects have acquired from the stimulus presentation. To further eliminate bias, signal detection theory is typically applied to the analysis of the data (Macmillan and Creelman 1991). This results in an objective measure of stimulus detection/discrimination sensitivity known as d', which is independent of the subject's choice of criterion or strategy (i.e. liberal or conservative). In other words, d' objectively characterizes a system's capacity or effectiveness to capture the information regarding the stimulus. Using d' as a measure of performance is the gold standard in psychophysical experiments, including studies on consciousness.

However, because of how objective and straightforward the analysis is, signal detection theory could actually also be applied to characterizing the sensitivity of a photodiode and other similarly simple electronic devices. In fact, signal detection theory was partly developed for such purposes (Green and Swets 1966). The realization that task performance essentially reflects any system's basic information processing effectiveness allows for an alternative and trivializing interpretation of current NCC findings. Typically, when the neural activity for a condition where subjects consciously perceive a stimulus is compared against an 'unconscious' condition, there is more activity in many regions throughout the cortex (including the 'frontoparietal network'). However, few have reported results for the reverse comparison. One way to look at the findings is to shift our focus away from the term 'consciousness', but to simply consider the fact that in the 'conscious' condition the brain is processing a lot of information, whereas in the 'unconscious' condition there is little to be processed. This is reflected by the difference in objectively measured task-performance levels, which captures the basic effectiveness of information processing in the brain. The frontal and parietal lobes contain the important association cortical areas, which are highly connected with each other as well as with many areas in the rest of the brain. It is not surprising that when the brain is processing information in a highly effective and sophisticated fashion, as compared to no processing at all, the frontal and parietal areas are employed.

Therefore, according to this alternative explanation, in studying the NCC what we are actually looking at is the basic mechanism of how the brain processes information and produces useful responses. A suitable analogy would be the comparison between a normal working computer that is receiving input and producing output in an effective and productive fashion, and a malfunctioning computer which is neither responding to inputs nor doing any useful work. It would not be surprising to find out that there is more electrical activity in the major components of the computer when it is working normally.

Is this what we are interested in looking at? I suspect that a good number of NCC researchers are prepared to give a blunt 'yes' to this question. Consciousness, according to them, is the basic mechanism by which the brain processes information. So studying the latter is the same as studying the former. But the reason that there is so much interest in the NCC project is that we have the powerful intuition that consciousness is more than simple information processing. Consciousness, that is the phenomenal and qualitative character of perception, seems to be beyond our normal understanding of mechanical information processing systems. This is why consciousness is considered to be such a mystery. Indeed, it has even been suggested that to fully understand consciousness we need to revise our framework of cognitive science and even the foundation of physics (Chalmers 1996). The 'hard problem' of explaining the subjective character of consciousness is the reason why many scientists, like myself, have entered the field in the first place. We would not like to concede that in trying to study consciousness objectively, we end up just studying the basic mechanism of information processing in the brain—for this is exactly what we suspect consciousness outstrips. At the very least, consciousness has to be a specific form of information processing that is not universal to all neural processes (Lau 2007).

This is not to say that the basic mechanism of neural information processing is unimportant to understanding consciousness. In fact, outside the context of the NCC project, the term 'consciousness' is often used to refer to our level of wakefulness (Laureys 2005), and this is closely related to the basic capacity with which the brain can process information regarding the external world. Even if we keep our focus on perceptual consciousness, it is reasonable to assume that the brain has to be able to process or represent the information regarding a stimulus in order for us to be conscious of it. But similarly, in the long run we would also need a healthy respiratory system for us to consciously *see*, assuming that dead people do not enjoy perceptual consciousness. The question here is not whether the task performance is *relevant* to the study of consciousness. But, given the fact that task performance is a poor measure of perceptual consciousness, the question is whether we should consider subjective reports as a more proper measure. After all, it is the subjective character of consciousness that makes the problem so intriguing. This choice between objective and subjective measures has to be made because these measures do not always agree.

One classic example of a subjective measure of consciousness is the commentary key introduced by Weiskrantz (1999). In studying blindsight patients, after a forced-choice question concerning the stimulus, a subjective question is often asked as to whether the patients consciously see the stimulus event or are just guessing. Given that the patients can perform well, though reporting that they are just guessing, one would sensibly consider the subjective report to be more reliable. When blindsight patients claim that they are guessing, we consider them to be perceptually unconscious even though their forced-choice performance is high.

Many have already argued, on theoretical grounds, that subjective measures are the more valid and relevant measures of consciousness (Chalmers 1994; Dienes and Perner 1994). Recently, new subjective measures are also being developed for NCC studies (Sergent and Dehaene 2004; Sergent *et al.* 2005). The 'Seen/Guesses' report in the commentary key paradigm could be reduced to a single button press, and is not substantially more difficult to analyse than other psychophysical data. There are concerns about personal bias in subjective measures, but this could be circumvented by comparing the measure within the same subject in different conditions (Lau and Passingham 2006). The undeniable merit of subjective measures is their face validity: they are reports about the conscious experience itself, rather than reports about something else, that is the stimulus.

One might think that the choice between subjective and objective measures is not so important, because the two correlate most of the time anyway. However, as we have seen in the examples from the last section, they do not *always* correlate. And precisely because they correlate most but not all of the time, we need to apply extra caution in disentangling the two, so as to find out whether our NCC results reflect one measure or the other.

Just acknowledging the importance of subjective measures is not enough. It is important to note we do not solve the problem of performance confound by simply collecting subjective reports as an additional procedure, or, worse, as an afterthought. Nor is it enough to just replace objective measures with subjective measures. In searching for the NCC, we need to take extra steps to ensure that any difference of neural activity in our comparison between conditions reflects the difference in subjective reports, *but not* the difference in performance. One way to achieve this is to collect data for both measures, and set up conditions where performance levels are matched, and try to look for a difference in the subject report.

8.5 Performance matching

The idea of matching performance in studying perceptual consciousness is not new. Weiskrantz *et al.* (1995) have suggested that the studies of blindsight patients offer such an opportunity, because performance level for the 'blind' visual field could be as high as that afforded by the unimpaired visual field, if we choose the stimulus strengths for the two differently and carefully. Here I present an example of how this can also be done in healthy subjects in neuroimaging experiments. I have presented similar stimuli to the same visual location to the same subject in the same experiment, under conditions in which performance levels are matched, but the degree of subjectively reported consciousness differed. I show that this is a useful and practical strategy for uncovering the NCC.

In the experiment (Lau and Passingham 2006), subjects were presented with either a square or a diamond in every trial, and they had to indicate which was presented to them (forced-choice performance). This visual target, however, was masked by a metacontrast stimulus (Fig. 8.1). Here the masking stimulus surrounds but does not overlap with the masked stimulus. One interesting property of metacontrast masking is that the masking function is U-shaped: as one increases the temporal gap between the target and the mask (the stimulus onset asynchrony, SOA), the forced-choice performance decreases and then later

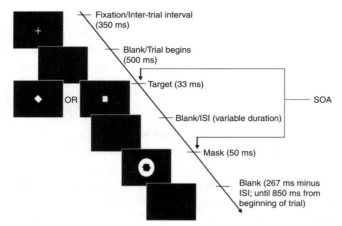

Fig. 8.1. Visual discrimination task with meta-contrast masking. After the presentation of the target and the mask, the participants were first asked to decide whether a diamond or a square was presented. Then, they had to indicate whether they actually saw the target or simply guessed the answer. Shown in the brackets are the durations of each stimulus.

increases again (Fig. 8.2). This means that there will always be pairs of SOA points at which the performance level is matched (33 ms and 100 ms in the data shown). As in the 'commentary key' paradigm, on each trial, after the forced choice, we collected subjective reports of whether the subjects saw the identity of the target or just guessed what it was. This subjective measure of perceptual consciousness

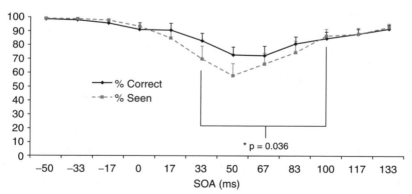

Fig. 8.2. Consciousness and performance. These results were obtained using the procedure described in Fig. 8.1, except that it also included trials where the mask was presented before the target (paracontrast masking). Note that at the SOAs where the performance levels (% correct) were the same (e.g. 33ms and 100ms), the awareness levels (% seen) differed significantly.

differed significantly when the SOAs were 33 ms and 100 ms, even though the performance levels were matched.

We then used fMRI to scan subjects while they were performing this task. For each subject we focused on the two SOAs for which performance was matched but the proportion of guesses differed. The higher level of subjectively reported perceptual consciousness in the longer SOA condition was associated with a higher intensity of neural activity in the mid-dorsolateral prefrontal cortex (mid-DLPFC). This region was the only area where we found a difference in the whole brain. In other words, we did not find a difference in either parietal cortex or the ventral visual stream.

Our result differs substantially from the results of previous NCC studies, since typically these report widespread activity throughout the cortex (Rees *et al.* 2002). In the context of the debate as to whether the dorsal visual stream contributes to visual consciousness, it is important to note that the mid-DLPFC has been formally characterized as a final converging point of the dorsal and ventral visual streams (Young 1992). I have elsewhere proposed that perceptual consciousness depends on the higher-order statistical learning of the behaviour of early sensory signals, and have suggested that the mid-DLPFC is likely to be playing precisely this role (Lau 2007).

8.6 Closing remarks

The point of this chapter is not to argue for any particular notion of NCC. There are many alternative notions, such as 40 Hz synchrony (Singer 2001) and recurrent processing (Lamme 2003), and these have not been discussed. However, though these notions differ, the same formal argument can be made, that is that one must eliminate a performance confound when studying these too.

The main point here is to raise certain concerns about the behavioural paradigms we use in the search for the NCC. The arguments do not depend on the new data I presented. Rather they depend on more 'classic' findings such as blindsight, and seminal experiments by Sperling (1960) and Marcel (1993). These results, some reported well before the popularity of the NCC project, gave early warnings about the validity of using forced-choice task performance as a measure of consciousness. In particular it was the study of blindsight that led to the development of the 'commentary key' paradigm, and to the important idea that we could study consciousness while having performance levels matched (Weiskrantz *et al.* 1995).

The fact that the foregoing arguments depend essentially on old findings suggests that the answer to the awkward question of 'Are we studying consciousness yet?' could be equally awkward: 'We had been! We were almost there!' Of course, as a neuroimager himself the author is by no means implying that neuroimaging is a retrograde step. Without doubt neuroimaging has shed great light on the nature of consciousness and its associated neural mechanisms. However, we should not take for granted that powerful machines or sophisticated data-analysis algorithms will guarantee conceptually interesting results. The dramatic progress in technology in neuroscience has given us a great sense of optimism, but this should not mean that we ignore important conceptual issues in experimental research.

Acknowledgements

I thank Chris Frith, Tom Schofield, Dick Passingham, Uriah Kriegel, Dave Chalmers, David Rosenthal, and the editors for reading earlier drafts of this chapter. This final version has benefited greatly from their valuable comments.

References

Azzopardi, P. and Cowey, A. (1997). Is blindsight like normal, near-threshold vision? *Proceedings of the National Academy of Sciences of the USA* **94**, 14190–14194.

Baars, B.J. (1988). *A Cognitive Theory of Consciousness*. Cambridge: Cambridge University Press.

Beck, D.M., Rees, G., Frith, C.D., and Lavie, N. (2001). Neural correlates of change detection and change blindness. *Nature Neuroscience* **4**, 645–50.

Blake, R. and Camisa, J. (1979). On the inhibitory nature of binocular rivalry suppression. *Journal of Experimental Psychology. Human Perception and Performance* **5**, 315–323.

Blake, R. and Logothetis, N.K. (2002). Visual competition. *Nature Reviews. Neuroscience* **3**, 13–21.

Block, N. (2007). Consciousness, accessibility, and the mesh between psychology and neuroscience. *Behavioral and Brain Sciences* **30**, 481–499.

Cabeza, R. and Nyberg, L. (2000). Imaging cognition II: an empirical review of 275 PET and fMRI studies. *Journal of Cognitive Neuroscience* **12**, 1–47.

Chalmers, D. (1994). How can we construct a science of consciousness? In Gazzaniga, M (ed.) *The Cognitive Neurosciences*, III. Cambridge, MA: MIT Press.

Chalmers, D. (1996). *The Conscious Mind: In Search of a Fundamental Theory*. Oxford: Oxford University Press.

Cowey, A. (2004). The 30th Sir Frederick Bartlett lecture. Fact, artefact, and myth about blindsight. *Quarterly Journal of Experimental Psychology A* **57**, 577–609.

Crick, F. and Koch, C. (1990). Some reflections on visual awareness. *Cold Spring Harbor Symposium on Quantitative Biology* **55**, 953–962.

Crick, F. and Koch, C. (1998). Consciousness and neuroscience. *Cerebral Cortex* **8**, 97–107.

Crick, F. and Koch, C. (2003). A framework for consciousness. *Nature Neuroscience* **6**, 119–26.

Dehaene, S., Naccache, L., Cohen, L., Bihan, D.L., Mangin, J.F., Poline, J.B., and Riviere, D. (2001). Cerebral mechanisms of word masking and unconscious repetition priming. *Nature Neuroscience* **4**, 752–758.

Dehaene, S., Sergent, C., and Changeux, J.P. (2003). A neuronal network model linking subjective reports and objective physiological data during conscious perception. *Proceedings of the National Academy of Sciences of the USA* **100**, 8520–5.

Dienes, Z. and Perner, J. (1994). Assumptions of subjective measures of unconscious mental states: three mappings. In Gennaro, R (ed.) *Higher-order Theories of Consciousness*, pp. 173–199. Amsterdam: John Benjamins.

Duncan, J. and Owen, A.M. (2000). Common regions of the human frontal lobe recruited by diverse cognitive demands. *Trends in Neuroscience* **23**, 475–83.

Göbel, S.M., Johansen-Berg, H., Behrens, T., and Rushworth, M.F. (2004). Response-selection-related parietal activation during number comparison. *Journal of Cognitive Neuroscience* **16**, 1536–1551.

Goodale, M.A. and Milner, A.D. (1992). Separate visual pathways for perception and action. *Trends in Neuroscience* **15**, 20–25.

Green, D. and Swets, S. (1966). *Signal Detection Theory and Psychophysics*. New York: Wiley.

Hon, N., Epstein, R.A., Owen, A.M., and Duncan, J. (2006). Frontoparietal activity with minimal decision and control. *Journal of Neuroscience* **26**, 9805–9809.

Lamme, V.A. (2003). Why visual attention and awareness are different. *Trends in Cognitive Science* **7**, 12–18.

Lau, H.C. (2007). A higher-order Bayesian decision theory of consciousness. *Progress in Brain Research* **168**, 35–48.

Lau, H.C. and Passingham, R.E. (2006). Relative blindsight in normal observers and the neural correlate of visual consciousness. *Proceedings of the National Academy of Sciences of the USA* **103**, 18763–18768.

Laureys, S. (2005). The neural correlate of (un)awareness: lessons from the vegetative state. *Trends in Cognitive Science* **9**, 556–559.

Macmillan, N. and Creelman C (1991). Detection theory: a user's guide. Cambridge: Cambridge University Press.

Marcel, A.J. (1993). Slippage in the unity of consciousness. *Ciba Foundation Symposium* **174**, 168–180.

Rees, G., Kreiman, G., and Koch, C. (2002). Neural correlates of consciousness in humans. *Nature Reviews. Neuroscience* **3**, 261–270.

Sergent, C. and Dehaene, S. (2004). Is consciousness a gradual phenomenon? Evidence for an all-or-none bifurcation during the attentional blink. *Psychological Science* **15**, 720–728.

Sergent, C., Baillet, S., and Dehaene, S. (2005). Timing of the brain events underlying access to consciousness during the attentional blink. *Nature Neuroscience* **8**, 1391–1400.

Simons, D.J. and Rensink, R.A. (2005). Change blindness: past, present, and future. *Trends in Cognitive Science* **9**, 16–20.

Singer, W. (2001). Consciousness and the binding problem. *Annals of the New York Academy of Science* **929**, 123–146.

Sperling, G. (1960). The information available in brief visual presentations. *Psychological Monographs: General and Applied* **74**(11), 1–30.

Takada, T. and Miyamoto, T. (2004). A fronto-parietal network for chewing of gum: a study on human subjects with functional magnetic resonance imaging. *Neuroscience Letters* **360**, 137–140.

Ungerleider, L.G. and Haxby, J.V. (1994). 'What' and 'where' in the human brain. *Current Opinions in Neurobiology* **4**, 157–165.

Weiskrantz, L. (1986). *Blindsight. A Case Study and Implications*. Oxford: Oxford University Press.

Weiskrantz, L. (1999). *Consciousness Lost and Found*. Oxford: Oxford University Press.

Weiskrantz, L., Barbur, J.L., and Sahraie, A. (1995). Parameters affecting conscious versus unconscious visual discrimination with damage to the visual cortex (V1). *Proceedings of the National Academy of Sciences of the USA* **92**, 6122–6126.

Young, M.P. (1992). Objective analysis of the topological organization of the primate cortical visual system. *Nature* **358**, 152–155.

Chapter 9

Beast machines? Questions of animal consciousness

Cecilia Heyes

When the philosopher-scientist René Descartes coined the term 'beast machines' he is understood to have been denying that non-human animals have souls, minds, or, in contemporary language, 'consciousness' (Descartes 1637; Huxley 1874). His views fuelled a debate that not only continues to the present day, but may be more lively, more heated, and apparently more tantalizingly close to resolution, than at any other time.

This debate is often said to relate to 'the question of animal consciousness', but naturally it comprises many questions (Carruthers 2005a; Griffin 1976). For example: What, in this context, is meant by 'consciousness'? Why is the question of animal consciousness a focus of such broad interest? Can it be addressed through scientific studies of animal behaviour? What has been shown by scientific studies that appear to address the question of animal consciousness? This chapter examines these questions, focusing on the claim that animal consciousness is subject to scientific investigation.

9.1 What and why

Three senses of the terms 'conscious' and 'consciousness' are commonly distinguished. First, we often say that a person is conscious meaning that they are awake rather than asleep or in a coma. Consciousness of this kind, 'creature consciousness' (Rosenthal 2002) is an intriguing phenomenon (Zeman, this volume), but it is not the focus of the animal consciousness controversy. No one doubts that the vast majority of animals can be conscious in this sense. Second, we sometimes say that a person is conscious of an object or event when we mean that they can detect, or are able to react to, the object or event. This has been described as 'transitive consciousness' (Rosenthal 2002) and, again, there is little doubt that the term conscious can be applied to animals in this second sense. Most people would agree that a mouse sometimes is, and sometimes is not, transitively conscious of a cat sneaking up on him.

The third sense of 'consciousness' is the one at the core of the animal consciousness debate, and it is famously difficult to capture or define. In this third sense, consciousness is a property, not of an individual, person or animal, but of a mental state—a percept, thought, or feeling. A mental state that is conscious in this third sense, 'state conscious' (Rosenthal 2002), or 'phenomenally conscious', is said to have 'subjective' properties. If a creature has phenomenally conscious mental states, then there is, in Nagel's (1974) words, 'something that it is like to be' that creature (Davies, this volume).

A number of factors contribute to contemporary interest in the question of whether animals are phenomenally conscious. At the most general level, a powerful motivation for much contemporary enquiry about phenomenal consciousness, in human and non-human animals, is the desire to demonstrate that the scientific, naturalistic world view is comprehensive—that there is nothing that, in principle, eludes its explanatory reach. In this respect, phenomenal consciousness has the character of a final frontier. Investigating animal consciousness is part of this frontier project, and, in this area as in many others, it is thought that the similarities and differences between humans and other animals can tell us about 'man's place in nature' (Huxley 1873).

More specifically, the question of animal consciousness is relevant to biomedical research. For example, it is questionable whether animals that are not phenomenally conscious can provide adequate substitutes for humans in research leading to the development of analgesic drugs. Turning from pragmatics to ethics, many philosophers and members of the public regard the question of phenomenal consciousness in animals as having substantial moral significance; strong implications regarding the way in which animals should and should not be treated, not only in biomedical research, but also in food production, as sources of labour, and as pets or 'companion animals'. Questions regarding animal welfare policy do not depend solely on the assumption that many animals are conscious, but it is undoubtedly an important consideration (Carruthers 2005a, Shriver and Allen 2005).

Reference to companion animals reminds us that concern about animal consciousness is not only academic and political in nature, it is also deeply personal. We engage in a great deal of what Carruthers (2005a) has called 'imaginative identification' with animals throughout our lives. In the style of Aesop, we ascribe to them human thoughts and feelings, and derive from our observations and engagements with them both moral lessons and emotional support. The possibility that, like Sony's robotic dog, Aibo, they are automata—impressively complex machines without phenomenal consciousness—puts these sources of companionship under threat. It raises the spectre of a profound, species-wide, human loneliness.

9.2 Methods

The question whether animal consciousness can be investigated scientifically seems to be answered with a resounding 'yes' in the literature on animal behaviour. Many books and journals in comparative psychology, ethology, and behavioural biology report empirical work probing the nature and distribution of consciousness in the animal kingdom. Broadly speaking, this work is of four kinds, distinguishable by the combination of its empirical approach and underlying theory, or assumptions, about consciousness.

9.2.1 Empirical approaches

The majority of empirical studies of animal consciousness are based on analogical reasoning. They start with an assumption that humans are phenomenally conscious when they exhibit a specified behaviour, B. The focal behaviour is sometimes simple, such as withdrawing from a potentially damaging stimulus, and sometimes more complex and protracted, such as learning from experience. The study then demonstrates, typically through observation of spontaneously occurring behaviour rather than by experiment, that animals of a particular species also exhibit behaviour B. It is then concluded that, when they exhibit behaviour B, members of the studied species are phenomenally conscious.

The other empirical approach, which is adopted more rarely in research concerned specifically with animal consciousness, uses experimental methods to test against one another two or more alternative explanations for a focal behaviour, B. One of these hypotheses suggests that the behaviour is a product of phenomenally conscious states or processes, similar to those that produce the behaviour in humans, whereas the other suggests that B is produced by alternative, non-conscious states and processes. The method of testing alternative hypotheses is a pervasive feature of scientific investigation (Sober 2000).

Both of these empirical approaches, analogical reasoning and testing alternative hypotheses, depend on assumptions about the kinds of behaviour that are associated with phenomenal consciousness in humans. This is obvious in the case of reasoning by analogy, but it is also true of studies that test alternative hypotheses. Typically the alternative hypotheses each postulate functional states and processes, i.e. processes defined in terms of what they do within the information-processing system, but not in terms of how it feels to be in the states or engaging in the processes. The functionally defined states and processes are then classified as consciousness-related or consciousness-unrelated on the basis of assumptions about which functional states and processes are and are not associated with phenomenal consciousness in humans.

9.2.2 Theories of consciousness

There is no consensus, among scientists or philosophers, about the nature of the behaviour, or the functional states and processes, that are associated with phenomenal consciousness in humans. However, there are, broadly speaking, two schools of thought on this issue.

The first school asserts that, in humans, consciousness is associated with first-order representation; that we are conscious of those sensations, emotions, and thoughts that contribute, or have the potential to contribute, to the rational control of action (Dretske 1995; Tye 1997; LeDoux, this volume). Since rational control of action is traditionally understood to result from the interaction of beliefs and desires, this view implies that, when looking for evidence of animal consciousness, we should be looking for behaviour indicating that animals have beliefs and desires.

The other school of thought is more demanding. Commonly known as higher-order thought theory, it suggests that we humans are phenomenally conscious only of those sensations, emotions, and thoughts that we think about, or conceptualize (Carruthers 2000, 2005b; Rosenthal 2005). According to this view, it is not enough for sensations and emotions to be capable of influencing beliefs and desires, they must themselves be objects of thought. Thus, for this camp, behaviour indicative of animal consciousness is behaviour indicative of higher-order thought; of thinking about mental states. Higher-order thought theory is consistent with the approaches to the investigation of human consciousness pursued by Frith (this volume) and Milner (this volume).

9.2.3 Four methods

Although students of animal behaviour rarely make explicit reference to any theory of consciousness, their assumptions tend to conform to one of the two positions outlined above. When these are combined with the two empirical approaches, four methods of investigating animal consciousness are discernible: analogical reasoning based on the assumption that phenomenal consciousness requires only first-order representation; testing alternative hypotheses while assuming that first-order representation is sufficient for consciousness; analogical reasoning based on the assumption that phenomenal consciousness is a property of the content of higher-order thought; and testing alternative hypotheses while assuming that conscious content is the content of higher-order thought.

Many studies using the first of these methods—analogical reasoning with first-order representation—are reviewed in Griffin's early work on 'animal awareness' (1976, 1984) and in Tye's more recent analysis of 'simple minds' (1997).

Taking one example, Tye refers to a study by Russell (1934) in which he observed the responses of sticklebacks to food placed in a jar at the bottom of their tank. Initially the fish attempted to seize the food through the glass, but after a while they nearly all learned to solve the detour problem: to move away from the food, up the side of the jar, and over the rim to the inside. This study, among others, is taken by Tye to indicate that fish can learn and 'make cognitive classifications', which, according to Tye's first-order theory, is sufficient to qualify them as creatures with phenomenal consciousness.

Studies of the first kind are scientific investigations of animal consciousness according to institutional criteria; for example, they have been conducted by people with scientific training, and published in scientific journals. But it is strange to regard them in this way because scientific institutions have been largely incidental to their genesis and development. One does not need scientific training, or a laboratory, or the peer-review processes of scientific publication, to observe animal behaviour suggesting, via analogical reasoning, the presence of phenomenally consciousness mental states. The fourteenth-century Scottish king, Robert the Bruce, needed none of these to attribute perseverance to the spider that tried repeatedly to cast her thread from one beam to another (Barton 1912). Arguably, analogical reasoning based on first-order theory is the method that we all use spontaneously in our day-to-day dealings with animals, in an attempt to understand and to anticipate their behaviour.

Because the first method is not distinctively scientific, and the focus of the present discussion is the question whether animal consciousness can be investigated scientifically, research using this method will not be discussed further. Examples of each of the other three methods are considered below.

9.3 Examples

9.3.1 Alternative hypotheses / first-order representation

In everyday life, we commonly assume that non-human animals have beliefs and desires. When we see a cat prowling around its food bowl at feeding time, we naturally assume that the cat wants (or desires) some food, and that he believes that prowling in the vicinity where food is delivered will, if not make it come sooner, then ensure that he can tuck-in as soon as possible after it arrives. But this is not the only plausible explanation for the cat's prowling behaviour. Instead, he may simply have developed a blind habit of approaching his dish when he is in a physiological state of hunger. According to this hypothesis, the cat does not believe that approaching the dish will hasten the moment when he can eat, or indeed have any thoughts about what the outcome of his behaviour might be. He is just drawn towards the dish when hungry because that behaviour has been rewarded in the past; a stimulus–response link, or habit, has been established (Heyes and Dickinson 1990).

Dickinson and colleagues have tested the belief–desire hypothesis against the habit hypothesis in experiments, not with cats approaching dishes for food, but with rats pressing levers for food (Balleine and Dickinson 1998). The seminal experiment of this kind, by Adams and Dickinson (1981), had three stages. In the first, rats pressed a lever and received two kinds of food, A and B. A pellet of food A was delivered whenever the rat pressed the lever. It was the outcome of that action. Food B was delivered equally frequently but free of charge; its delivery did not depend on what the rat was doing. Over time, the frequency of lever presses went up. Was this because the rats had formed a belief that lever pressing caused the delivery of food A, and desired food A?

To find out, in the second stage of the experiment, Adams and Dickinson gave half of the rats an experience that would devalue food A. They allowed them to eat some of A and then injected them with lithium chloride, an emetic. Rats typically avoid eating foods that they have previously eaten just before administration of lithium chloride. The other half of the rats, the controls, ate food B, the free food, before being injected with lithium chloride. In the final stage, several days later when they were no longer nauseous, all rats were given a test in which they could press the lever but this action had no outcome; neither food A nor food B was delivered.

If stage 1 experience had established lever pressing as a blind habit, one would not expect devaluation to affect the rats' propensity to press the lever. If the behaviour was based on a habit rather than a belief about the outcome, then changing the value of the outcome, should not change the behaviour. In contrast, if the rats had formed a belief in stage 1—the belief that lever pressing causes the delivery of food A—then devaluation of food A should reduce their propensity to press the lever. Believing that lever pressing causes A, and *not* desiring A, they should press the lever on test less than the controls who, by hypothesis, believe that lever pressing causes A and *do* desire A. In accordance with the second hypothesis, the results showed that the rats averted to A in stage 2 pressed the lever less often in stage 3 than the controls, who had been averted to food B in stage 2.

This is one example of many experiments by Dickinson and colleagues testing belief–desire hypotheses against habit–drive hypotheses. Others, focusing on the desire vs drive distinction, show something remarkable: that animals are not merely driven to seek what their body needs—food when hungry, fluids when thirsty—but must experience the commodity, the food or drink, in the relevant physiological need state before it can function as an incentive (Balleine and Dickinson 1998).

In all of these experiments, the belief–desire hypothesis provides a better explanation for the rats' behaviour, it renders the observations more probable, than the alternative habit–drive explanation. Therefore these experiments may be said to provide evidence of beliefs and desires in rats. There is a certain irony in this. In the minds of many people, lever-pressing rats represent behaviourism, the school of psychology that sought to deny mental life of any kind to all animals, human and non-human. Yet it is this 'preparation', this animal/procedure combination, that has yielded some of the strongest evidence of mentality, and specifically of first-order representation, in animals.

9.3.2 Analogical reasoning / higher-order thought

In 1970 evidence began to emerge that chimpanzees are able to use mirrors to detect marks on their bodies. The most compelling evidence of this capacity, which is known as 'mirror self-recognition', has come from experiments using a mark test (Gallup 1970). In this procedure, chimpanzees are first exposed to a mirror for a period of days. Then they are anaesthetized and, while under the anaesthetic, marked with an odourless, non-irritant dye, typically on an eyebrow ridge and on the opposite ear. After recovery from the anaesthetic, the chimpanzees touch the marks on their heads more often when a mirror is present than when it is absent. The results of many mark tests could be due to an anaesthetic artefact; being sleepier in the first, mirror absent, observation period, the chimpanzees may have been less active in general than during the second, mirror present, observation period (Heyes 1994). However, this problem was overcome in a more recent study (Povinelli *et al.* 1997), and now there is no reason to doubt that chimpanzees can detect marks on their bodies using a mirror.

It is often claimed that mirror self-recognition is indicative of higher-order thought, and the leading figure in the field, Gordon Gallup, claims explicitly that it implies *conscious* higher-order thought (Gallup 1970; Keenan *et al.* 2003). The reasoning behind these claims has never been articulated, but it seems to be roughly as follows. (1) When I (a human) use my mirror image, I understand the image to represent my 'self', and I understand my self to be an entity with thoughts and feelings. (2) This chimpanzee uses his mirror image. (3) Therefore this chimpanzee understands his mirror image to represent his 'self', an entity with thoughts and feelings

There is something seductive about this line of reasoning, perhaps because the mirror is a recurrent metaphor for the soul. However, on reflection, the argument is not compelling. Humans indicate by what they say that they understand a mirror image to represent or 'stand for' their body; to be in some sense a symbol. It is not clear that such a rich understanding of the relationship between object and image is necessary for use of a mirror image as a source of information about the body. For example, it may be sufficient to know that changes in the appearance of the image regularly co-occur with changes in the body. One could learn this by performing exploratory movements while looking at one's mirror image, and exploit the information to guide body-directed movement, without conceptualizing the image as a representation. Therefore, it is quite possible that, although chimpanzees can use a mirror to detect marks on their bodies, they either do not understand the image to be a representation at all, or they understand it to be a representation of a body—a body they care about—but not necessarily a body that thinks. In other words, mirror self-recognition need not involve higher-order thought.

Research on mirror self-recognition is based on analogical reasoning, and therefore the hypothesis that it depends on higher-order thought has not been explicitly tested against leaner alternatives. However, indirect evidence in support of a leaner interpretation comes from studies showing that the probability that a chimpanzee will pass the mark test declines with age (Heyes 1995, de Veer *et al.* 2002). Nearly 90% of infant chimpanzees (1–5 years old) pass the test, but the proportion declines in childhood (6–7 years) until, in adolescence and early maturity (8–24 years), only 40–50% of chimpanzees are successful. If mirror self-recognition were based on conceptualization of the mirror image as a representation, and the 'self' as a thinker, one would expect success on the mark test to be low in infancy and to rise with age, as chimpanzees undergo cognitive development. The observed pattern suggests, in contrast, that mark test performance depends on how much a chimpanzee makes spontaneous movements when looking in the mirror, and thereby gives itself the opportunity to learn that felt changes in the body co-occur with viewed changes in the image.

9.3.3 Alternative hypotheses / higher-order thought

Studies of meta-memory seek evidence that animals, typically monkeys, know when they remember, i.e. that they are capable of higher-order thought about their memory states. Most studies of this kind (see Smith *et al.* 2003, for a review) use variants of the 'commentary key' method, proposed by Weiskrantz (1986, 1995), and pursued by Cowey and Stoerig (1995).[1] Two experiments on meta-memory will be described. The first is not very successful because it relies on analogical reasoning, but its weaknesses highlight the very considerable strengths of the second experiment, which tests alternative hypotheses.

---

[1] The seminal experiments of Cowey and Stoerig (1995) are not discussed here because they addressed a subtly but importantly different question. Here we are concerned with

On each trial in the first experiment (Hampton *et al.* 2004), food was placed in one of four opaque, horizontal tubes. On some trials the monkey could see this baiting operation, and on other trials his view of baiting was occluded by a screen. After baiting, the monkey was given access to the tubes and allowed to select one of them to search for food. If the selected tube contained the food, the monkey was allowed to eat it, but if the selected tube did not contain the food, the monkey ended the trial empty handed. The researchers recorded when and how often the monkeys looked down the tubes before selecting one to search, and found that seven out of nine monkeys looked less often when they had seen baiting than when their view of the baiting operation had been occluded.

Analogical reasoning tempts one to interpret this result as evidence of metamemory. For example, the monkeys' behaviour may be seen as analogous to that of a person who wants a telephone number (Hampton 2005). In this situation, we often ask ourselves whether we can remember the number and, if we cannot summon a clear mental image of the sequence of digits, look it up in a directory. This way of interpreting the monkeys' behaviour is enticing, but it is not justified by the data because they can be explained by several alternative hypotheses. For example, the monkeys could have been using the occluding screen (a public perceptual cue) or their own hesitant behaviour (a public motor cue) as the basis for the decision whether or not to look. In other words, each monkey may have learned, during the experiment or previously, that in choice situations it is a good idea to look before you leap *if* you were recently behind a screen, or find your arm hovering between the alternatives. These hypotheses account for the experimental findings while assuming that monkeys monitor, not private mental states, but states of the world that are observable to all.

The second experiment (Hampton 2001, Experiment 3) overcame the public/private problem using a procedure that, in effect, asks a monkey to report whether he remembers a picture he saw recently. At the beginning of each trial, the monkey was shown one of four pictures on a computer screen. (A new set of four pictures was used in each test session.) After picture presentation, there was a delay, of variable duration, in which the screen was blank. After the delay,

---

whether scientific research can show that animals are phenomenally conscious. In contrast, Cowey and Stoerig *assumed*, plausibly and for sound methodological reasons, that monkeys are typically conscious of their perceptual states, and asked whether this capacity is impaired in blindsight. Similarly, Leopold and Logothetis (1996) assumed that monkeys are phenomenally conscious of the perceptual states that comprise binocular rivalry in order to investigate the neurological correlates of such states. They did not test this assumption.

the monkey was usually required to touch one of two flags on the screen. Touching the 'test flag' resulted in the monkey being presented with a display containing all four pictures. If he selected from this array the picture he had seen at the beginning of the trial, he was rewarded with a peanut, a preferred food. Touching the other, 'escape flag', resulted in the monkey being given a lesser reward, a pellet of ordinary primate diet, but without him taking the test.

Hampton (2001) found that the frequency with which the monkey chose the escape key over the test key increased with the duration of the delay since the original picture was presented. If he had been reasoning by analogy, Hampton may have stopped there, but in reality he recognized that this pattern of behaviour need not be due to the monkey reflecting on the quality of his memory. The target image was not present when the monkey made his choice between the flags, so his tendency to escape after longer delays could not have been due to use of a public motor cue; he could not have based his decision on his own hesitant behaviour. However, the monkey could in principle have used a public perceptual cue, i.e. the duration of the delay since the original image was presented. He may have learned that test taking has a happy result when the trial started a short time ago, and an unhappy result when it started a long time ago. He may have no idea why this is the case, that delay influences memory quality.

To test this lean hypothesis against the meta-memory hypothesis, Hampton compared the accuracy of the monkey's performance on trials, like those described above, when he chose to take the test, and on other trials when he was forced to take the test. At each delay duration, approximately one-third of trials were forced rather than chosen. If, in tests he has chosen, the monkey decides to take the test by assessing the quality of his memory, by asking himself whether he has a clear mental image of the picture, then he should be less accurate on forced than on chosen test trials because the forced trials will sometimes be trials in which he does not have a clear picture. In contrast, if the monkey decides whether or not to take the test just by remembering how long it was since the trial started, he should be equally accurate on chosen and forced test trials because the two types are likely to include a similar mix of trials in which he does and does not have a strong memory. In fact, the accuracy data were consistent with the predictions of the meta-memory hypothesis. When tests were chosen, accuracy was high and fairly constant across delay durations, but when the tests were forced, accuracy declined with delay. This is an intriguing result. In providing evidence that favours the meta-memory hypothesis over an alternative, public cue hypothesis, this study provides scientific evidence of meta-memory, of thinking about memory, in a monkey.

9.4 Conclusions

At least two conclusions can be drawn from the examples discussed above. First, studies based on analogical reasoning tell us about animal *behaviour*, but provide very little information about animals' *minds*, about the psychological states and processes, conscious or otherwise, that generate behaviour. For example, they show that chimpanzees can use mirrors to detect marks on their bodies (Povinelli *et al.* 1997), and that monkeys look for food when it is in their interests to do so (Hampton *et al.* 2004). However, research based on analogical reasoning is no better, and no worse, than common sense or intuition in telling us what kind of thinking is behind these behavioural achievements. Second, the method of testing alternative hypotheses *can* provide information, not only about behaviour, but also about the psychology of animals. In the examples above, this method showed that rats have beliefs and desires (Balleine and Dickinson 1998; Dickinson, this volume), and that monkeys can think about their memories (Hampton 2001).[2]

Does the second of these conclusions imply that animal consciousness can be investigated scientifically, and that the products of this research have shown that rats and monkeys have phenomenally conscious mental states? Many psychologists and biologists would argue that it does, but to secure this conclusion one must make at least three major assumptions.

The first assumption is that further research will not undermine the consciousness-related hypotheses favoured by research to date. For example, the studies by Dickinson and his colleagues have shown that the belief–desire hypothesis (consciousness-related) provides a better explanation than the habit–drive hypothesis (consciousness-unrelated) for the instrumental behaviour of rats. Further experiments are very unlikely to require revision of this conclusion, but they may well favour a new consciousness-unrelated hypothesis over the belief–desire hypothesis. Indeed, Dickinson's (1994) own associative-cybernetic model of learning is a rich potential source of such a rival hypothesis. Thus, the first assumption is tenuous, but it is not unusual.

---

[2] This chapter does not favour higher-order thought theories over first-order theories of consciousness, but it does suggest that testing alternative hypotheses is a more scientific, and a more reliable, method of investigating mentality and consciousness than analogical reasoning. The latter suggestion is intended to apply to both human and non-human animals; I am not setting the bar higher for people than for other animals. In everyday life we use analogical reasoning to make inferences about the conscious states of fellow humans, but effective scientific research on human consciousness is theory-guided; it does not rely solely on the observer's experience of correlations between his or her conscious states and behaviour.

Arguably all conclusions of scientific research rest on the same kind of assumption; they are all, in this sense, provisional.

The second assumption is more specific to the investigation of consciousness. To interpret the studies reviewed above as providing evidence of consciousness in animals, one must assume that one of the major theories of consciousness, the first-order representational theory or higher-order thought theory, is broadly correct. Scientific conclusions typically depend on assumptions about the validity of background theories, but those theories are seldom as contentious as are theories of consciousness. In addition to controversy among supporters of first-order and higher-order theories, there are many scientists and philosophers who deny the coherence and validity of the whole project that these theories represent. They argue that consciousness cannot be identified with *any* functionally defined states, or that such identification completely fails to explain the core features of consciousness, its subjective properties (Nagel 1974; Chalmers 2002; see Hurley and Nöe 2003 for a contrasting view). This 'explanatory gap' (see Davies, this volume; Zeman, this volume), or the suspicion of such a gap, does not accompany other scientific ventures.

The third assumption is related to the second, and is perhaps the most interesting because it concerns the investigation of animal consciousness specifically. Even if we disregard worries about the explanatory gap, and assume that one or other of the functionalist theories of consciousness is correct *for humans*, the studies reviewed above provide evidence of animal consciousness only if we assume that it is also correct for non-human animals. In other words, we must assume that consciousness is a property of the same functionally defined states in humans and in other animals.

The presence and pervasive influence of language in humans, but not in other animals (Pinker 1994), is a major obstacle to this third assumption. The absence of language in other animals is sometimes regarded merely as an inconvenience, as a factor that makes it hard to find out about animal consciousness, but which does not impact on the likelihood that animals are, in fact, conscious. But the absence of language in non-human animals does not only create a measurement problem. It also raises the fundamental question of whether creatures that do not have the potential to use language, and have not acquired, through language, the knowledge and cognitive skills of human culture, can have mental states that are phenomenally conscious (MacPhail 1998).

Some theories of consciousness claim explicitly that it is dependent on language (Dennett 1991). Others, such as higher-order thought theory, are more cryptic. They are formulated primarily to account for human consciousness, and they describe the higher-order thoughts that render mental content phenomenally conscious as if those thoughts were propositions or 'sentences in the head' (e.g. Carruthers 2002; Rosenthal 2005). However, they do not consider in any detail what kind of non-linguistic, non-propositional mental structure, if any, could fulfill the same higher-order function.

Consider, for example, the monkey that passed Hampton's (2001) test of meta-memory. In spite of lacking natural language, he may have a 'language of thought', sometimes known as 'mentalese' (Fodor 1975). If so, the monkey could mentally represent his memory state as a component of a sentence in mentalese. If the monkey does not have a language of thought, then his test performance could be based on an unusual kind of contingency knowledge. It could be that he has encoded his task experience as pairs of mental images of his state, linked to mental images of trial outcomes. For example, an image of himself in state X (a state that we humans would characterize as having a weak memory), combined with an image of himself touching the test flag, is linked with an image of an unhappy outcome, no food at the end of the trial. In contrast, an image of himself in state Y (a state that we would characterize as having a strong memory), combined with an image of himself touching the test flag, is linked to a representation of a happy outcome, the arrival of a peanut. This is a non-linguistic, functionalist account of what was going on in the monkey's head during the meta-memory test. If it were correct, would the monkey be engaging in higher-order thought of the kind that has been associated with phenomenal consciousness in humans? Currently, higher-order theories of consciousness do not offer a clear answer to this question.

To summarize: When psychological processes are functionally defined—in terms of what they do, rather than how they feel—there is good reason to believe that at least some research, that which tests alternative hypotheses, provides reliable, scientific information about the psychological processes that generate animal behaviour. To conclude from this that animal consciousness is subject to scientific investigation, one must assume (1) that further research will not favour consciousness-unrelated functional hypotheses over the consciousness-related hypotheses that are currently plausible, (2) that a functionalist theory of consciousness—first-order or higher-order—is broadly correct for humans, and (3) that the same functionally defined processes are phenomenally conscious in humans and in other animals.

The second and third of these assumptions require one to ignore the explanatory gap, and to take it largely on faith that consciousness does not depend on language. Personally, I find these demands too great, and therefore doubt that animal consciousness is currently the subject of scientific investigation. However, I assume on non-scientific grounds that many animals experience phenomenally conscious states—that they are not 'beast machines'—and I find it plausible that, at some time in the future, the presence and character of these states will be discoverable by scientific methods. To make that possible, we need stronger theories of consciousness; theories that close the explanatory gap, elucidate the relationship between language and consciousness, and are grounded in the kind of empirical work on human consciousness reported elsewhere in this volume.

Acknowledgements

I am grateful to the Warden and Fellows of All Souls College, and to the McDonnell Centre for Cognitive Neuroscience, for their hospitality and support during the period in which this chapter was written. I would like also to thank my fellow Chichele lecturers for valuable discussion during that period, and Martin Davies, David Milner, Nick Shea, and Larry Weiskrantz for their comments on an earlier draft of the manuscript. I shall always be grateful to Susan Hurley both for her intellectual gifts and her friendship.

References

Adams, C.D. and Dickinson, A. (1981). Instrumental responding following reinforcer devaluation. *Quarterly Journal of Experimental Psychology* **33**B, 109–122.

Balleine, B. and Dickinson, A. (1998). Consciousness: the interface between affect and cognition. In Cornwell, J. (ed.) *Consciousness and Human Identity*. Oxford: Oxford University Press.

Barton, B. (1912). Bruce and the spider. Reprinted from Holland, R.S. (ed.). *Historic Ballads and Poems*. Philadelphia: George W. Jacobs & Co.

Carruthers, P. (2000). *Phenomenal Consciousness: A Naturalistic Theory*. Cambridge: Cambridge University Press.

Carruthers, P. (2002). The cognitive functions of language. *Behavioral and Brain Sciences* **25**, 657–726.

Carruthers, P. (2005a). Why the question of animal consciousness might not matter very much. *Philosophical Psychology*, **18**, 83–102.

Carruthers, P. (2005b). *Consciousness: Essays from a Higher-Order Perspective*. Oxford: Oxford University Press.

Chalmers, D. (2002). Consciousness and its place in nature. In Stich, S. and Warfield, T. (eds), *Blackwell Guide to the Philosophy of Mind*. Oxford: Blackwell.

Cowey, A. and Stoerig, P. (1995). Blindsight in monkeys. *Nature* **373**, 247–249.

Dennett, D.C. (1991). *Consciousness Explained*. London: Penguin.

Descartes, R. (1637) *Discourse on Method*. Vol. XXXIV, Part 1. The Harvard Classics. New York: P.F. Collier & Son, 1909–14; Bartleby.com, 2001.

de Veer, M.W., Gallup, G.G., Theall, L.A., van den Bos, R., and Povinelli, D.J. (2002). An 8-year longitudinal study of mirror self-recognition in chimpanzees. *Neuropsychologia* **1493**, 1–6.

Dickinson, A. (1994). Instrumental conditioning. In Mackintosh, N.J. (ed.) *Animal Learning and Cognition. Handbook of Learning and Cognition*, 2nd edn. San Diego, CA: Academic Press.

Dretske, F. (1995). *Naturalizing the Mind*. Cambridge, MA: MIT Press.

Fodor, J.A. (1975). *The Language of Thought*. Cambridge, MA: Harvard University Press.

Gallup, G.G. (1970). Chimpanzees: self-recognition. *Science* **167**, 86–87.

Griffin, D.R. (1976). *The Question of Animal Awareness*. New York: Rockefeller University Press.

Griffin, D.R. (1984). *Animal Thinking*. Cambridge, MA: Harvard University Press.

Hampton, R.R. (2001). Rhesus monkeys know when they remember. *Proceedings of the National Academy of Sciences of the USA* **98**, 5359–5362.

Hampton, R.R. (2005). Can rhesus monkeys discriminate between remembering and forgetting? In Terrace, H.S. and Metcalfe, J. (eds), *The Missing Link in Cognition*. Oxford: Oxford University Press.

Hampton, R.R., Zivin, A., and Murray, E.A. (2004). Rhesus monkeys discriminate between knowing and not knowing and collect information as needed before acting. *Animal Cognition* **7**, 239–246.

Heyes, C.M. (1994). Reflections on self-recognition in primates. *Animal Behaviour* **47**, 909–919.

Heyes, C.M. (1995). Self-recognition in primates: further reflections create a hall of mirrors. *Animal Behaviour* **50**, 1533–1542.

Heyes, C.M. and Dickinson, A. (1990). The intentionality of animal action. *Mind and Language* **5**, 87–104.

Hurley, S. and Nöe, A. (2003). Neural plasticity and consciousness. *Biology and Philosophy* **18**, 131–168.

Huxley, T.H. (1873). *Man's Place in Nature*. New York: D. Appleton & Co.

Huxley, T.H. (1874) On the hypothesis that animals are automata, and its history. *Fortnightly Review* **16**, 555–580.

Keenan, J.P., Gallup, G.G., and Falk, D. (2003). *The Face in the Mirror: The Search for the Origins of Consciousness*. London: Ecco/HarperCollins.

Leopold, D.A. and Logothetis, N.K. (1996). Activity changes in early visual cortext reflect monkeys' percepts during binocular rivalry. *Nature* **379**, 549–553.

Macphail, E.M. (1998). *The Evolution of Consciousness*. Oxford: Oxford University Press.

Nagel, T. (1974). What is it like to be a bat? *Philosophical Review* **83**, 435–450.

Pinker, S. (1994). *The Language Instinct*. London: Penguin.

Povinelli, D.J., Gallup, G.G., Eddy, T.J., Bierschwale, D.T., Engstrom, M.C., Perilloux, H.K., and Toxopeus, I.B. (1997). Chimpanzees recognize themselves in mirrors. *Animal Behaviour* **53**, 1083–1088.

Rosenthal, D. (2002). Explaining consciousness. In Chalmers, D.J. (ed.) *Philosophy of Mind: Classical and Contemporary Readings*. Oxford: Oxford University Press.

Rosenthal, D. (2005). *Consciousness and Mind*. Oxford: Oxford University Press.

Russell, E. (1934). *The Behaviour of Animals*. London: Edward Arnold.

Shriver, A. and Allen, C. (2005). Consciousness might matter very much. *Philosophical Psychology* **18**, 103–111.

Smith, J.D., Shields, W.E., and Washburn, D.A. (2003). The comparative psychology of uncertainty monitoring and metacognition. *Behavioral and Brain Sciences* **26**, 317–373.

Sober, E. (2000). Evolution and the problem of other minds. *Journal of Philosophy* XCVII, 365–386.

Tye, M. (1997). The problem of simple minds: is there anything it is like to be a honey bee? *Philosophical Studies* **88**, 289–317.

Weiskrantz, L. (1986). *Blindsight: A Case Study and Implications*. Oxford: Clarendon Press.

Weiskrantz, L. (1995). The problem of animal consciousness in relation to neuropsychology. *Behavioural Brain Research* **71**, 171–175.

Chapter 10
Why a rat is not a beast machine
Anthony Dickinson

With characteristic analytic incision, Cecilia Heyes has dissected the problem of animal consciousness with a two-by-two factorial analysis. The theoretical factor contrasts theories that attribute consciousness to either first-order or higher-order representations, whereas the empirical variable distinguishes between two sources of evidence for animal consciousness, the analogical and the experimental evaluation of alternative hypotheses. Within this analysis, Heyes gives most weight to the alternative hypotheses/higher-order thought cell by discussing Hampton's (2001) demonstration that a monkey was capable of reporting on the state of its memory using Weiskrantz's (1995) 'commentary response' procedure. Although there is further evidence from this cell that other animals are capable of higher-order thought, especially in the domain of social cognition and experience projection (Emery and Clayton 2001), whether or not my chosen animal, the humble laboratory rat, would pass this criterion by 'commenting' on its first-order mental states is uncertain (Smith *et al.* 2003; Foote and Crystal 2007). However, my argument is that the higher-order representation cell is not the critical one for the ascription of phenomenal consciousness.

While endorsing Heyes' emphasis on experimental rather than analogical evidence, I argue that first-order mentality can be phenomenally conscious in animals incapable of higher-order thought. The argument is functional, and indeed Heyes also concludes her analysis in doubt about whether experimental demonstrations of higher-order thought warrant the inference to phenomenal consciousness on functional grounds. I share her scepticism. As she points out, the concern turns on the assumption that 'functionally defined processes are phenomenally conscious in humans and other animals' (p. 271). But what is the warrant for this assumption? From proponents of higher-order accounts of phenomenal consciousness, we find claims to the effect that 'if a system were doing this type of processing (thinking about its own thoughts), it would be very plausible that it should feel like something to be doing this' (Rolls 2005). But why is this assumption plausible—surely higher-order, and for that matter first-order cognitive processes, such as planning, decision-making, or error correction, can all be conducted perfectly effectively without it feeling like something to do it? As in the case of all purely cognitive theories, conscious experience is epiphenomenal with no psychological function within the processes of meta-cognition. For those of us who view mentality in general, and phenomenal conscious in particular, as evolved capacities, any account that renders conscious experience functionless is deeply unsatisfactory.

Some years ago, Bernard Balleine and I (Balleine and Dickinson 1998a; Dickinson and Balleine 2000) argued that the function of phenomenal conscious is to act as a motivational interface between cognition and response systems, and I shall spend the remainder of this commentary unpacking this idea and describing the experimental tests that it generated. To do so, however, I shall have to re-tell a personal anecdote, which motivated the experiments testing our claim.

10.1 **Watermelons and desires**

Many years ago during a Sicilian holiday, my wife and I spent a few days camping by a beach near Palermo. One day, feeling thirsty after too many hours on the beach, she suggested that we go in search of watermelons, which we had seen on stalls in a market square near the centre of town. After a bit of exploration, we discovered the route from the beach to the square and I had my first watermelon—delicious it was too in slaking my thirst. Unfortunately, however, the day ended in hedonic distress because later that evening I staggered back to the camp site drunk and nauseous after imbibing far too much Sicilian red wine in a local taverna.

So what had I learned from that day's experience? I had acquired an instrumental belief about how to get from the beach to the market square and a desire for watermelons when thirsty (and a respect for the local red wine, but this is incidental to the purpose of my story), and so it is not surprising that a couple of days later it was I who suggested that we again sought out watermelons to quench our thirst. Having successfully retraced the route from the beach to the square, we rounded the corner to be confronted with the fruit stalls, and I can still remember feeling a slight rise in my gorge at the sight of the rosy segments of melon, a fleeting feeling that I dismissed with the recollection of how delicious they had been a couple of days ago. So, after purchasing a segment each, I had my last ever bite of watermelon—it was disgusting and, overcome by feeling of nausea, I gagged and spat it out. I have never since knowingly tasted watermelon.

Although at the time I did not understand why my desire had changed so capriciously, much has become clear since I became a student of learning. It is well known that a strong flavour aversion can be conditioned by following a single experience of a flavour with sickness even though a relatively long interval intervenes between the two events. So it would appear that the sickness induced by the red wine had conditioned an aversion to the watermelon flavour. You may well ask why, if this was so, was not an even stronger aversion conditioned to red wine. The answer lies with the phenomenon of so-called latent inhibition. Aversions are conditioned much more readily to novel flavours than to familiar ones and, not surprisingly, I was highly familiar with red wine but had never before tasted watermelon.

Although this analysis answers a number of the more superficial puzzles, the whole experience raises more profound implications for the nature of our psychology. The overarching one is that we are beings with a dual psychology in that we are both cognitive creatures and beast machines. Presumably, in retracing my steps to the Palermo square in search of watermelons, I was acting as a cognitive creature in that this behaviour was a rational and intentional action controlled by a practical inference from my belief about the route to the square and my desire for watermelons. In contrast, the disgust that I manifested (or, more strictly speaking, my orofacial reactions: gagging and spitting out) on re-tasting the melon was a manifestation of my non-representational, beast machine psychology because, prior to this second tasting, I thought that they had a fresh and delicious flavour. Consequently, during the intervening days between my two trips to the market square, these two psychologies appeared to remain radically disconnected and at variance with each other, at least with respect to the melons. As a cognitive creature I represented the watermelons as highly desirable and valued (especially when thirsty), whereas latent within my beast machine psychology was an aversion of which my cognitive psychology was totally ignorant.

Importantly, what connected these two psychologies and allowed them to interact in the control of my subsequent behaviour was my phenomenal experience on the second exposure to the melon. It was the experience of a powerful negative affect in the form of nausea and disgust, in conjunction with a perceptual representation of the melon, that led to the loss of my desire. If I had not experienced nor cared about that the feeling of disgust phenomenally, and I can assure you that there was something it was like to experience that nausea, I would probably still love and seek out watermelons on hot summer days.

Discussions with Bernard Balleine about this episode led us to the conclusion that a possible function for phenomenal consciousness is to act as a motivational interface between the psychologies of the cognitive creature and the beast machine. The function of this interface is to ground intentional desires, or in other words cognitive representations of goal values, in the biological responses of the beast machine to motivationally relevant variables, such as nutritional and fluid depletion, poisoning, hormonal states, or body temperature. This grounding occurs through the contiguous experience in phenomenal consciousness of the perception (or thought) of the target object or event (the melon) and the affect that it engenders (disgust) with the perception (or thought) being mediated by the cognitive psychology and the affect by the beast machine psychology. Because my loss of desire for watermelons in the presence of the nausea was immediate and did not require any reflection on the cause of the nausea, we assumed that the assignment of goal or incentive value on the basis of affective experience is a function of first-order rather than higher-order mentality. I did not have to represent to myself that I was eating a watermelon and that I was experiencing disgust at the same time—rather, it was just disgusting.

10.2 The Palermo experiments

An implication of this hypothesis is that the capacity for cognitively mediated goal-directed action must always accompanied by the capacity for phenomenal or experiential consciousness if the value of a goal is to be grounded in biological motivational states and variables. I have long argued, on a variety of grounds (Dickinson 1980; Heyes and Dickinson 1990), that the goal-directed actions of the rat are mediated by intentional causal beliefs about the consequences or outcomes of its actions, a claim that has received recent support from a demonstration that the instrumental behaviour of the rat manifests a form of causal inference that lies outside the scope of simple associative processes (Blaisdell *et al.* 2006; Clayton and Dickinson 2006). Given these claims, Balleine and I entertained the idea that the humble rat is phenomenally conscious, a line of reasoning that we decided to evaluate by running our rats through my Palermo experience with the watermelon.

10.2.1 The analogical demonstration

The first study sought to determine whether our rats also acted in apparent ignorance of a latent aversion (Balleine and Dickinson 1991). The rats were thirsty, just as I had been, but instead of learning the route from the beach to the square, we taught them to press a lever in a single session, not for watermelon, but for a novel sugar water solution. Following this session, we made them mildly ill, not by intoxication with red wine, but by injecting a mild toxin that induces gastric malaise. In spite of these procedural variations, the basic structure of the experience was the same as mine on the day of my first experience of watermelons. Consequently, if the rats were anything like me, as cognitive creatures, what they should have initially learned is how to get sugar water by lever pressing, and that sugar water is delicious when thirsty and therefore has value in this state. But, as beast machines, they should also have subsequently acquired a latent aversion to the sugar water, an aversion that should have been unknown to their cognitive psychology controlling their purposive actions. Consequently, when once again given the opportunity to seek sugar water by lever pressing, our rats should have readily done so. And, just like me, this is what they did—their propensity to press the lever was unaffected by the aversion treatment. It is important to note that this test was conducted in the absence of the sugar water because it was designed to assess the rats' motivation to seek out the solution on the basis of their representation of its value. However, as soon as they earned the sugar water and tasted it, thereby discovering that it was now noxious, they immediately stopped pressing. In all significant ways, the behaviour of our rats was analogous to my own all those years ago in Palermo. They too appeared to act in ignorance of their latent aversion to the sugar water.

10.2.2 Engaging the cognitive-motivational interface

As Heyes so cogently points out, a case for animal consciousness cannot be based on analogies alone; what is required are predictions from our account of the function of phenomenal consciousness. The first prediction arose from considering what would have happened if on the day following my initial experience with watermelon, I had been presented with a slice in a different context, for example, at dinner on the following evening. This episode would have allowed me to experience how disgusting the melon had become, thereby transforming my cognitive desire into an aversion and short-circuiting any future propensity to seek it out. So our initial prediction was that allowing our rats to re-taste the sugar water following their initial experience with the solution during training should have reduced any propensity to seek it again.

To test this prediction, we gave a second group of rats the full Palmero treatment as above. The only difference was that between the two opportunities to seek the sugar water by lever pressing, we gave these rats a taste of the solution in a separate drinking cage. Figure 10.1 illustrates what we expect to happen during such a re-tasting. At the outset of the re-tasting the rats have a desire for sugar water in their cognitive psychology and a learned associative connection between a stimulus unit activated by sugar water and response unit generating sickness reactions in their beast machine stimulus–response (S-R) psychology. Our proposal is that re-tasting the sugar water has two effects. First, the sugar water is represented by the cognitive system as a consciously experienced perception of this solution. Secondly, the re-tasting also activates the sickness response unit in the S-R psychology via the stimulus unit, which is experienced consciously as nausea. The perception of the sugar water conjointly with the strong aversive motivational imperative of the experienced nausea

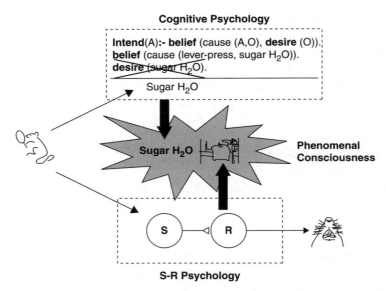

Fig. 10.1 Schematic representation of the interaction between the cognitive and S-R psychologies through the motivational interface of phenomenal consciousness during the re-tasting of the sugar water in the modified Palermo experiment (see text). The contents of the cognitive psychology represent (in PROLOG) the instrumental causal belief and the desire for the sugar water acquired during the initial training and the rule of practical inference that takes this belief and desire as arguments to generate an intention to lever press (see Heyes and Dickinson, 1990). The S-R psychology contains an associative connection between a stimulus unit (S) activated by the tasting of the sugar water, illustrated in the form of a rat licking an ice lolly, and a response unit (R) generating the sickness responses, such as gaping. Re-tasting the sugar water leads to a concurrent cognizance in phenomenal consciousness of this solution as a perception and the experience of nausea as an affective reaction with motivational imperative, illustrated by a rat ill in bed. The consequence of this experience is the removal of the desire for sugar water from the cognitive psychology.

then negates the cognitive desire for this solution. Therefore, our prediction was that these animals, unlike the standard Palermo rats, should have had no desire to seek the sugar water on the second occasion, and indeed they did not do so (Balleine and Dickinson 1991).

10.2.3 Manipulating affective experience

According to our account, the crucial feature of this re-tasting episode was the experience of disgust. If we could attenuate this experience, our rats should have retained their sugar-water desire in spite of the re-tasting, and one way to do so is by the administration of an antiemetic drug known to minimize the experience of nausea. Consequently, in a further study with another group of rats (Balleine *et al.* 1995), we re-ran the Palermo scenario with the interpolated re-tasting opportunity except in this case the rats were injected with an antiemetic just prior to the re-tasting experience. Therefore, these rats should have perceived the solution in conjunction with an attenuated experience of nausea, which at the very least should have reduced the loss of desire. In accord with this prediction, when subsequently given the opportunity to perform the instrumental action that had previously yielded the solution, they sought it more than one re-tasted without the antiemetic.

10.2.4 The autonomy of the cognitive

The final prediction, although the most subtle, illustrates an important feature of our theory, namely that the cognitive control of action is itself affect free. Consider again my current lack of any desire for watermelons. Today my avoidance of this fruit is not accompanied or modulated by any concurrent nausea—I just have no desire for the melon. In fact, I have no idea whether or not I still have the aversion latent in my beast psychology and would have to discover the fact of the matter by a re-tasting test. I have not done so yet. Therefore, in our final Palermo experiments, we set out to see whether rodent action is similarly dissociated from concurrent affect at the time of performance.

The rationale for this experiment is based upon a thought experiment in which I am offered a choice between seeking watermelon and some other thirst-quenching fruit while under the influence of the antiemetic drug. If my instrumental choice is controlled online by affective feedback, as suggested for example by Damasio's somatic marker hypothesis (Damasio 1996), then the antiemetic should reduce my avoidance of the melon choice. According to this hypothesis, without the drug, I avoid choosing the melon because contemplating making this choice produces a concurrent feeling of disgust, which in turn leads me to avoid choosing the melon. Consequently, attenuating this feedback with the antiemetic should make me more likely to choose the melon. At variance with this prediction, what we found was that the presence of the antiemetic had no effect on the instrumental behaviour of our rats at the time of choice (Balleine *et al.* 1995, and see Balleine 2005). So, although manipulating their affective experience at the time when they were assigning value or desirability during re-tasting modulated their subsequently instrumental choices, the same manipulation was ineffective at the time of the choice itself. In this sense, performing intentional action can be independent of the affective experiences that grounded the goals in the first place and does not require a concurrent representation of the goal in phenomenal consciousness. In other words, goal-directed action can be purely cognitive once the incentive or goal values have been acquired.

I have explicated the Palermo experiments in some detail to illustrate evidence for animal consciousness from the alternative hypothesis/first-order representation cell of Heyes' analysis by making a case for our theory of the function of first-order phenomenal consciousness. The primary function, we argue, is to ground cognitive representations of value, or in other words desires, in basic biological needs and motivational states through the conjoint experience in phenomenal consciousness of a cognitive representation, usually but not necessarily perceptual, of the object of the desire and the affective experience elicited by that object through the basic biological systems. This is a process that Balleine and I refer to more prosaically as *instrumental incentive learning*. But it may be thought that this is far too grand a claim to be mounted on the basis of a relatively trivial life experience and a single set of rat experiments. We now know, however, that the control of goal-directed action by a range of motivational states—hunger, thirst, sexual, and thermoregulation in addition to disgust—all depend upon instrumental incentive learning (Dickinson and Balleine 1994, 2002).

10.3 The evolutionary just-so story

As well as attempting to specify the primary function of phenomenal consciousness, Balleine and I also sketched an evolutionary just-so story for its origin (Dickinson and Balleine 2000). It goes like this. When neural plasticity first evolved to support learning, it endowed creatures with an adaptive but a non-cognitive, non-representational learning system, possibly Thorndike's (1911) classic S-R/reinforcement mechanism.[1] The core idea of the S-R mechanism is that pairing a stimulus or response with a biological potent event strengthens or reinforces an association or connection between a current stimulus and an adaptive response so that the re-presentation of that stimulus is more likely to elicit the response in future. This simple learning mechanism enables the animal to respond adaptively to predictive Pavlovian signals of biologically important reinforcers and to control the occurrence of such events through acquisition of instrumental habits, especially once it is recognized that the associated response can vary all the way from a change in the gain of the spinal stretch reflex in motor learning (Wolpaw 1997) to the activation of complex motivational systems (Balleine and Killcross 2006). Indeed, this form of

---

[1] Although Thorndike (1911) originally defined a reinforcer in terms of its hedonic or affective properties, such a definition is not necessary. Reinforcers usually generate an affective experience in the phenomenal consciousness of cognitive creatures, but there is no reason to believe that this experience is necessary for their capacity to reinforce S-R habits.

learning has been deployed to great effect in the sophisticated 'reinforcement' learning algorithms of artificial intelligence (Sutton and Barto 1998).

A model S-R system is that mediating the orofacial ingestive and rejection reactions elicited by biologically significant flavours in many different mammalian species (Berridge 2000), including human infants (Steiner et al. 2001). For example, a sweet sugar solution elicits a set of ingestive reactions, whereas aversive, rejection reactions follow an intra-oral infusion of a bitter quinine solution. There are two points to note about this response system. First, the responses can be modified by learning. Pairing a sugar solution with gastric illness, as in the Palermo experiments, changes the reactions elicited by the solution from ingestive to aversive (Breslin et al. 1992). Secondly, these response systems are directly modulated by primary biological needs. Perhaps the most compelling example of this direct modulation is the response of rats to a strong salt solution. Normally, this solution elicits pronounced aversive reactions, but the very first time the solution is tasted after salt depletion, these reactions are transformed into the ingestive pattern (Berridge et al. 1984). The fact that this shift occurs on very first exposure to the salt after depletion shows the modulation by the need state is direct and does not depend upon learning.

Although some have claimed that phenomenal consciousness is an intrinsic property of such basic biological response systems (Panksepp 2006), Balleine and I argue that any animal endowed only with response systems of this type is a true beast machine. As cognitive creatures, we have phenomenal experiences when tasting these solutions, both perceptual and hedonic, but these conscious experiences have no functional role within the psychological economy of the beast machine. Such an animal can react appropriately to different flavours in a way that reflects its current needs and learn adaptive responses to the contingencies of the environment without any experience of sweetness or bitterness, of palatability or disgust, nor any belief about the consequences of ingesting the solutions. Such mental states simply have no function within the mechanisms controlling the adaptive behaviour of an S-R beast machine, and in the absence of function there is no warrant for the ascription of phenomenal consciousness.

The major constraint on a beast machine is its inability to adapt immediately to changes in goal value and therefore to plan a course of action in light of its current goals. This limitation can again be illustrated by the Palermo experiments. Recall that having taught our rats how get sugar water by lever pressing, we allowed them learn about their aversion to this reward by tasting it outside the context of the lever pressing. When once again given the opportunity to press the lever, they were now very reluctant to do so. In contrast to our rats, however, the change in value of the sugar water would have no direct impact on lever pressing by a beast machine. All that the beast machine has acquired is a compulsion to press whenever it sees the lever, but it has no knowledge that this response gives access to sugar water. Therefore, it cannot adapt its behaviour directly to changes in the values of outcomes or reinforcers; rather it has to re-experience the outcome contingent upon the response for a change in its reinforcing properties to impact on responding.

Our argument is that the capacity for goal-directed action could not be supported by an elaboration of the basic S-R psychology of the beast machine[2] but rather required the evolution of a capacity for some form of practical inference that takes as its arguments beliefs about the causal outcomes of its actions and the desirability or value of the outcomes (Dickinson 1980; Heyes and Dickinson 1990). However, this adaptation leaves the animal with two incommensurate psychologies. The S-R process is a psychological *mechanism* in that its processes, excitation, inhibition, and connection strengthening, are explanatory by analogy with physical processes. In contrast, the cognitive inference process and its arguments, the animal's beliefs and desires, are intentional or representational with the process itself constrained by canons of rationality rather than by those of psychological mechanistic interactions.

This distinction at the psychological level is manifest by the nature of the implementation of the psychologies in terms of brain processes. Being mechanistic, S-R learning is, at least in principle, transparent with respect to neural processes. The psychological process can be implemented directly in terms of experience-dependent synaptic plasticity of the connections between sensory and motor brain systems. By contrast, the cognitive inference process, being intentional and rational, is more opaque with respect to the underlying brain processes just as the processes operating on symbolic variables of a high-level computer language are opaque with respect to the machine's hardware processes.

The evolution of a dual psychology in the service of goal-directed action inevitably brought with it the problem of connecting up or interfacing these two systems, a problem that is particularly acute in the case of motivation. As we have already noted for ingestive/rejection behaviour, it is basic response systems that are directly sensitive to the biological motivational states and variables. By contrast, the values of or desires for goals are represented

---

[2] As Heyes points out (p. 264), Balleine and I have considered an elaboration of the S-R mechanisms in the form of an associative–cybernetic model (Dickinson and Balleine 1993; Dickinson 1994). However, there are a number of problems with this mechanism that are too technical and detailed to consider in the present context.

abstractly in the cognitive system where their function is both to cause and rationalize behavioural decisions. Therefore, in order to be effective, the cognitive control of goal-directed action must have co-evolved with an interface between the two psychologies so that the values and desires of the cognitive system could be grounded in the biologically relevant variables of the S-R psychology.

In principle, the interface or transducer between the two psychologies could have taken many forms. For example, in a computer-controlled industrial plant the interface between the pressure within a vessel and the symbolically encoded procedures for pressure control takes the form of an analogue-to-digital pressure-sensitive transducer. By analogy, it possible that the evolution could have 'solved' the interface problem by a non-conscious transducer. However, our claim is that Nature did, as a matter of fact, solve the interface problem by the co-evolution of the capacity for phenomenal, first-order consciousness along with the cognitive control of goal-directed action. And our scientific strategy for evaluating this hypothesis has been to generate a number of predictions about the interactions of the two psychologies, which we have in turn tested in the Palermo and other experiments.

Within the interface, the motor commands elicited with the S-R mechanisms, for example the orofacial reactions to a flavour, are transformed by some mysterious neural process into the phenomenal experience of disgust or delight,[3] which in turn leads to assignment of an appropriate value or desirability to any flavour that is concurrently perceived. The crucial point about this transduction is that the affective or hedonic imperative of the experience is not epiphenomenal to the function of the interface—it is the fact that the flavour is so disgusting, the toothache so painful, the caress so erotic that grounds the very assignment of value.[4]

---

[3] Of course, I have nothing to say about how to solve this 'hard problem' (Chalmers 1996).

[4] Our view of the structure and evolution of phenomenal consciousness bears certain similarities to that developed by Humphrey (1992, 2006). He too draws a distinction between two psychologies, in his case between perceptual knowledge and sensations with the latter providing the contents of phenomenal consciousness. Moreover, like us, Humphrey argues that phenomenal experience evolved through internalization of responses to biologically important stimuli. Where the accounts seem to differ is in the primary function attributed to phenomenal consciousness. Whereas for us the function is the transmission of value through affect between the cognitive and mechanistic psychologies in the service of goal-directed and planned action, Humphrey gives sensation the more general function of, as he puts it, keeping perception honest: 'Sensation lends a here-ness and now-ness and a me-ness to the experience of the world, of which pure perception in the absence of sensation is bereft' (Humphrey 1992, p. 73).

10.4 Which animals are conscious?

A minimal demand of any theory of animal consciousness is that it provides a principled behavioural criterion for the ascription of phenomenal consciousness. In her chapter, Heyes focused on the Weiskrantz 'commentary response' procedure derived from the higher-order thought theory. In this commentary, I have attempted to complement her focus by arguing for the 'alternative hypotheses/first-order representation' cell of her analysis in the form of our theory of the motivational interface between cognitive and S-R psychologies. According to this theory, the criterion for ascribing phenomenal consciousness is that the animal should be capable of goal-directed action. However, this claim clearly needs further unpacking. According to this theory, goal-directed actions are behaviours mediated by a causal belief about the outcome of the action and the desire for or value of the outcome, and therefore the behavioural criteria are twofold: performance of the action should be sensitive to both the causal relationship with the outcome and the current value of the outcome. Without going into details, the first is assessed by an action–outcome contingency manipulation, which maintains the temporal pairing of the action and outcome while varying the causal relationship by altering the likelihood that the outcome occurs in the absence of the action. The second criterion is sensitivity to goal value as assessed by an outcome re-valuation test, such as that implemented in the Palermo experiments.

To the best of my knowledge, only the humble laboratory rat has fulfilled both these criteria (Balleine and Dickinson 1998b). However, if we allow the outcome re-valuation test to be sufficient, monkeys (Izquierdo *et al.* 2004) and at least some birds (Clayton and Dickinson 1999) join the ranks of the conscious creatures, and a generous generalization would encompass all mammals and birds. But whether fish, reptiles, and amphibians, let alone invertebrates, are pure beast machines or conscious, cognitive creatures remains unknown within the framework of our theory.

Acknowledgements

I should like to thank Bernard Balleine, Nicola Clayton, Cecilia Heyes, Susan Hurley, and Kristjan Laane for their comments on an earlier draft of the manuscript. Although these ideas were developed and tested in collaboration with Bernard Balleine, this should not be taken to imply that he endorses the present exposition.

References

Balleine, B.W. (2005). Neural bases of food-seeking: Affect, arousal and reward in corticostriatolimbic circuits. *Physiology and Behavior* **86**, 717–730.

Balleine, B. and Dickinson, A. (1991). Instrumental performance following reinforcer devaluation depends upon incentive learning. *Quarterly Journal of Experimental Psychology* **43B**, 279–296.

Balleine, B.W. and Dickinson, A. (1998a). Consciousness—the interface between affect and cognition. In Cornwall, J. (ed.) *Consciousness and Human Identity,* pp. 57–85. Oxford: Oxford University Press.

Balleine, B.W. and Dickinson, A. (1998b). Goal-directed instrumental action: contingency and incentive learning and their cortical substrates. *Neuropharmacology* **37**, 407–419.

Balleine, B.W. and Killcross, S. (2006). Parallel incentive processing: an intergrated view of amygdala function. *Trends in Neuroscience* **29**, 272–279.

Balleine, B.W., Garner, C., and Dickinson, A. (1995). Instrumental outcome devaluation is attenuated by the anti-emetic ondansetron. *Quarterly Journal of Experimental Psychology* **48B**, 235–251.

Berridge, K.C. (2000). Measuring hedonic impact in animals and infants: microstructure of affective taste reactivity patterns. *Neuroscience and Biobehavioral Reviews* **24**, 173–198.

Berridge, K.C., Flynn, F.W., Schulkin, J., and Grill, H.J. (1984). Sodium depletion enhances salt palatability in rats. *Behavioral Neuroscience* **98**, 652–660.

Blaisdell, A.P., Sawa, K., Leising, K.J., and Waldmann, M.R. (2006). Causal reasoning in rats. *Science* **311**, 1020–1022.

Breslin, P.A., Spector, A.C., and Grill, H.J. (1992). A quanative comparison of taste reactivity behaviors to sucrose before and after lithium chloride pairings: a unidimensional account of palatability. *Behavioral Neuroscience* **106**, 820–836.

Chalmers, D. (1996). *The Conscious Mind.* Oxford: Oxford University Press.

Clayton, N.S. and Dickinson, A. (1999). Memory for the contents of caches by scrub jays (*Aphelocoma coerulescens*). *Journal of Experimental Psychology: Animal Behavior Processes* **25**, 82–91.

Clayton, N. and Dickinson, A. (2006). Rational rats. *Nature Neuroscience* **9**, 472–474.

Damasio, A.R. (1996). The somatic marker hypothesis and the possble functions of the prefrontal cortex. *Philosophical Transactions of the Royal Society of London Series B Biological Sciences* **351**, 1413–1420.

Dickinson, A. (1980). *Contemporary Animal Learning Theory.* Cambridge: Cambridge University Press.

Dickinson, A. (1994). Instrumental conditioning. In Mackintosh, N.J. (ed.) *Animal Cognition and Learning,* pp. 45–79. London: Academic Press.

Dickinson, A. and Balleine, B. (1993). Actions and responses; the dual psychology of behaviour. In Eilan, N., McCarthy, R. and Brewer, B. (eds), *Spatial Representation,* pp. 277–293. Oxford: Blackwell.

Dickinson, A. and Balleine, B. (1994). Motivational control of goal-directed action. *Animal Learning and Behavior* **22**, 1–18.

Dickinson, A. and Balleine, B.W. (2000). Causal cognition and goal-directed action. In Heyes, C. and Huber, L. (eds), *The Evolution of Cognition*, pp. 185–204. Cambridge, MA: MIT Press.

Dickinson, A. and Balleine, B. (2002). The role of learning in the operation of motivational systems. In Pashler, H. and Gallistel, R. (eds), *Stevens' Handbook of Experimental Psychology*, 3rd ed., Vol. 3, *Learning, Motivation, and Emotion*, pp. 497–533. New York: Wiley.

Emery, N.J. and Clayton, N.S. (2001). Effects of experience and social context on prospective caching strategies in scrub jays. *Nature* **414**, 443–446.

Foote, A.L. and Crystal, J.D. (2007). Metacognition in the rat. *Current Biology* **17**, 551–555.

Hampton, R.R. (2001). Rhesus monkeys know when they remember. *Proceedings of the National Academy of Sciences of the USA* **98**, 5359–5362.

Heyes, C. and Dickinson, A. (1990). The intentionality of animal action. *Mind and Language* **5**, 87–104.

Humphrey, N. (1992). *A History of the Mind*. London: Chatto & Windus.

Humphrey, N. (2006). *Seeing Red*. Cambridge, MA: Harvard University Press.

Izquierdo, A., Suda, R.K., and Murray, E.A. (2004). Bilateral orbital prefrontal cortex lesions in rhesus monkeys disrupt choices guided by both reward value and reward contingency. *Journal of Neuroscience* **24**, 7540–7548.

Panksepp, J. (2006). Affective consciousness: core emotional feelings in animals and humans. *Consciousness and Cognition* **14**, 30–80.

Rolls, E.T. (2005). *Emotion Explained*. Oxford: Oxford University Press.

Smith, J.D., Shields, W.E., and Washburn, D.A. (2003). The comparative psychology of uncertainty monitoring and metacognition. *Behavioral and Brain Sciences* **26**, 317–373.

Steiner, J.E., Glaser, D., Hawilo, M.E., and Berridge, K.C. (2001). Comparative expression of hedonic impact: affective reactions to taste by human infants and other primates. *Neuroscience and Biobehavioral Reviews* **25**, 53–74.

Sutton, R.S. and Barto, A.G. (1998). *Reinforcement Learning*. Cambridge, MA: MIT Press.

Thorndike, E.L. (1911). *Animal intelligence*. London: Macmillan.

Weiskrantz, L. (1995). The problem of animal consciousness in relation to neuropsychology. *Behavioural Brain Research* **71**, 171–175.

Wolpaw, J.R. (1997). The complex structure of memory. *Trends in Neuroscience* **20**, 588–594.

Chapter 11

Does consciousness spring from the brain? Dilemmas of awareness in practice and in theory

Adam Zeman

11.1 When were you conscious last?

When were you conscious last?—Evidently, just a moment ago. But we all have our favourite examples of the state. Mine tend to be moments of varied perceptual experience—standing last weekend in the garden of an English pub, enjoying a sip of its bitter-sweet beer, gazing across a river estuary studded with boats and birds, toward a facing strip of land, just a few degrees of field and forest at that distance, a slender arc of terra firma suspended between reaches of water beneath and sky above—but others will single out quite different species of awareness: the purer experience of a simpler and more intense sensation, a physical exhilaration like descending a black run fast, a meditative emptying of the mind, a fruitful conversation, a period of absorption in a task, a reflective act of memory, a flight of the imagination, a feat of will, a surge of desire or emotion. Wherever you encounter consciousness most keenly, you will surely acknowledge the force of the other examples: painful or pleasurable, perception, action, thought, attention, memory, imagination, will, desire, emotion provide the data, the 'givens', that any science of consciousness must explain.

We tend to look for their explanation to the brain, for we are pretty much persuaded, nowadays, that the brain is the 'organ of mind', the seat of the soul, the wellspring of consciousness. And yet, and yet—many commentators report deep puzzlement at this idea. The brain, they grant, is relevant to consciousness. But they question whether we have any inkling of how a physical process occurring in an object, even one as complex as the brain, could give rise to something that seems as different in kind as an experience in a subject—you or me. The 'explanatory gap', David Chalmer's 'hard question', Colin McGinn's 'transformation of the water of the brain into the wine of consciousness' confront us: how can what happens in our brains possibly give a satisfying explanation for what passes through our minds?

This is of course a contemporary version of the ancient 'mind–brain problem': the 'problem of consciousness' is its modern disguise. Several related distinctions, deeply embedded in our theory and practice, echo this central dichotomy. In our hospitals, neurologists care for disorders of the brain, while psychiatrists care for disorders of the mind, often at opposite ends of the city. In schools and universities, students have to choose between studying the arts, which emphasize the experiences of subjects, and the sciences, focused on processes in objects. We tend to assume that the terms in these twin pairs, mind–brain and subject–object, are well-defined, polar, opposites.

This chapter will examine these distinctions from several perspectives—not in the hope of solving the problem of consciousness, but with the more modest ambition of assuaging it. The problem is often thought to lie entirely in the relatively primitive state of our understanding of the brain. This certainly does not help, but I believe that two other factors play a part: the history of our thinking about mind, which has implanted powerful, but potentially misleading, assumptions that we can only begin to question once we realize we possess them, and the ambiguity of our terms. Once we have taken account of these, the metaphysical problem of consciousness may look less daunting. We might even glimpse some prospect of *rapprochement* between those ancient antagonists, subject and object, mind and brain.

This chapter originated in a seminar emphasizing clinical aspects of the relationship between mind and brain, and a lecture addressing more theoretical questions. These beginnings remain visible. In sections 11.2–11.4 I consider mainly clinical material: section 11.2 discusses evidence on attitudes to mind and brain among patients with 'medically unexplained symptoms', doctors and the general public; section 11.3 illustrates the artificiality of any rigid separation between biology and psychology in the care of patients with disorders of the central nervous system; section 11.4 introduces a 'biopsychosocial' approach that can help doctors to integrate their approach to illness in general. Sections 11.5–11.9 consider the implications of the intimate interdependence of the mental and the physical for theories of consciousness. Section 11.5 discusses the meanings and etymology of consciousness and related terms. Section 11.6 briefly reviews key findings from the 'science of consciousness'. Section 11.7 sketches the remarkably wide range of theories of consciousness on offer. Section 11.8 considers the preconditions for even simple conscious experiences, as you and I enjoy them. Section 11.9 asks whether the exclusive emphasis on brain processes in many theories of consciousness may be misplaced—one of the sources of our puzzlement over consciousness.

11.2 Attitudes to mind and brain

Clinical neurology is the field of medicine that deals with disorders of brain, spinal cord, peripheral nerve, and muscle, or, more accurately, the field that deals with patients whose symptoms suggest that they might have such disorders. The everyday work of neurologists constantly illustrates the traffic between mind and brain (Butler and Zeman 2005):

- Neurologists see patients with disorders of experience ranging from pins and needles in the feet, through visual hallucinations, to atypical panic attacks and loss of memory, reaching such diagnoses as, respectively, diabetes affecting peripheral nerves, stroke affecting visual cortices, epilepsy due to a tumour in the temporal lobe, and Alzheimer's disease. These are examples of 'structural' pathologies in the nervous system impinging on sensory, perceptual, emotional or cognitive functioning—all bona fide functions of 'mind'.

- Conversely, about 30% of the work of a neurologist is with patients whose neurological complaints are 'poorly or not at all explained' by neurological disease. Such patients, with 'medically unexplained symptoms' are just as disabled as patients with neurological disease, and more likely to have serious mood disturbance, but require an entirely different approach to management (Stone 2006). These patients have psychological difficulties giving rise to physical symptoms.

- Neurological disease, like disease generally, frequently gives rise to a secondary psychological reaction—depression, anxiety, panic, hypochondriasis, and so on—and this reaction, rather than the disease itself, may be the major cause of disability in a sufferer's life.

- Psychological states have important effects upon the nervous system: for example, high levels of circulating steroid hormones, associated with chronic stress, cause shrinkage of the hippocampus, a region of the brain closely involved with memory, with resulting memory impairment (Sapolsky 2000).

Despite this busy traffic between mind and brain, many patients, doctors, and lay people are inclined to draw sharp distinctions between matters mental and physical. I shall illustrate each tendency in turn.

11.2.1 Hysteria

The term 'hysteria' derives from the ancient notion that the *hystera*, the womb, could cause disease by wandering like an animal around the female body. Plato subscribed to this theory in the *Timaeus*: 'The womb is an animal which longs to generate children. When it remains barren too long after puberty, it is distressed and sorely disturbed, and straying about the body and cutting off the passages of the breath, it impedes respiration and brings the sufferer into the extremest anguish and provokes all manner of diseases besides' (cited in Veith 1965). Although the theory of the wandering womb was refuted by Galen among others, the term and concept of hysteria survived, because they referred to a persistent if varied set of clinical phenomena. As the physician, Thomas Sydenham, wrote in the seventeenth century: 'The frequency of hysteria is no less remarkable than the multiformity of the shapes which it puts on. Few of the maladies of miserable mortality are not imitated by it.' (Veith 1965). Sydenham was echoing Galen's words from the second century AD—hysteria 'is just one name; varied and innumerable, however, are the forms which it encompasses' (Veith 1965).

The term 'hysteria' is now felt to be tactless, but the problem it denotes is still very much alive. Its less perjorative successors, like 'medically unexplained symptoms, and 'somatoform' or 'functional' disorder, refer to the same frequently encountered cluster of symptoms. Broadly speaking, these are symptoms—like tingling, weakness, loss of speech, or episodes of loss of consciousness—that initially suggest neurological disease, but occur in patients in whom there turns out to be no evidence of conventional neuropathology, while psychosocial factors—such as stress and depression—seem likely to be a crucial part of the explanation.

In recent decades this common problem has tended to fall into a no man's land between neurologists, defining their role as the care of patients with diseases of the nervous system which patients with hysteria by definition do not have, and psychiatrists, confining their work to patients presenting with overtly psychological symptoms, which patients with hysteria are reluctant to admit to. Happily there has been a recent resurgence of interest in this interdisciplinary problem, not least because it is so common. In a recent comparison of 107 patients with significant medically unexplained weakness with 46 whose weakness was explicable neurologically, Stone and colleagues found that patients in the first group were just as disabled, more likely to have given up work, described a larger number of symptoms, and were more likely to be suffering from depression or panic (Stone 2006). Yet despite clear evidence that these patients were ill, and that a key component of their illness was psychological, they were less likely than patients with orthodox neurological disease to 'agree that stress was a possible cause for their symptoms' or to 'wish that the doctor would ask about emotional symptoms'. This suggests a certain incapacity, or at least reluctance, to 'think psychologically' as most of us tend to do. Such a trait might help to explain why psychological distress is channelled—or 'converted' in the vocabulary of psychoanalysis—into physical symptoms in these patients.

Thus, whereas most of us expect some physical fall-out from our emotions, the group of patients presenting with medically unexplained symptoms appears to be reluctant to entertain psychological explanations for physical going-on. They might be described as excessively 'dualistic'. But these patients are surely a special case. Doctors are much more enlightened: or are they?

11.2.2 Medicine and the mind

Many good doctors are, and always have been, sensitive to the kinds of traffic between body and mind described earlier. But there are strong institutional forces in medicine that work against the integration of physical and psychological approaches. Medicine and psychiatry are usually practised in different places. It is interesting that investment in the physical environment tends to flow into the former. The dinginess of many psychiatric institutions in the UK is quite remarkable, although from first principles one might have thought that the quality of one's surroundings would be an especially important aspect of the care of people with mental distress. Psychiatry is usually taught to undergraduates quite separately from other aspects of medicine, and, although psychological problems are extremely common among patients on medical wards, 'liaison psychiatry' is a scarce and fragile resource, one of the first victims of cost-cutting drives. Astonishingly, psychiatry and psychological medicine are usually omitted entirely from the postgraduate training of physicians, who are expected to rotate through a wide range of medical specialities before they make their own choice. This is astonishing because the one certainty for prospective physicians is that, whichever specialty they eventually make their own, around 30% of their patients will present for primarily psychological reasons—which they may have difficulty acknowledging. It is a pity that medical education should compound the problem by making this acknowledgement difficult for physicians too, by depriving them of the relevant training.

Another unfortunate outcome of the segregation of medicine from psychiatry is that doctors sometimes assume that people can be segregated in a similar fashion. Much time is spent in clinical meetings in neurology debating whether a particular patient's clinical problem is 'functional' or 'organic'. This distinction, which has a magnetic attraction for physicians, is self-evidently nonsensical: we are all organisms, and all that we do is 'function'—or sometimes dysfunction. Epilepsy is a prime example of a disorder of function, yet no neurologist would question its 'organic' status. What is really at issue in these discussions is whether the explanation for a patient's problem is most appropriately framed in terms of psychology—experience, behaviour—or in terms of neural mechanism. This is a reasonable, pragmatic, distinction to be trying to draw—even accepting that, in some sense, all our experience and behaviour depend on neural mechanisms. But this reasonable distinction easily shades into a related but unreasonable one, implying that some patients can be thought of purely as sick bodies and others purely as sick minds. We are all composed of body, brain, and mind: the full understanding and treatment of a medical disorder requires due attention to each.

Doctors, blinded by medical science, may be slow to take this simple point, but surely the lay person will see sense?

11.2.3 The public understanding of the mind

Granted that the relationship between body and mind remains a vexed topic within science and philosophy, the non-expert could be forgiven some puzzlement about the matter. But it is of interest to know the assumptions, even if they are inconsistent and part-formed, that the public bring to the area. It should be helpful for doctors to know about these, as they may influence patients' responses, for example, to suggestions that the mind may be playing a part in creating physical symptoms. In the more theoretical context of consciousness research, given that consciousness is not in fact primarily a term of science, widely shared background beliefs about the mind may exert some influence on the kinds of questions that researchers pose—and, for sure, these beliefs will affect the reception of their theories by the public.

In a recent survey of 250 Edinburgh undergraduates, 168/250 (67%) of responders agreed that 'mind and brain are two separate things', while 158/248 (64%) disputed the statement that 'the mind is fundamentally physical'. 161/246 (65%) agreed that 'each of us has a soul which is separate from the body', 174/248 (70%) that some spiritual part of us survives after death, and 150/239 (63%) believed in the existence of a God or gods. Presented to academic and medical audiences, these figures often meet with surprise. But they probably underestimate the frequency of belief in 'mental substance' and supernatural beings in the developed world. A recent American survey of religious belief revealed that 66% of the population 'has no doubt that God exists' (Time 2006). In our survey, belief in God was strongly associated with belief in the soul and spiritual survival (n = 139/148 vs 29/89; χ^2 $p < 0.0001$), and with disagreement with the view that the mind is fundamentally physical (n =112/155 vs 44/91, χ^2 $p < 0.0001$).

More extensive work along these lines would surely reveal a broad range of distinct but overlapping conceptions of the mental and the physical. But the beliefs that mind and matter are separable, and that mind—or soul—can potentially survive without the body, are clearly widespread. They are linked, I suspect, at least in our culture, to another group of beliefs which has a more philosophical flavour but develops the same underlying theme of the radical independence of mind from body (Bering 2006; Kirmayer 2001). These beliefs include:

- transparency and privacy—we cannot but know the contents of our minds, and only we can know them
- global rationality—these contents are all potentially connected by chains of reasoning
- autonomy—our minds freely give rise to considered decisions that we implement through acts of will and these govern our actions
- responsibility—we are responsible for the reasonings and decisions.

Taken together, these two groups of beliefs, in the independence of mind from matter and in the wholly rational, transparent yet private nature of the mind, identify many of our distinctively human characteristics—self-knowledge, rationality, autonomy, morality—with the supposedly immaterial, inaccessible realm of mind.

Prevalent, though of course not universal, these beliefs are in harmony with the medical tendency to separate disorders of the body from those of the mind, and with the reluctance of patients with 'functional disorders' to admit the possibility that their symptoms might have a psychological basis: to do so would be step from the realms of disease and blind, irresponsible matter into the realms of mind and culpability. For those with an interest in a science of consciousness, the relevance of these beliefs is that, almost without our noticing, they influence our expectations of a theory of awareness.

11.3 Building bridges between neurology, psychology, and psychiatry

Despite the intellectual and institutional forces that tend to separate the care of the body from the care of the mind, the practice of medicine keeps bringing them back together. This is especially clear in my field, neurology, as disorders of the brain typically affect cognitive abilities, like thought and memory, and psychological functions such as mood, personality, and behaviour as well as basic neurological functions like movement and sensation. I shall illustrate this point briefly with some published case examples and research findings from my own work, not because these are unique, but precisely because their general orientation is typical of the field.

S.P., an unemployed man of 50, was compulsorily admitted to a psychiatric ward because of bizarre and disinhibited behaviour (Zeman et al. 2005). This included going naked in his garden, filling his house with odds and ends collected in the streets for which he had no use, and, finally, dismantling the power supply to the tenement block in which he lived. He had a lifelong tendency to fidget, which had gradually increased so that he was now unable to keep still for more than a few seconds at a time. Formal neuropsychological assessment revealed evidence of a 'dysexecutive syndrome'—difficulty with the organization of thought and behaviour. On neurological examination S.P. displayed 'chorea', constant involuntary movements with a semi-purposeful appearance. The clue to diagnosis was finally given by the discovery of spiky red blood cells, acanthocytes, on microscopy of a blood film. S.P. proved to be suffering from a genetic disorder, the McLeod syndrome, due to a previously undescribed mutation of the responsible *XK* gene. This gene is expressed both in the membrane of red blood cells and in the brain. Does S.P. suffer from a neurological, psychological, psychiatric, or, for that matter, haematological disorder? Expertise from each of these directions was required to reach a diagnosis, but shoehorning S.P.'s disorder into one or other category would be a futile exercise.

A.J., a 51-year-old right-handed man, was admitted to a medical ward after the police brought him to hospital because of bizarre behaviour in a shopping centre (Zeman *et al.* 2006). This is a sample of his speech soon after admission:

> 'It's unther ah excuse me it's garvo it's gungle black it's clim it's clung cleetly aatly clung clu clat clee-artly and danzai duver cloutly fouchy dill debs dill doot dilartly . . . clexus, clexus, clexus'.

His bizarre speech was due to a fluent or 'jargon' dysphasia, a language disturbance usually caused by damage to Wernicke's area in the left superior temporal lobe. It emerged that A.J. had a long-standing history of epilepsy, and that following his seizures his speech always tended to become circumlocutory, thought not as a rule to this degree. A brain scan revealed an abnormal tangle of blood vessels, an arteriovenous malformation, in the left middle and inferior temporal lobes. The rapid flow through this vascular short circuit appeared to be 'stealing' blood from Wernicke's area just above it.

Over the following weeks and months, A.J.'s language disorder improved and then resolved, but, as it did so, an intermittent disorder of thought took its place:

> 'What is meant by 'rolling stones gather no moss'?'—'stock that moves is healthy and doesn't accumulate explosive characteristics.'

> 'What are you up to these days?'—'There's great potential for wastepaper, mainly because sodium chlorate, put by furnaces into 2 year old or three year old timber, when it enters the repulping process, gives rise to swayback or DCN [the initials of our department] in cows'.

Disorders of language are traditionally distinguished from disorders of thought. But recent research on thought disorder in schizophrenia has shown that it is in fact associated with shrinkage and underactivity of the brain in or close to Wernicke's area. It seems likely that both A.J.'s language and his thought disorder are twin manifestations of a single underlying structural pathology in his left temporal lobe. Once again, it would be fruitless to debate whether his illness is neurological, cognitive, or psychiatric.

A 34-year-old woman was admitted to a psychiatry ward with a persistent sense of déjà vu, the sense that she could predict what was about to happen, and odd bodily sensations (Wright et al. 2006). In view of a past history of epilepsy, she was investigated for possible continuous partial seizure activity: this was confirmed on EEG. Following appropriate antiepileptic treatment her symptoms, which had by now been present for about a fortnight, promptly resolved. However immediately after this episode she complained first that she was unable to see, and then that she had become unable to recognize familiar faces. The symptom, prosopagnosia, has persisted over 40 years. Her epilepsy proved to be due to an abnormal blood vessel—probably a venous angioma—compressing her left fusiform gyrus. The fusiform gyrus in now known to contain a critical node in the network of areas responsible for face recognition. Presumably the protracted seizure activity in this area caused permanent damage in the left fusiform gyrus and perhaps bilaterally. In this case a neurological disorder left a psychological deficit in its wake.

The most familiar manifestation of epilepsy is of course the 'grand mal' or 'tonic-clonic' convulsion, with loss of consciousness, stiffening, shaking, tongue biting, and incontinence—a frightening, embarrassing, and disruptive loss of self-control. But, as the last case illustrates, 'focal' epilepsy, which involves abnormally synchronized electrical activity confined to a small region of the cerebral cortex, can have much more subtle manifestations. One such manifestation is a short period of transient amnesia.

Sufferers from transient epileptic amnesia, who are generally middle-aged, experience brief spells of amnesia, lasting around half an hour, during which they have difficulty in recalling recent events and in laying down memories of current ones (Butler et al. 2007). Attacks occur repeatedly, at an average rate of one a month, and often start on waking. Cognitive abilities apart from memory are substantially preserved, but other more familiar manifestations of epilepsy—such as olfactory hallucinations or brief loss of contact—can occur. Diagnosis is generally delayed—as transient amnesia is an unfamiliar cognitive presentation of epilepsy to most doctors—but once diagnosed, the attacks almost always cease on antiepileptic drug treatment. Around two-thirds of sufferers report one or both of two relatively unexplored forms of memory impairment—accelerated forgetting, the unexpectedly rapid loss of recently acquired memories, and autobiographical amnesia, a patchy but dense loss of recall for salient personal episodes from the recent or remote past. These cognitive phenomena are, again, psychological manifestations of a core neurological disorder.

The cerebellum has traditionally been regarded as a motor control system, a kind of onboard slave computer, necessary for precise coordination of movement but of limited cognitive or psychological interest. Thomas Willis, a seventeenth-century Oxford physician, gave an early formulation of this view, describing the cerebellum as the 'Mistress of the Involuntary Function': 'The spirits inhabiting the Cerebel perform unperceivedly and silently their works of Nature without our knowledge or care'. But recent work has thrown this tidy idea into doubt. The cerebellum has anatomical connections with regions of prefrontal cortex associated with cognition and behaviour, rather than the immediate control of movement; functional imaging studies often reveal selective cerebellar activation by cognitive tasks; clinical studies have suggested that cerebellar damage can cause a 'cerebellar cognitive affective syndrome'. We have obtained evidence that an inherited cerebellar syndrome, spinocerebellar ataxia type 8, is associated with a high rate of cognitive and affective symptoms and with impairment of the control of thought and behaviour, a 'dysexecutive' syndrome (Zeman *et al.* 2004; Torrens *et al.* 2008). In this case, disorder in a region of the brain traditionally associated with motor control proves to have important cognitive and behavioural ramifications.

A final study emphasizes that there are revealable neural correlates for even the most subtle disorders of experience, such as hallucinations, déjà vu, shivers down the spine, and, in this case, the capacity for visual imagery (Zeman *et al.* 2007). A recently retired surveyor M.X. reported the abrupt loss of the capacity to summon up visual imagery—which he had previously relished. At the same time his dreams had lost their visual content. His IQ and general memory quotient were both in the very superior range. There was no evidence of neurological or psychiatric disorder, and standard brain imaging was normal. Although we had no doubt that the symptom was genuine—and similar cases have been reported before, though rarely—it proved extremely difficult to demonstrate any associated neuropsychological deficit. We therefore used functional imaging to explore his brain activity during perception and imagination. On viewing famous faces, his brain activity was indistinguishable from that of controls. However, while controls were able to activate posterior, predominantly 'visual' brain regions, when asked to imagine famous faces, M.X. was unable to do so. In this case a disorder of brain function, later confirmed by functional imaging, had presented with an unusual, subtle, but subjectively significant aberration of experience.

11.4 Treating Cartesian ills

I will summarize the line of thought so far. Consciousness, and mentality in general, are clearly associated with the activity of the brain, yet many commentators express deep puzzlement about how awareness or subjectivity could arise from a physical organ. In the background of this puzzlement lie a set of assumptions about the nature of the mental and the physical. At the risk of caricature, the mental is associated with immateriality, immortality, rationality, autonomy, responsibility, while the physical is associated with mechanistic, mindless, objective process. Mental disorder, the province of psychiatry, is stigmatized, as it seems to imply a defect in those moral and intellectual qualities that make us human; physical disorder, the domain of medicine, is immune from moral disapproval, as it is disconnected from the sphere of human reasoning and choice. But the idea that the mental and physical are insulated one from the other is systematically belied by the discoveries of psychology and neuroscience, as well as by everyday clinical experience—not to speak of common sense! It is certainly impossible to practise medicine well by concentrating exclusively on physical or mental aspects of disease, or without paying close attention to the constant traffic between them.

Without even beginning to try to solve the problem of consciousness, there is a simple practical remedy for the Cartesian tendencies to which doctors are prone. This is to use—as psychiatrists tend to, but physicians tend not to—a 'bio-psycho-social' model of illness: such a model envisages biological, psychological, and social predisposing, provoking, and maintaining factors for any and all presentations. It provides a salutary reminder that patients combine these dimensions, and that all three are often relevant to the diagnosis and management of illness. Given the confusing tendency for 'biological' disorder to present with 'psychological' symptoms, and vice versa, the model can be of real practical help.

Turning to the problem of consciousness itself, I believe that the ground we have travelled so far, remote though it may seem, is highly relevant. Consciousness is not a straightforward scientific variable. The concept is embedded in a rich cultural, religious, philosophical context. We have seen its close relative, 'mind', at work in practice, among patients with hysteria, doctors, the public at large—and have concluded that it often stands at one pole of an unsustainable dichotomy between mind and body. In pursuing a science of consciousness with all the technological resources of cognitive neuroscience, there is a real risk that we may embark on an impossible 'neurology of the soul'. As Alva Noë has written, instead of demystifying the mind, we may end up with a mysterious conception of the body (Noë 2004). I suggested at the start of the chapter that we are familiar with the data, the 'givens', of consciousness. Certainly, something is given to us. But in the light of our tendency to make debatable assumptions about the mind, we may need to reconceptualize just what it is before we can solve the problem—of mind and matter—that first caught our attention.

11.5 The senses and etymology of consciousness—and its cousins

Although seekers after the neural basis of experience tend to choose 'consciousness' as their watchword, several alternative terms crop up repeatedly—these include awareness, cognition, and experience. One of the 'problems of consciousness' is that these terms are all mired in ambiguity. It is therefore worth pausing to reflect on the senses of consciousness and its cousins (Zeman 2002). Their etymology provides a useful clue to the basis of their family relationship: all these words referred at the outset to the getting and having of knowledge.

'Consciousness' derives from the Latin *cum-scio/cum-scire*, literally 'to know together with'. The knowledge in question was often guilty, of the kind one might share with a co-conspirator. Knowledge that can be shared with another can also be shared with oneself—in this case one does more than simply know: one knows that one knows. C.S. Lewis called this the 'strong' sense of consciousness (Lewis 1960). But even in Latin the word had a tendency to slide back into a more basic or 'weak' sense, in which it referred simply to knowledge. 'Awareness' also has connotations of knowledge—of the kind we garner when we are put on our guard: 'wary' and 'beware' are other English relatives of *vereri*, the Latin for fear. 'Cognition' derives from another Latin word for knowing—*cognoscere*—which shares its Indo-European root with 'gnosis'—knowledge of spiritual mysteries—and with our 'knowledge' pure and simple. 'Experience' is the getting of knowledge at first hand, 'perilously', as we subject our prior assumptions to trial—or experiment—by the senses.

Of course, the origins of words do not compel us to maintain their early uses, but their shared point of departure is of interest: not least because knowledge extends from very simple instances—like knowing how to walk—to highly sophisticated ones—like knowing how to do calculus, knowing beauty, knowing ourselves. We tend to draw sharp lines of demarcation between varieties of knowledge. For example we often distinguish procedural knowledge—know-how—from perceptual knowledge, on the grounds that the former is a quintessentially implicit, or unconscious, form of knowledge, while vision yields 'qualia', 'subjectivity', 'phenomenology'—the 'qualities of visual experience' that are often regarded as a prime target of the science of consciousness. The etymology of these 'consciousness-words' quietly reminds us that while there are distinctions that need marking here, we should not lose sight of important continuities between knowledge of all kinds.

Turning from origins to current usage, 'consciousness' today is used in two key senses. The first equates roughly to 'wakefulness' or 'alertness' or 'vigilance', and refers to a state of consciousness. The second, often picked out by the word 'awareness', refers to the content of consciousness. The helpful description of the vegetative state, for example, as a condition of 'wakefulness without awareness' appeals to this distinction: it is possible to exist with open eyes, and some degree of basic reactivity, in a 'state of wakefulness', but to lack awareness of self or surroundings (Zeman 1997).

The contents of consciousness can be supplied by any of our psychological capacities—as we saw at the start of this essay: we can be conscious of perceptions, imaginings, memories, emotions, desires and intentions for example. At least in typical instances of mature human consciousness, if we are so conscious, then we know that we are seeing, imagining, etc., and can report on the fact by a variety of means: the knowledge that is 'in the system' has become 'knowledge for the system' (Cleeremans 2005).

The senses of self-consciousness also amount to a series of increasingly sophisticated forms of knowledge, in this case self-knowledge: these range from the primitive knowledge that a stimulus is directly impinging upon us, to the metacognitive ability that enables us to predict our likely success in a cognitive task, through the knowledge that enables us to recognize ourselves in a mirror, to the ability to appreciate that others also are knowers like us, who form beliefs on the basis of evidence and are vulnerable to misinformation, culminating in our poignant knowledge of ourselves as heroes, or villains, of a finite personal narrative, Tulving's 'autonoetic' awareness, the capacity for mental time-travel in both past and future.

I have emphasized the close relationship between 'knowledge' and 'consciousness' as an antidote to the strong tendency in many writings on consciousness (including mine) to demarcate a certain kind of knowledge, 'phenomenal' or 'experiential' knowledge, and then to treat this, supposedly mysterious, form of knowledge as constituting the problem of consciousness. Perhaps this kind of knowledge is not so special after all: perhaps there is no single, critical threshold at which the light of 'consciousness' first dawns, but rather a series of increasingly sophisticated ways of discovering a world in which the sun already shines, a nested series of ever more powerful forms of cognition. We shall return to this idea.

In the following section (11.6) I will supply a very brief reminder of what we know already about the science of consciousness in each of the senses just distinguished—a good deal by now. I will then turn in section 11.7 to consider several theories of consciousness, each of which locates the source of the real McCoy, the 'miracle of phenomenal experience', in a different kind of neural or psychological process. In section 11.8 I will ask what is presupposed by even very simple passages of experience, like the experience of seeing red, and whether, once we understand all this, there will be any residual 'mystery' to explain. This will lead to a reconsideration, in the final section, of something we took for granted at the outset, but perhaps too readily—the contemporary assumption that consciousness springs from the brain.

11.6 The science of consciousness

Much has been written recently about the 'neural correlates' of consciousness and self-consciousness in the senses distinguished in the preceding section. Although the link is often left implicit, a great deal of work in cognitive neuroscience, and neurology more generally, is relevant to the scientific understanding of consciousness. This is not the place for an extensive review (see Zeman 2001): I shall briefly summarize the key themes and findings.

The neurological basis of wakefulness, the sleep–wake cycle, and of states of impaired consciousness, such as coma and the vegetative state, has been the target of productive physiological, anatomical and pharmacological research over the past two centuries. The demonstration by the German neuropsychiatrist Hans Berger in 1929 that it was possible to record the brain's electrical activity from the scalp provided a tool—the electroencephalogram or EEG—with which to track the concerted shifts in cerebral activity that accompany changes in conscious state (Berger 1929). Berger and others soon described the fundamental rhythms of the EEG: beta, at >13 Hz, which accompanies mental effort; alpha, at 8–13 Hz, the signature of relaxed wakefulness; theta (4–7 Hz) and delta (<4 Hz) which predominate in deep sleep (Fig. 11.1). In the 1950s Kleitman and his co-workers in Chicago discovered that sleep itself has an internal architecture: over the first hour of sleep, the sleeper descends through a series of deepening stages into stage III and IV sleep in which slow waves predominate (slow wave sleep, SWS) only to ascend back through these stages into a state resembling wakefulness in its EEG appearance, accompanied by rapid eye movements, profound muscular atonia, autonomic arousal, and vivid mentation—dreaming, paradoxical, or rapid eye movement sleep (REM) (Dement and Kleitman 1957). This cycle repeats itself four or five times in the course of the night, with decreasing amounts of SWS and increasing amounts of REM as the night proceeds (Kelly 1991; Fig. 11.2). Recent work on the brain's electrical rhythms has highlighted the potential importance of rapid, widely synchronized, high frequency gamma oscillations (25–100 Hz) in wakefulness and REM, although their true significance is not yet clear (Llinas and Ribary 1993) (Fig. 11.3).

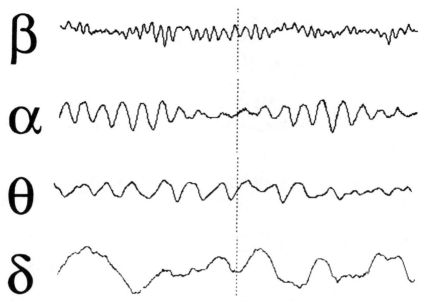

Fig. 11.1. The rhythms of the EEG: records from diagnostic encephalograms performed in four different patients, showing beta rhythm (>14 Hz); alpha rhythm (8–13 Hz); theta rhythm (4–7 Hz); delta rhythm (4 Hz). In each case the dotted line bisects a 2 second sample.

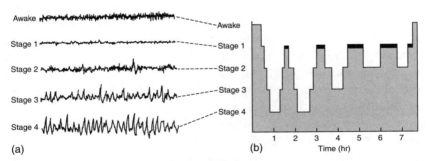

Fig. 11.2. The architecture of sleep: an example of sleep staging over the course of a single night. The sleeper passes from wakefulness to deep sleep and then ascends to REM sleep. Five similar cycles occur in the course of the night. The EEG tracings to the left show the EEG appearances associated with the stages of sleep; the EEG in REM resembles the 'awake' trace.

304 | DOES CONSCIOUSNESS SPRING FROM THE BRAIN?

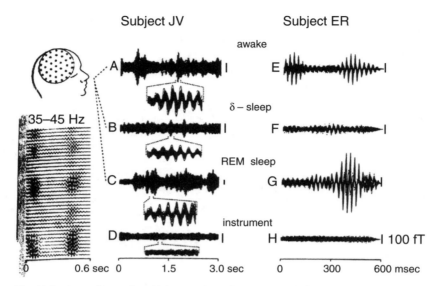

Fig. 11.3. Recordings of rapid (gamma) oscillations in wakefulness, delta or slow wave sleep and rapid eye movement (REM) sleep made using magnetoencephalography (Llinas and Ribary 1993). The diagram at top left indicates distribution of sensors over the head; recordings from these sensors, filtered to pass signals at 35–45 Hz, are shown below. The figures at right show superimpositions of these oscillations in two subjects during wakefulness, slow wave sleep and REM sleep. Note the differing time bases of the two recordings. The amplitude of synchronized gamma oscillations is markedly diminished in slow wave sleep in comparison to wakefulness and REM sleep.

The anatomical and pharmacological mechanisms which control these cycling states have also been clarified over the past 100 years. Moruzzi and Magoun's proposal that the brain stem and thalamus are home to an 'activating system' which maintains arousal in the hemispheres has stood the test of time(Moruzzi and Magoun 1949) (Fig. 11.4). However, the notion of a single monolithic system has given way to a pharmacologically complex picture of multiple interacting systems innervating the cerebral hemispheres widely from the brain stem and diencephalon (Robbins and Everitt 1995) (Fig. 11.5). These systems are defined by their neurotransmitters, which include acetylcholine, serotonin, noradrenaline, dopamine, histamine, hypocretin, and glutamate. The normal succession of conscious states is regulated by these systems: for example, in SWS all these systems become relatively quiescent; in REM periods the ascending cholinergic system becomes disproportionately active; REM periods are eventually brought to an end by rising levels of activity in noradrenergic and serotonergic neuronal groups which had fallen silent at REM onset. These brain chemicals are of course associated with the

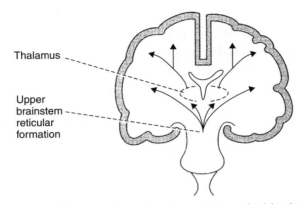

Fig. 11.4. A cartoon of the ascending activating system, emphasizing its origins in the upper brain stem and thalamus, and its wide-reaching control of the state of arousal of the cerebral hemispheres.

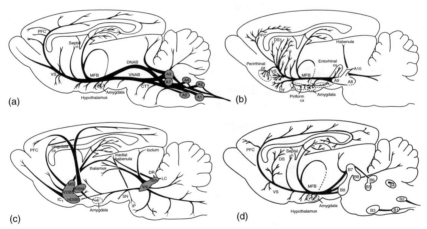

Fig. 11.5. The pharmacology of the brainstem activating systems: (a) shows the origin and distribution of the central noradrenergic pathways in the rat brain; (b) the dopaminergic pathways; (c) the cholinergic pathways; (d) the serotonergic pathways. CTT = central tegmental tract; dltn = dorsolateral tegmental nucleus; DNAB = dorsal noradrenergic ascending bundle; DR = dorsal raphe; DS = dorsal striatum; HDBB = horizontal limb nucleus of the diagonal band of Broca; Icj = islands of Calleja; IP = interpeduncular nucleus; LC = locus ceruleus; MFB = medial forebrain bundle; MS = medial septum; NBM = nucleus basalis magnocellularis (Meynert in primates); OT = olfactory tubercle; PFC = prefrontal cortex; SN = substantia nigra; tpp = tegmental pedunculopontine nucleus; VDBB = vertical limb nucleus of the diagonal band of Broca; VNAB = ventral noradrenergic ascending bundle; VS = ventral striatum (Robbins and Everitt 1995).

regulation of mood, motivation and movement as well as consciousness. Hobson's AIM model attempts to integrate several lines of evidence on the genesis and nature of conscious states (Hobson and Pace-Schott 2002).

The practical upshot of these advances is a useful taxonomy of states of healthy and disordered consciousness. In health, we cycle between wakefulness, SWS, and REM (Fig. 11.6). Pathological states include coma (Glasgow Coma Score <7, eyes closed), the vegetative state mentioned, brain death and the locked-in syndrome (Table 11.1) (adapted from Royal College of Physicians Working Party 2003). While wakefulness, SWS, and REM are as a rule mutually exclusive, overlaps between these states occasionally occur (Mahowald and Schenck 1992). For example, sleepwalking reflects motor activation of the kind seen during wakefulness occurring at a time when much of the brain is deactivated by SWS (Bassetti et al. 2000); REM sleep behaviour disorder, in which sufferers enact their dreams, results from a failure of the normal atonia of REM sleep, allowing dream mentation to give rise to behaviour, like self-defence, of a kind which would normally be confined to wakefulness (Schenck et al. 1986; Schenck and Mahowald 2002).

Knowledge of the neural basis of awareness, the second key sense of consciousness, has also been transformed over the past century by groundbreaking work on the biology of cognition, exploring the neurology of perception, language, memory, emotion, and action. Work on these psychological processes, and their disruption by disease, is demonstrating increasingly fine grained correlations between features of our experience and details of neural processes. Much cited examples include the key roles of visual area V4

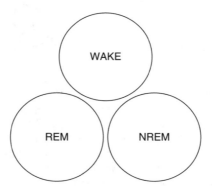

Fig. 11.6. The three principal states of wakefulness, REM, and NREM sleep are normally distinct. Many parasomnias can be understood as the result of overlap between two or more states: for example, overlap between the phenomena of REM sleep and wakefulness gives rise to REM sleep behaviour disorder; overlap between NREM sleep and wakefulness occurs during sleepwalking and night terrors.

Table 11.1 Disorders of conscious state, and related conditions

Condition	Vegetative state	Minimally conscious state	Locked-in syndrome	Coma	Death confirmed by brain stem tests
Awareness	Absent	Present	Present	Absent	Absent
Sleep–wake cycle	Present	Present	Present	Absent	Absent
Response to pain	+/–	Present	Present (in eyes only)	+/–	Absent
Glasgow coma score	E4, M1–4, V1–2	E4, M1–5, V1–4	E4, M1, V1	E1, M1–4, V1–2	E1, M1–3, V1
Motor function	No purposeful movement	Some verbal or purposeful motor behaviour	Volitional vertical eye movements or eyeblink preserved	No purposeful movement	None or only reflex spinal movement
Respiratory function	Typically preserved	Typically preserved	Typically preserved	Variable	Absent
EEG activity	Typically slow wave activity	Insufficient data	Typically normal	Typically slow wave activity	Typically absent
Cerebral metabolism (PET)	Severely reduced	Insufficient data	Mildly reduced	Moderately–severely reduced	Severely reduced or absent
Prognosis	Variable: if permanent, continued vegetative state or death	Variable	Depends on cause but full recovery unlikely	Recovery, vegetative state or death within weeks	Already dead

in the conscious perception of colour, damage to this area causing central achromatopsia (Zeki 1990) (loss of colour vision); of area V5 in the perception of visual motion, damage here giving rise to central akinetopsia (loss of the perception of visual movement) (Zeki 1991); and of the fusiform face area in the perception of faces, damage here causing prosopagnosia (inability to recognize familiar faces) (Kanwisher 2001; Kanwisher et al. 1997).

Recent work in this area has explored two complementary approaches to delineating the neurology of awareness. The first is to investigate how cerebral activity changes when our experience changes without any corresponding change in the world: examples include studies of imagery (Ishai et al. 2000), hallucinations (Ffytche et al. 1998) and the modulation of awareness by attention or during binocular rivalry (Moutoussis and Zeki 2002). The rationale is that changes in brain activity occurring when material enters consciousness in these circumstances will be closely linked with the experience itself.

The second strategy is to approach awareness by stealth, so to speak, by studying the neurology of unconscious processes. Blindsight is an especially well-known example of such processes: patients with blindsight are blind in part of their visual fields, in the sense that they deny the occurrence of any visual experience of the affected part of visual space, yet, when asked to guess about the features of visual stimuli in the affected region, they perform well above chance (Weiskrantz 1998). But blindsight is just one among many examples. Unconscious processes are ubiquitous, and can be studied in health as well in disease, for example in subjects exposed to 'unperceived' stimuli that nevertheless exert an influence on their behaviour: such stimuli may be unperceived because they are too weak, too brief, or masked by preceding or succeeding stimuli (Dehaene and Naccache 2003; Dehaene et al. 1998). Work of this kind need not be restricted to perception: a parallel approach can be taken in the study of the neurology of memory and of action (Frith et al. 1999).

These approaches open up the possibility of a 'contrastive analysis', comparing the neural correlates of conscious and unconscious processes (Baars 2002). From first principles one might propose several criteria likely to characterize the neural activity associated with consciousness: its 'quantity' (related to its amplitude and duration); its quality (related, for example, to the degree of synchronization in the activity of the neurons involved); its site (for example cortical versus subcortical); and its 'reach' (partly a function of the connectivity of the neurons concerned). The results summarized in Table 11.2 provide some support for each of these criteria. The dominant view in contemporary theorizing, the 'global work-space model', is that awareness is linked to a step change in the nature of cerebral processing (Baars 2002; Dehaene and Naccache 2003). When this step change occurs, information that is, at other times, processed locally within independent neural modules, becomes widely available throughout much of the brain, making the information accessible for report and providing it with a key role in the control of action. In Dan Dennett's memorable phrase, the proposal is that the appearance of information in awareness is a function of its 'cerebral celebrity'. The finding that selective

Table 11.2 'Contrastive analysis': studies comparing conscious and unconscious brain activity

Study	Context	Comparison	Results
Laureys et al. 2000	Vegetative state	Vegetative state vs recovery	Increase in cortical metabolic rate and restoration of connectivity with recovery
John et al. 2001	Anaesthesia	Anaesthesia vs awareness	Loss of gamma band activity and cross-cortical coherence under anaesthesia
Sahraie et al. 1997	Blindsight	Aware vs unaware mode of perception in blindsight patient G.Y.	Aware mode associated with DLPF and PS activity, unaware with medial F and subcortical
Dehaene et al. 1998	Backward masking	Perceived numbers vs backward masked but processed numbers	Unreported numbers underwent perceptual, semantic and motor processing similar but less intense to reported numbers
Kanwisher 2000	Binocular rivalry	Attention to 'face' or 'place' when stimuli of both kind are simultaneously in view, or perception of face or place during binocular rivalry	Activity in FFA and PPA locked to presence or absence of awareness of face and place
Moutoussis and Zeki 2002	Invisible stimuli	Perceived vs 'invisible' but processed faces/houses	Similar but less intense activation of FFA and PPA by invisible stimuli
Engel et al. 1999	Binocular rivalry	Perception of one or other of a pair of rivalrous stimuli	Firing of cells processing currently perceived stimulus better synchronized than firing of cells processing suppressed stimulus
Tononi and Edelman 1998	Binocular rivalry	Perception of high vs low frequency flicker during binocular rivalry	More widespread and intense activation by perceived stimulus
Petersen et al. 1998	Task automatization	Effortful verb generation task vs performance after training	LPF, ant cing and cerebellar activation shifts to left perisylvian activation with training

Key: ant cing = anterior cingulate; DLPF = dorsolateral prefrontal cortex; FFA = fusiform face area: LPF = lateral prefrontal cortex; medial F = medial frontal cortex; PPA = parahippocampal place area; PS = prestriate.

depression of metabolic activity in a network of frontoparietal brain regions is a common feature of sleep, coma, general anaesthesia, and the vegetative state hints at a possible neural basis for the global work space (Fig. 11.7).

We have briefly surveyed knowledge of the neural basis of wakefulness and awareness, the two key senses of consciousness. What of the various senses of self-consciousness? Suffice it to say that these, too, have become legitimate topics for study in cognitive neuroscience, exemplified by interesting recent work on the neural basis of empathy (Rankin *et al.* 2006), theory of mind (Baron-Cohen 1995; Abu-Akel 2003) and autonoetic awareness (Tulving 1985).

11.7 **Theories of consciousness**

The much discussed global workspace model is probably the leading current theory of the neural basis of consciousness (Baars 2002; Baars *et al.* 2003;

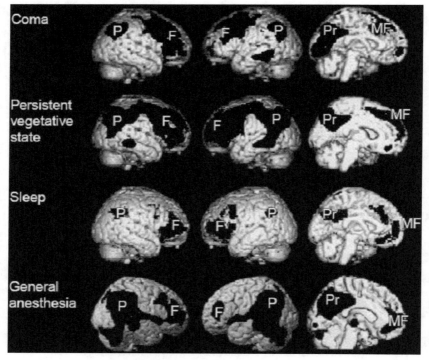

Fig. 11.7. Areas in which brain metabolism is decreased by comparison with levels in conscious control subjects in four states of diminished awareness. Column 1 shows the right lateral aspect of the brain, column 2 the left lateral aspect, column 3 a medial view of the left hemisphere. F, prefrontal cortex; MF mesiofrontal; P, posterior parietal; Pr posterior cingulate/precuneus.

Dehaene and Naccache 2003). But it by no means the only one. In this section I will take a step back from the detail of recent empirical findings, to place this proposal in the context of the broad range of theories of consciousness. Mapping these theories along an evolutionary spectrum that runs from inorganic matter to the most sophisticated human cultural achievements, reveals that these theories variously identify the origin of consciousness with processes at points distributed very widely along this 'consciousness line' (Table 11.3).

All naturalistic, scientific, theories of consciousness regard consciousness as a property of matter, at some level of organization. Panpsychism regards it as a property of matter *per se*, in its simplest atomic, or subatomic forms. Panpsychism represents an understandable reaction to the puzzlement engendered when we try to understand how complex systems of matter can conjure consciousness from unconscious elements: to obviate this puzzlement, panpsychists propose that consciousness is there from the start, in the minutest particles of matter. Panpsychism continues to attract supporters. Animism, which identifies consciousness with life, is less popular these days, although most if not all pre-modern societies were animistic, as are young children. Our continuing enjoyment of Walt Disney's films, well into adulthood, suggests that many of us retain this tendency life-long.

But of course most contemporary theories of consciousness regard it as a property of nervous systems. The most left-wing, or bottom-up, of such theories

Table 11.3 The entities and functions to which consciousness is linked by a variety of contemporary theories

Postulated source of consciousness:							
Matter	Life	Complex neuronal architecture	Working memory, novel solutions, voluntary control	Commentary	Theory of mind	Language	Culture
Proponents:							
Panpsychists	Animists	Zeki Block Chalmers	Baars Dehaene	Weiskrantz LeDoux	Frith Humphrey	Heyes	Jaynes
Bottom-up theories->					<-Top-down theories		
Consciousness line							

envisage consciousness as a property of the first-order representations provided by neuronal systems—such as the network of areas that processes information about colour in the human visual system (by 'first-order' I have in mind a representation that maps some feature of the world: a 'second-order' representation would map some feature of a first-order representation). To help yourself decide whether you are sympathetic to this kind of theory, imagine the entirely hypothetical case of an isolated visual system, teased apart from the rest of the brain by the most expert of neurosurgeons, and activated as your visual system is when you see an expanse of red: could such a system enjoy an experience of redness? If you are inclined to think so, then you are sympathetic to the notion of 'microconsciousness'—like Semir Zeki, in neuroscience (Zeki and Bartels 1998), and, I suspect, Ned Block (Block 2005) and David Chalmers in philosophy (Chalmers 1996). On this theory consciousness is a kind of inner glow, an invisible phosphorescence, possessed by certain neural processes. Consciousness of this kind would not, in itself, have any function.

By contrast, the theories that follow envisage some intrinsic function for consciousness. In the broadest terms, the function is to free the organism from control by its immediate environment, linking consciousness, in the majority of such theories, with voluntary action, developing the intuitively appealing view of Sherrington and Lloyd Morgan that 'the primary aim, object and purpose of consciousness is control' (Sherrington 1953).

Most contemporary theories, like the global workspace theory, suggest that consciousness arises when first-order representations are made available for further processing—for example in the service of problem solving, decision making, or report. The link between consciousness and report, or commentary, is particularly strongly emphasized by Larry Weiskrantz, who has proposed, with examples like blindsight in mind, that it is that ability to 'render a parallel acknowledged commentary' on an otherwise unconscious stream of sensorimotor processing that endows us with consciousness (Weiskrantz 1997). Joseph LeDoux has made a rather similar suggestion about the awareness of emotion (LeDoux 1998). These ideas sound a neuropsychological echo of David Rosenthal's philosophical 'higher-order theory' of consciousness, the view that a mental state become conscious when another mental state is directed upon it (Rosenthal 1986).

Heading off to the right wing we encounter more decidedly top-down theories like Nick Humphrey's and Chris Frith's, that envisage consciousness arising late in evolution, as a product of evolutionary pressure created by social interactions among hominids (Humphrey 1978). On such views consciousness came about when we began to share, and to conceal, our states of mind from one another. Celia Heyes has suggested (this volume) that language may be a prerequisite for consciousness. On the far right wing lie theories, like Julian Jaynes', that locate the emergence of consciousness in a particular phase of human culture (Jaynes 1976). One encounters creative variations on these themes in contemporary fiction: in the Northern Lights trilogy, Philip Pullman locates the origin of consciousness in an influx of elementary particles of consciousness that streamed into our ancestors' heads at a critical moment in human development some 30 000 years ago (Pullman 1995).

Each of these theories maps some intrinsically interesting and important aspect of human cognition. But it seems extraordinary that views can differ so markedly on the phase of cognitive evolution that is relevant to the emergence of consciousness. One has to wonder whether a question that receives quite such varied answers might be misconceived: are we absolutely sure that we know what we are trying to explain? Could we be pursuing an imaginary prey? This possibility is worth considering—especially as all these contrasting proposals can arouse the same uneasy dissatisfaction: none of them quite seems to do the trick, to conjure experience from mechanism. Undoubtedly, the thought goes, all the processes identified in these theories are relevant to human cognition. But why should any of them give rise to *consciousness*?

These difficulties prompt a counter-thought: could it be that there is in fact no need for any trick—no miracle to perform? Perhaps both the ceaseless effort to locate the source of consciousness, and our lingering unease with each and every proposal, can be traced to the subliminal effects of those questionable background assumptions about consciousness that we encountered earlier. If so, the appropriate antidote is not to produce yet another theory of the origin of consciousness—but to adjust our assumptions.

11.8 Going conscious

Imagine a really simple experience, a pure 'mental event', like seeing a large red square, as we look perhaps at one of Rothko's great expanses of coloured canvas. This apparently simple, atomistic experience is, of course, we know now, extraordinarily complex.

It presupposes, first, and fundamentally, life—especially the kind of life possessed by animals, typified, as Aristotle recognized, by movement, movement that, as it happens, is almost always organized into cycles of activity, like our cycles of sleep and wakefulness. This kind of life underwent a process of evolution, over several hundred million years, that enabled eyes like ours to detect light, form an image, distinguish different wavelengths, and send on the news of their arrival. The resulting mechanism is wonderfully honed for its purpose. Darwin came close to admitting defeat when confronted by the exquisite mechanism of the eye, writing that 'to this day it gives me a cold shudder'. But evolution, of course, operates on the basis of reproductive outcomes that eyes can't achieve all on their own: they need bodies. Sensation was embodied from the start.

Darwinian evolution is not the only kind of history involved in seeing red. Our experience is the outcome of a long, almost wholly forgotten, process of individual development. Over the first few years of infancy, during the period of maximum 'plasticity' in the nervous system, the responses of the cells in our visual system are gradually tuned to the occurrence of features in the world: our brains are literally shaped by our environments. Simultaneously, through our active exploration of the world, we accumulate a huge stock of sensorimotor knowledge which becomes the bedrock of all our subsequent understanding, providing a second key sense in which human cognition is embodied. To see the square before us is to know that if we move our eyes directly to the left from the top right corner we will follow the upper boundary; drop them by a few degrees, and we will lose the line. People who have their sight restored in adulthood complain that they 'can't see', despite the undeniable flood of information arriving from their eyes, because they lack the stock of sensorimotor knowledge that usually enables us to negotiate the visual world so effortlessly. As Herschel wrote: 'seeing is an art that must be learned: we cannot see at sight'.

All the while, as human infants, we are exposed to language, which we gradually map onto our ever-enlarging sensorimotor knowledge of the world. Language provides a rapid entrée into the cultural realm that soon comes to pervade experience—as gradually the pigment on the canvas becomes a cipher for passion, bull rings, communism. Simultaneously we build our own idiosyncratic, personal, repertoire of associations. By the time we become aware of our awareness, it is firmly embedded in a dense cultural and personal context, suffusing and transforming consciousness.

Finally, the process of 'seeing red' in our encounter with Rothko is just that—a process, rather than an instantaneous atom of experience. To see is always to explore, to interrogate, to navigate: experience is, therefore, typically extended in space and time. In Kevin O'Regan's phrase, the world serves as our 'outside memory' (O'Regan and Noë 2001): the richness of our experience derives in large part from the richness of the world itself, which we are equipped to sample with our skilful, hard won, techniques of exploration.

Any passage of mature human experience, however simple it appears to us at first glance, is therefore 'embodied, embedded, and extended'. Once we have explained all this—the innate properties of our sensory systems, their tuning by the environment, the sensorimotor skills that enable us to navigate our surroundings, the culturally transmitted skills through which we articulate and share what we perceive, the cumulative personal associations we bring to every moment of perception—might we have explained experience? Are we *sure* there will be further work to do?

11.9 **Does consciousness spring from the brain?**

During the series of lectures that gave rise to this book, a quizzical philosophical Fellow of All Souls College, which hosted our series, asked me for my title. On hearing that it was 'Does consciousness spring from the brain', he grinned: 'Can't think where else it springs from!'. Fair comment: but there is a serious point. We may make an important error in restricting our hunt for the source of consciousness to the brain, and consciousness may not be the kind of the thing that springs from anywhere. When we ask, in a similar vein, how the brain can possibly 'generate' consciousness, we may be resorting to the wrong kind of metaphor for mind.

The 'quest for consciousness' is often described as one of the great intellectual challenges of the age. As we now have a reasonably satisfying explanation of how matter can give rise to life, the 'mystery of consciousness' has replaced the 'mystery of life' as the great unanswered question. For anyone working within a broadly biological tradition, it is natural to look to life and its workings for an explanation of consciousness—and therefore, ultimately, to look to matter. This seems a reasonable project, on the face of it: few of us would quarrel with the idea that organisms can gain knowledge of the world, at least, through natural processes. We can also see that it might be advantageous to such an organism to gain knowledge of its knowing, by modelling its own cognition. The series of 'theories of consciousness' we reviewed in section 11.7 are mostly made in this naturalistic spirit, but as we have seen it is possible to survey all these proposals and come away dissatisfied.

I have floated the idea that our dissatisfaction stems partly from unreasonable expectations: we catch ourselves looking for a 'neurology of the soul' that can never be delivered. But if we reject the Cartesian conception of mind without matter from which this expectation stems, we should not feel committed, either, to a Cartesian conception of matter without mind. The classical problem of consciousness begins with precisely this polar opposition—of mindless matter to matterless mind. But the opposition is surely false. The relationship of mind and matter is not oppositional but circular: mind emerges from matter, and matter is conceived by mind. The stark opposition between subject and object is also false. We can never attain absolute objectivity: knowledge is always shaped by the knower, the subject is always present in the object that the subject conceives. Nor can we ever hope to enjoy pure subjectivity: knowledge always contains something of the world and arises from it. Much the same can be said of the opposition between the arts and the sciences: science seeks objectivity, yet the ultimate 'view from nowhere' is unattainable—science depends upon human powers of conception and imagination. Conversely, the arts place human experience at the centre stage, but our experience is always embedded in the world.

We find ourselves in the same pickle whether we approach the problem of consciousness from a conviction in the rock-bottom reality of experience—or from a conviction in the rock-bottom reality of the physical world. Starting from the second point of departure, for example, we ask: how can experience arise from this physical object, the brain? And such is our enthusiasm for science that, in our puzzlement, we can be tempted to doubt the reality of experience. Yet this outcome is odd: for experience is where we start. Even our knowledge of the brain is given to us in experience—and through experience we come to realize that experience itself is given through the brain. The world gives us experience, and is given through experience.

If we, subjects of consciousness, are in and of the world from the start—if the problem of consciousness is at least partly the product of a false opposition between our minds and the world they inhabit—then how should we view the activity of the brain? Surely not, like many contemporary accounts, as a kind of magic lamp, which, once rubbed the right way, gives rise to a stream of invisible, immaterial 'mind events' to parallel the visible, material, stream of brain events, nor as a neural power station, generating 'awareness' much as a gas-fired one might generate electricity. Instead, we should regard the brain as an enabler—an instrument that brings us into contact with the world, enabling us to interact with its extraordinary richness. Of course, what we are able to find in the world depends on the nature of the instrument we use: but what we find really is in the world, not in our heads.

This kind of approach to the problem of consciousness, sometimes deemed 'enactive', offers one promising escape route from the ancient dilemma surrounding the relationship between mind and body, by denying its usual premises. It accepts that our knowledge of the world is in a certain sense 'groundless'. If subject and object are in an inescapably dynamic, circular, relationship we cannot look to either to 'ground' the other: neither idealism nor materialism is believable. Though appealing, and challenging, this approach is not without its problems.

First, it not easy for it to make sense of the kinds of experience that occur entirely independently of the external world—imagery, hallucinations, dreams, experience during brain stimulation. One might respond that these experiences flow from prior interactions between organism and world, and simply could not occur without a long history of exploration and learning; that they tend to involve simulated action; that we can be mistaken about the nature of these experiences: they may be less rich than we suppose. Though we are sometimes tricked into believing we are experiencing the real thing, we are seldom tricked for long: without access to the world, experience is radically impoverished.

A second problem for the alternative view I am exploring is where to 'place' the appearances. We need to put them somewhere in our scheme of things, but we have mostly come to accept, following Locke, that 'feels' and looks, like warmth or colour, are not in the world but reside instead in the mental realm of experience: these secondary qualities are 'phantoms of the brain' even if they track physical variables. But if one follows through the earlier line of argument, then appearances like these *are* in the world: they are our world. I understand that this idea and its difficulties, sometimes referred to as the 'body–colour problem', by analogy with the mind–brain problem, are a focus of attention in philosophy. The help of philosophy is certainly needed here.

If at least some of the difficulty of the 'problem of consciousness' flows from questionable features of our concept of consciousness, what aspects of the concept will we have to give up before the problem can be solved? The following, I suspect: the notion that experience is an invisible process in an immaterial medium; the notion that mind can be detached from matter; the notion of an absolutely private inner space; the belief that our knowledge of the contents and the underlying nature of our own experience is infallible. You may never seriously have entertained any of these thoughts about the mind—but many of us do.

One reason, I suspect, that many of us cling to these assumptions, is the sense that to relinquish them would be to abandon mystery: to accept that no part of our lives or of the world is safe from the prying eye and reductive theorizing of science. But there is surely mystery enough, in the fact of existence itself, for us to do without a replication of that same mystery under the guise of consciousness. As Darwin wrote: 'There is grandeur in this view of life'. Indeed, there is no reason to suppose that explaining consciousness will prevent us from continuing to be astonished by it: the phenomena of life—whether created by nature or in a test-tube—are no less awe-inspiring for the discoveries of modern science.

So, in the end: Does consciousness spring from the brain? Beyond doubt the brain takes a leading part in the processes work that enable our wakefulness, awareness and self-knowledge. But as we learn more about what the brain does, and as old intellectual assumptions loosen their grip on our thinking, our understanding of what it means to be conscious is likely to change. As this happens, we will gradually change the metaphors we use in thinking about the functions of the brain. We may come to see it less as the source of a mysterious emanation than as an intermediary, a subtle instrument, affording us a range of forms of knowledge of the world. We may come to acknowledge that the richness of experience derives as much from the physical world we inhabit, the cultural world we inherit, and the processes of development through which we become experts in navigating both, as from the intricacies of the workings of the brain.

The paradox of consciousness is that in one sense we know it so intimately, yet, in another sense, we seem scarcely to know it at all:

> '... the end of all our exploring
> Will be to arrive where we started
> And to know the place for the first time'.

(Eliot 1944)

References

Abu-Akel, A. (2003). A neurobiological mapping of theory of mind. *Brain Research. Brain Research Review* **43**(1), 29–40.

Baars, B.J. (2002). The conscious access hypothesis: origins and recent evidence. *Trends in Cognitive Sciences* **6**, 47–52.

Baars, B.J., Ramsoy, T.Z., and Laureys, S. (2003). Brain, conscious experience and the observing self. *Trends in Neurosciences* **26**, 671–675.

Baron-Cohen, S. (1995). *Mindblindness*. Cambridge, MA: MIT Press.

Bassetti, C., Vella, S., Donati, F., Wielepp, P., and Weder, B. (2000). SPECT during sleepwalking. *Lancet* **356**, 484–485.

Berger, H. (1929). Uber das elektrenkephalogramm des Menschen. *Archiv für Psychiatrie* **87**, 527–570.

Bering, J.M. (2006). The folk psychology of souls. *Behavioral and Brain Sciences* **29**, 453–462.

Block, N. (2005). Two neural correlates of consciousness. *Trends in Cognitive Science* **9**(2), 46–52.

Butler, C. and Zeman, A. (2005). Neurological syndromes which can be mistaken for psychiatric conditions. *Journal of Neurology, Neurosurgery and Psychiatry (Neurology in Practice Supplement)* **76**, supplement I, 31–38.

Butler, C., Hodges, J., Wardlaw, J., Kapur, N., Graham, K., and Zeman, A. (2007). The syndrome of transient epileptic amnesia. *Annals of Neurology* **61**(6), 587–598.

Chalmers, D.J. (1996). *The Conscious Mind*. Oxford: Oxford University Press.

Cleeremans, A. (2005). Computational correlates of consciousness. *Progress in Brain Research* **150**, 81–98.

Dehaene, S. and Naccache, L. (2003). Towards a cognitive neuroscience of consciousness: basic evidence and workspace framework. *Cognition* **79**, 1–37.

Dehaene, S., Naccache, L., Le Clec, H.G., Koechlin, E., Mueller, M., and Dehaene-Lambertz, G. (1998). Imaging unconscious semantic priming. *Nature* **395**, 595–600.

Dement, W. and Kleitman, N. (1957). Cyclic variations in EEG during sleep and their relation to eye movements, body motility, and dreaming. *Electroencephalography and Clinical Neurophysiology* **9**, 673–690.

Eliot, T.S. (1944). Little Gidding. In *The Four Quartets*. London: Faber & Faber.

Engel, A.K., Fries, P., König, P., Brecht, M., and Singer, W. (1999). Temporal binding, binocular rivalry, and consciousness. *Consciousness and Cognition* **8**, 128–151.

Ffytche, D.H., Howard, R.J., Brammer, M.J., David, A., Woodruff, P., and Williams, S. (1998). The anatomy of conscious vision: an fMRI study of visual hallucinations. *Nature Neuroscience* **1**(8), 738–742.

Frith, C., Perry, R., and Lumer, E. (1999). The neural correlates of conscious experience: an experimental framework. *Trends in Cognitive Science* **3**(3), 105–114.

Hobson, J.A. and Pace-Schott, E.F. (2002). The cognitive neuroscience of sleep: neuronal systems, consciousness and learning 5. *Nature Reviews. Neuroscience* **3**(9), 679–693.

Humphrey, N. (1978). Nature's psychologists. *New Scientist* **78**, 900–903.

Ishai A., Ungerleider L., and Haxby J.V. (2000). Distributed neural systems for the generation of visual images. *Neuron* **28**, 979–990.

Jaynes, J. (1976). *The Origin of Consciousness in the Breakdown of the Bicameral Mind*. London: Penguin.

John, E.R., Prichep, L.S., Kox, W., Valdés-Sosa, P., Bosch-Bayard, J., Aubert, E., Tom, M., diMichele, F., and Gugino, L.D. (2001). Invariant reversible QEEG effects of anesthetics. *Consciousness and Cognition* **10**, 165–183.

Kanwisher, N. (2000). Neural correlates of changes in perceptual awareness in the absence of changes in the stimulus. *Towards a Science of Consciousness* 2000 Abstr. No. 164.

Kanwisher, N. (2001). Neural events and perceptual awareness. *Cognition* **79**(1–2), 89–113.

Kanwisher, N., McDermott, J., and Chun, M.M. (1997). The fusiform face area: a module in human extrastriate cortex specialized for face perception. *Journal of Neuroscience* **17**(11), 4302–4311.

Kelly, D.D. (1991). Sleep and dreaming. In Kandel, E.R. et al. (eds) *Principles of Neural Science*, 3rd edn, pp. 792–804. NewYork: Elsevier.

Kirmayer, L.J. (2001). The anthropology of hysteria. In Halligan P.W. et al. (eds) *Contemporary Approaches to the Study of Hysteria*, pp. 251–270. Oxford: Oxford University Press.

Laureys, S., Faymonville, M.E., Luxen, A., Lamy, M., Franck, G., and Maquet, P. (2000). Restoration of thalamocortical connectivity after recovery from persistent vegetative state. *Lancet* **355**, 1790–1791.

LeDoux, J. (1998). *The Emotional Brain*. London: Phoenix.

Lewis, C.S. (1960). *Studies in words*. Cambridge: Cambridge University Press.

Llinas, R. and Ribary, U. (1993). Coherent 40-Hz oscillation characterizes dream state in humans. *Proceedings of the National Academy of Sciences of the USA* **90**(5), 2078–2081.

Mahowald, M.W. and Schenck, C.H. (1992). Dissociated states of wakefulness and sleep. *Neurology* **42**(7) Suppl 6, 44–51.

Moruzzi, G. and Magoun, H.W. (1949). Brain stem reticular formation and the activation of the EEG. *Electroencephalography and Clinical Neurophysiology* **1**, 455–473.

Moutoussis, K. and Zeki, S. (2002). The relationship between cortical activation and perception investigated with invisible stimuli. *Proceedings of the National Academy of Sciences of the USA* **99**, 9527–9532.

Noë, A. (2004). *Action in Perception*. Cambridge, MA: MIT Press.

O'Regan, J.K. and Noe, A. (2001). A sensorimotor account of vision and visual consciousness, *Behavioral Brain Science* **24**(5), 939–973.

Petersen, S.E., van Mier, H., Fiez, J.A., and Raichle, M.E. (1998). The effects of practice on the functional anatomy of task performance. *Proceedings of the National Academy of Sciences of the USA* **95**, 853–860.

Pullman, P. (1995). *Northern Lights*. London: Hippo.

Rankin, K.P., Gorno-Tempini, M.L., Allison, S.C., Stanley, C.M., Glenn, S., Weiner, M.W., and Miller, B.L. (2006). Structural anatomy of empathy in neurodegenerative disease. *Brain* **129**, 2945–2956.

Robbins, T.W. and Everitt, B.J. (1995). Arousal systems and attention. In Gazzaniga, M.S. (ed.) *The Cognitive Neurosciences*, pp. 703–720. Cambridge, MA: MIT Press.

Rosenthal, D.M. (1986). Two concepts of consciousness. *Philosophical Studies* **49**, 329–359.

Royal College of Physicians Working Party (2003). *The Vegetative State: Guidance on Diagnosis and Management*. London: Royal College of Physicians.

Sahraie, A., Weiskrantz, L., Barbur, J.L., Simmons, A., Williams, S.C.R., and Brammer, M.J. (1997). Pattern of neuronal activity associated with conscious and unconscious processing of visual signals. *Proceedings of the National Academy of Sciences of the USA* **94**, 9406–9411.

Sapolsky, R.M. (2000). The possibility of neurotoxicity in the hippocampus in major depression: a primer on neuron death. *Biological Psychiatry* **48**, 755–765.

Schenck, C.H., Bundlie, S.R., Ettinger, M.G., and Mahowald, M.W. (1986). Chronic behavioural disorders of human REM sleep: a new category of parasomnia. *Sleep* **9**, 293–308.

Schenck, C.H. and Mahowald, M.W. (2002). REM sleep behavior disorder: clinical, developmental, and neuroscience perspectives 16 years after its formal identification in sleep. *Sleep* **25**(2), 120–138.

Sherrington, C. (1953). *Man on His Nature*. Cambridge: Cambridge University Press.

Stone, J. (2006). *Functional Weakness*. PhD thesis, University of Edinburgh.

Time (2006) Faith in figures. *Time* magazine, 6 November.

Tononi, G. and Edelman, G.M. (1998). Consciousness and complexity. *Science* **282**, 1846–1851.

Torrens L., Burns E., Stone J., Graham C., Wright H., Summers D., Sellar R., Porteous M., Warner J., and Zeman A. (2008). Spinocerebellar ataxia type 8 in Scotland: frequency, neurological, neuropsychological and neuropsychiatric findings. *Acta Neurologica Scandinavica* **117**, 41–48.

Tulving, E. (1985). Memory and consciousness. *Canadian Psychology* **26**, 1–12.

Veith, I. (1965). *Hysteria: The History of a Disease*, pp. 7, 39, 141. Chicago: University of Chicago Press.

Weiskrantz, L. (1998). *Blindsight—A Case Study and Implications*, 2nd edn. Oxford: Clarendon Press.

Weiskrantz, L. (1997). *Consciousness Lost and Found*. Oxford: Oxford University Press.

Wright, H., Wardlaw, J., Young, A.W. and Zeman, A. (2006). Prosopagnosia following nonconvulsive status epilepticus associated with a left fusiform gyrus malformation. *Epilepsy and Behavior* **9**, 197–203.

Zeki, S. (1990). A century of cerebral achromatopsia. *Brain* **113**(6), 1721–1777.

Zeki, S. (1991). Cerebral akinetopsia (visual motion blindness). A review. *Brain* **114**(2), 811–824.

Zeki, S. and Bartels, A. (1998). The asynchrony of consciousness. *Proceedings of the Royal Society of London Series B Biological Sciences* **265**(1405), 1583–1585.

Zeman, A. (1997). The persistent vegetative state (review). *Lancet* **350**, 795–799.

Zeman, A. (2001). Consciousness. *Brain* **124**, 1263–1289.

Zeman, A. (2002). *Consciousness: A User's Guide.* New Haven, CT: Yale University Press.

Zeman, A., Porteous, M., Stone, J., Burns, E., Barron, L., and Warner, J. (2004). Spinocerebellar ataxia type 8 in Scotland: genetic and clinical features in seven unrelated cases and a review of published reports. *Journal of Neurology, Neurosurgery and Psychiatry* **75**(3), 459–465.

Zeman, A., Daniels, G., Tilley, L., Dunn, M., Toplis, L., Bullock, T., Poole, J., and Blackwood, D. (2005). McLeod syndrome: life-long neuropsychiatric disorder due to a novel mutation of the XK gene. *Psychiatric Genetics* **15**, 291–293.

Zeman, A., Carson, A., Rivers, C., and Nath, U. (2006). A case of evolving post-ictal language disturbance secondary to a left temporal arteriovenous malformation: jargon aphasia or formal thought disorder? *Cognitive Neuropsychiatry* **11**(5), 465–479.

Zeman, A., McGonigle, D., Gountouna, E., Torrens, L., Della Sala, S., and Logie, R. (2007). Blind imagination: brain activation after loss of the mind's eye. *Journal of Neurology, Neurosurgery and Psychiatry* **78**, 209.

Chapter 12

On the ubiquity of conscious–unconscious dissociations in neuropsychology

Lawrence Weiskrantz

The interest of Adam Zeman and other neurologists in the topic of consciousness is something of a scene-change. I can remember the distinguished, but outspoken American neurologist, Norman Geschwind, a frequent visitor to the UK, complaining some 25 years ago that British neurologists had lost all interest in the mind—not entirely a fair accusation. However, Adam Zeman's recent invited review (2001) on the topic of consciousness—in *Brain*!—I am told reached a record citation index. The scene has changed in other ways also. We now have some neurologists familiar with philosophy, philosophers who know some neuroscience, and psychologists who study cognition and do brain scanning. The census has changed even if we may not have reached consensus. I have argued elsewhere that we actually need such a triangular approach—from psychology, neuroscience, and philosophy—for the scientific study of consciousness (Weiskrantz 1999). We need to know what our dependent variable is and what environmental events impinge on it, what brain events influence it or covary with it, and we need a theoretic base onto which to map the causal interactions of independent and dependent variables as well as the brain events. Zeman in his seminar has argued for another kind of broadening of the clinical perspective in neurology to a *bio-psycho-social* approach. It is perhaps a difficult aim for individual neurologists to achieve such a breadth—it is almost an argument for a committee consultation. And what kind of training would the practitioners of the future receive? But it is a powerful and welcome argument for a openness of mind not always seen in professionals—or in academics, for that matter.

Zeman has argued that one of the problems in our efforts to achieve a definition of consciousness stems from culture-soaked assumptions about it—from centuries of religious, philosophical, and intellectual tradition that lead us to have widely shared but questionable assumptions and sharply dualistic habits of thought.

But I wonder if the problem would go away in *any* culture that we could imagine unless it were one, perhaps, that was extremely animistic. I cannot imagine a culture in which one did not think, feel, imagine, on the one hand, and have physically enshrined material bodies and brains, on the other, and wonder about the explanatory gap. The problem is more enduring than could be solved by a culture shift. The matter is by no means simpler if we just start with the brain. As Horace Barlow wrote (1961), even if neurons were simply binary states, which they are not, there would be 2 to the power of 10 to the 10 possible brain states, more than there are particles in Eddington's universe. And so, even if we assigned a single particle to every possible state, we could not do it.

In fact, the important direction of flow is from outside the brain to the brain itself. One could dissect the brain as an anatomical organ as carefully as one wished and one would never conclude from that dissection alone that this slab of jelly-like matter is capable of controlling colour vision, speech, let alone thought, or memory, or conscious awareness. We are helped not by anatomy or physiology alone, but by starting with the knowledge that we do have speech and thought, and we can try, haltingly, to relate those to brain events. The point is that we cannot understand the brain without knowing what it is trying to understand.

Often the journey starts with the study of brain damage, such as the interesting cases reviewed by Zeman in his seminar. They illustrate just how wide a spectrum of knowledge is involved in their analysis and understanding, and how new discoveries and interdisciplinary links keep surfacing. The cases include an example of transient amnesia associated with epilepsy, another in which there is a thought disorder associated with epilepsy, another due to a novel genetic mutation. One of the more interesting cases, given the central topic of the series, was of *failure of imagination*. He described, illustrated by a video presentation, the loss in these patients of the capacity to visualize absent events and items—the subjects could not summon images to the *mind's eye*, although they were normal in formal tests of imagery. Such cases may turn out to be more common than usually reported. For earlier reports one goes back to Russell Brain in 1954. Zeman's promise of more complete fMRI analyses will be awaited keenly. He also discussed an interesting case of prosopagnosia (failure to recognize familiar faces) associated with a non-convulsive epileptic focus in the fusiform gyrus. Prosopagnosia offers me the opportunity to inject a lateral extension into a subject closely relevant to the main theme of this Chichele lecture series, consciousness and its loss, because it turns out, surprisingly, that in virtually all of the major cognitive categories that are disturbed by brain damage, prosopagnosia included, there can be remarkably preserved functioning without the patients themselves being aware of the residual function.

The best and most thoroughly studied example of this 'performance without awareness' is the *amnesic syndrome*, a severe memory disorder, for which experimentally proven evidence of residual function after damage emerged some 50 years ago. Amnesic syndrome patients are grossly impaired in remembering recent experiences even after an interval as short as a minute. The patients may have no impairment of short-term memory, for example in reciting back strings of digits, nor need they have any perceptual or intellectual impairments, but they have grave difficulty in acquiring and holding new information ('anterograde amnesia'). Memory for events from before the onset of the injury or brain disease is also typically affected ('retrograde amnesia'), especially for those events that occurred a few years before the brain damage. Older *knowledge* can be retained—the patients retain their vocabulary and acquired language skills, they know who they are, they may know where they went to school, although in some of the severe and densely amnesic cases patients may be vague even about such early knowledge, e.g. how many children they have, or whether their spouse is alive or dead. The striking aspect of the amnesic syndrome is that it can be relatively pure and in isolation from other cognitive, motor, and perceptual difficulties. The disorder is severely crippling, and such patients typically need constant custodial care. And yet there is good evidence of storage of new experiences.

Experimentally robust evidence was reported in the 1950s of the famous patient H.M., who became severely amnesic after bilateral surgery to structures in the medial portions of his temporal lobes for the relief of intractable epilepsy. It soon became apparent that H.M. was able to learn various perceptual and motor skills, such as mastering the path of a pursuit rotor in which one must learn to keep a stylus on a narrow track on a moving drum. He was also able to learn mirror drawing, that is, learning to copy from the reversed image of a pattern seen in a mirror (Corkin 1968). He was able to retain such skills excellently from session to session, but he demonstrated *no* awareness of having remembered the experimental situations nor could he recognize them—he claimed he had never seen them (or the experimenters) before.

But it is not only perceptual-motor skills that can be demonstrated to be intact. Amnesic subjects can also retain information about verbal material, but again without recognition or acknowledgement. The demonstration depended on showing the patients lists of pictures or words and, after an interval of some minutes, testing for recognition in a standard yes/no test, that is, asking them to say whether they did or did not recognize having seen the previously exposed words or pictures (together with new items that had not been exposed). Not surprisingly, the patients performed at the level of chance. But when asked to '*guess*' the *identity* of pictures or words from difficult fragmented drawings, however, they were much better able to do so for those items to which they had been exposed (Warrington and Weiskrantz 1968).

Another way of demonstrating this phenomenon was to present just some of the letters of the previously exposed word, for example, the initial pair or triplet of letters (Weiskrantz and Warrington 1970). The patients showed enhanced ability for finding the correct words to which they had been exposed earlier compared to control items to which they had not been exposed. This is a procedure that is called *priming*—the facilitation of retention induced by previous exposure. The demonstration of retention by amnesic patients has been repeatedly confirmed, and retention intervals can be as long as several months: in H.M, successful retention was reported for an interval of 4 months (Milner B. *et al.* 1968). Such patients, in fact, can learn a variety of novel types of tasks and information, such as new words, new meanings, new rules (McAndrews *et al.* 1987; Knowlton *et al.* 1992). Learning of novel pictures and diagrams has also been demonstrated by Schacter *et al.* (1991) and Gabrieli *et al.* (1990).

An early response by some memory researchers to the reports of positive memory retention by subjects who could not 'remember' was that they might be just like normal people but with 'weak' memory. But it gradually became clear that there are a number of different memory systems in the brain operating in parallel, although of course normally in interaction with each other. For example, it was found that lesions well removed from the medial temporal lobes, in the basal ganglia, produced loss of 'skills', such as learning a pursuit-rotor route, or riding a bicycle, and yet another lesion far removed from the temporal lobe (in the cerebellum) could interfere selectively with conditioned eyelid responses, in neither case causing any loss of 'recognition'. Yet other subjects had brain damage elsewhere that caused them to lose the meanings of words, but without behaving like amnesic subjects; that is, they could remember having been shown a word before, and remember that on that occasion they also did not know its *meaning*. Other brain-damaged subjects could have very impoverished short-term memory—being able to repeat back only one or two digits from a list—and yet were otherwise normal for remembering events and recognizing facts. They could *remember* that they could only repeat back two digits when tested the day before! (See reviews by Weiskrantz 1987, 1990b.)

Such sets of double and multiple dissociations thus demonstrated that 'memory' is not a term to be used as a singular concept The brain apparently has a variety of systems that deal with types of material according to their different demands and, at the very least, has a variety of independent processing modes. But the point here is that there is excellent capacity for acquisition and retention of material by amnesic patients.

Other examples of residual function following brain damage have now been found across virtually the whole spectrum of neuropsychological defects. *Unilateral neglect* is associated with lesions of the parietal lobe in humans, typically the right parietal lobe. The patients behave as though the left half of their visual (and sometimes also their tactile) world is missing. This striking syndrome is the subject of quite intense experimental research and attracts a number of theories. Even though the subjects neglect the left half of their visual world, it can be shown in some cases that 'missing' information is being processed by the brain. In a striking, classical experiment by John Marshall and Peter Halligan (1988) such a subject was shown two pictures of houses that were identical on their right halves, but different on the left. In one picture a fire was vigorously projecting from the chimney on the left side of the picture, and in the other no fire was shown. Because the subject neglected the left halves of the pictures, she repeatedly judged them as being identical and scored at chance in discriminating between them. But when asked which house she would prefer to live in she retorted that it was a silly question, because they were the *same*, but nevertheless she reliably chose the house not on fire. Edoardo Bisiach (1992) has carried out some variations on this theme. Other studies have asked whether an 'unseen' stimulus in the neglected left field can 'prime' a response by the subject to a stimulus in the right half-field. Thus, Làdavas *et al.* (1993) have shown that a word presented in the right visual field was processed faster when the word was preceded by a brief presentation of an associated word in the 'neglected' left field. When the subject was actually forced to respond directly to the word on the left, for example, by reading it aloud, he was not able to do so—it was genuinely 'neglected'. A similar demonstration was made by Berti and Rizzolatti (1992) using pictures of animals and fruit in the left neglected field as primes for pictures in the right which their patients had to classify as quickly as possible, either as animals or fruit, by pressing the appropriate key. The unseen picture shown on the left generated faster reaction times when it matched the category of the picture on the right. And so material, both pictorial and verbal, of which the subject has no awareness in the left visual field, and to which he or she cannot respond explicitly, nevertheless gets processed. The subject may not 'know' it, but some part of the brain does.

We have already taken as our touchstone the phenomenon of *prosopagnosia*, an impairment in the ability to recognize and identify familiar faces, a condition that obviously can be socially very embarrassing to the subject. The problem is not one of knowing that a face is a face, i.e. it is not a perceptual difficulty, but one of facial *memory*. The condition is associated with damage to the inferior posterior temporal lobe, especially in the right hemisphere. The condition can be so severe that patients do not recognize the faces of members of their own family. But it has been demonstrated clearly that the autonomic nervous system can tell the difference between familiar and unfamiliar faces. In one study, Bauer (1984) measured the skin conductance responses (SCR) of prosopagnosic patients when they were asked to read names when shown an individual face of a famous person or a family member. Some names were correct matches for the face, others were not. When the correct name was read, SCR responses were much larger than those to the incorrect names. Tranel and Damasio (1985) carried out a similar study in which patients were required to pick out familiar from unfamiliar faces. They scored at chance, but notwithstanding their SCRs were reliably larger for the familiar faces than for the unfamiliar faces. And so the patients do not 'know' the faces, but some part of their autonomic nervous system obviously does.

Even for that uniquely human cognitive achievement, the skilled use of language, when severely disturbed by brain damage, evidence exists of residual processing of which subjects remain unaware. One example comes from patients who cannot read whole words, although they can painfully extract the word by reading it 'letter by letter'. This form of *acquired dyslexia* is associated with damage to the left occipital lobe. In a study by Shallice and Saffran (1986), one such patient was tested with written words that he could not identify—he could neither read them aloud nor report their meanings. Nevertheless he performed reliably above chance on a lexical decision task, when asked to guess the difference between real words and nonsense words. Moreover, he could correctly categorize words at above chance levels according to their meanings, using forced-choice responding to one of two alternatives. For example, given the name of a country in writing he could say whether it was in Europe or not; he could say whether the name of a person was that of an author or politician; or, given the name of an object, whether that object was living or non-living. All this, despite his not being able to read or identify the word aloud or explicitly give its meaning.

An even more striking outcome emerges from studies of patients with severe loss of linguistic comprehension and production, *aphasia* (Frederici 1982; Linebarger et al. 1983). An informative study that tackled both grammatical and semantic aspects of comprehension in a severely impaired patient is that by Tyler (1988, 1992). She presented the subject with sentences that were degraded either semantically or syntactically. The subject succeeded in being able to follow the instruction to respond as quickly as possible whenever a particular target word was uttered. It is known that normal subjects are slower to respond to the target word when it is in a degraded context than when in a normal sentence. Even though Tyler's aphasic patient was severely impaired in his ability to judge whether a sentence was highly anomalous or normal, his pattern of reaction times to target words in a degraded context showed the same pattern of slowing as is characteristic of normal control subjects. By this means it was demonstrated that the patient retained an intact capacity to respond to both the semantic and grammatical structure of the sentences. But he could not use such a capacity, either in his comprehension or his use of speech. Tyler distinguishes between the 'online' use of linguistic information, which was preserved in her patient, and its exploitation 'offline'.

In all of the examples so far there has been, in some sense, a retained capacity in the absence of an acknowledged awareness—either of the information content itself or of the knowledge of the residual existence of the capacity. But there is a more subtle way in which *subparts* of perception can be shown to behave similarly, such that patients can be said not to perceive in one sense, but to perceive very well in another. In evidence garnered impressively by Milner A.D. and Goodale (1995), reviewed in another chapter in this volume by Milner, patients are described who are 'agnosic' for judging the shape of objects, or cannot even tell different orientations of lines apart. Yet such a patient might be able, without difficulty, when forced to *act*, to slot cards into a 'mail box' the orientation of which is set to various different angles.

Perhaps the most celebrated and dramatic evidence of function without awareness has come from the study of commissurotomized patients, colloquially known as *'split-brain'* patients (Sperry 1974). Because epileptic electrical outbursts in the brain tend to spread from one cerebral hemisphere to another (and, in fact, thereafter to become autonomous) surgeons have cut the massive connections between the hemispheres, the corpus callosum, in an effort to contain the electrical conflagration. Thus, at least at the cortical level, the two hemispheres are rendered independent, although the lower connecting structures of the midbrain still remain intact. Such subjects, for example, could respond correctly to visual stimuli presented to the non-speech right hemisphere, although they claimed they could not see them.

Thus, neuropsychology has exposed a large variety of examples in which, in some sense, awareness is disconnected from a capacity to discriminate or to remember or to attend or to read or to speak (Weiskrantz 1997). They are no longer surprising. Of course, there is nothing surprising in our performing without awareness—it could even be said that most of our bodily activity is unconscious, and that many of our interactions with the outside world are carried on 'automatically' and, in a sense, thoughtlessly. What is surprising about these examples from neuropsychology is that in all these cases the patients are unaware in precisely the situations in which we would normally expect someone to be very much aware.

Perhaps the most dramatic and most counter-intuitive, and perhaps still the most controversial, of all such examples is *blindsight*, the residual function that can be demonstrated after damage to the visual cortex. The story actually started with animal research, some of it more than a century old. The major target of the neural output from the eye lies in the occipital lobe (after a relay via the thalamus), in the 'striate cortex' ('V1'). But while this so-called geniculostriate pathway is the largest one from the eye destined for targets in the brain, it is not the only pathway. There are at least nine other pathways from the retina to targets in the brain that remain open after blockade or damage to the primary visual cortex (Cowey and Stoerig 1991). The important point is that, in the absence of striate cortex, there are a number of routes over which information from the eye can reach and be processed in various systems in the brain.

And so, given this state of affairs, it is perhaps not surprising that monkeys can still carry out visual discriminations even in the absence of V1. The paradox is that human patients in whom the striate cortex is damaged say that they are blind in that part of the visual field that maps onto the damaged V1. William James summarized the contemporary wisdom of his time in 1890, which has remained as current wisdom until quite recently: 'The literature is tedious *ad libitum* ... The occipital lobes are indispensable for vision in man. Hemiopic disturbance comes from lesion of either one of them, and total blindness, sensorial as well as psychic, from destruction of both.' (1890, p. 47).

Why are human patients apparently unable to make use of the parallel visual pathways from the retina that remain after damage to the occipital cortex and to V1? Why are they are blind? Or are they *really* blind? The answer to the question gradually emerged when such patients were tested in the way that one is forced to test animals (Humphrey and Weiskrantz 1967), i.e. without a dependence on a verbal response, revealing that human subjects can discriminate stimuli in their blind fields even though they are not aware of them. Hence, the term *blindsight* (Weiskrantz *et al.* 1974; Weiskrantz 1986).

When subjects were questioned as to how well they thought they had done, in many cases they said they were just guessing and lacked confidence, and thought they were not performing better than chance. There were conditions under which subjects do have *some* kind of awareness, and this has turned out to be of interest in its own right. That is, with some stimuli, especially those with very rapid movement or sudden onset, they 'knew' that something had moved in the blind field, even though they did not 'see' the movements as such (Barbur *et al.* 1994a, Weiskrantz *et al.* 1995), and this has been dubbed *blindsight type 2*. The distinction has allowed one to carry out functional imaging of the contrast between *aware* and *unaware* in blindsight (Sahraie *et al.* 1997). But other stimuli patients could discriminate without any awareness of them (*blindsight type 1*).

It is not only in the visual mode that examples of 'unaware' discrimination capacity have been reported. There are reports of 'blind touch' and also a report of 'deaf hearing'. The first case of 'blind touch' (Paillard *et al.* 1983) was closely similar to the early accounts of blindsight. A similar recent case has been described by Rossetti and his colleagues in Lyon (Rossetti *et al.* 1995, 1996), who have invented the splendid oxymoron, 'numbsense'.

Because many subjects resist answering questions about stimuli they cannot *see*, indirect methods of testing for residual visual processing have been developed that allow firm inferences to be drawn about its characteristics without forcing an instrumental response to an unseen stimulus (Torjussen 1976, 1978; Marzi *et al.* 1986; cf. Weiskrantz 1990a). For example, responses to stimuli in the intact hemifield can be shown to be influenced by stimuli in the blind hemifield, as with visual completion or by visual summation between the two hemifields. Of the various indirect methods, pupillometry offers a special opportunity because the pupil is surprisingly sensitive to spatial and temporal parameters of visual stimuli in a quantitatively precise way. Barbur and his colleagues have shown that, among other parameters, the pupil constricts sensitively to movement, to colour, and to contrast and spatial frequency of a grating, and that the acuity estimated by pupillometry correlates closely with that determined by conventional psychophysical methods in normal subjects (Barbur and Forsyth 1986; Barbur *et al.* 1992, 1994b; Barbur 1995). Such pupillary measurements can be obtained in the absence of verbal interchange about the effective visual stimuli. Therefore, obviously, the method is available not only for testing normal visual fields, but for the blind fields of patients, animals, or human infants, indeed in any situation where verbal interchange is impossible or is to be avoided. For an example of comparison between results of pupillometry and psychophysics in a human blindsight subject and pupillometry in hemianopic monkeys, see Weiskrantz *et al.* (1998).

The impetus for blindsight research actually started with results from monkeys with striate cortex lesions, in which it was shown that a considerably greater degree of residual visual function was available than was evident from gross observations, and it now seems that it may be possible to complete the circle. From the seminal experiments of Cowey and Stoerig (1995, 1997) it has been claimed that the monkey with a unilateral lesion of striate cortex also has blindsight in the sense that it treats visual stimuli in the affected half-field as *non-visual* events even though there is clear evidence that it can detect them very well. Mole and Kelly, in a recent critique (2006), have argued against this as a firm conclusion, but suggest that there might be an attentional bias towards the intact hemifield by the animals in these experiments. But if not the circle, then at least the ellipse is being brought nearer to closure.

And so, in summary of this evidence from the effects of brain damage, across the whole spectrum of cognitive neuropsychology there are residual functions of good capacity that continue in the absence of the subject's awareness. And by comparing aware with unaware modes, with matched performance levels, there is a route to brain imaging of these two modes by comparing brain states when a subject is aware, or is unaware but performing well. By now it is self-evident that the determination of whether the subject is aware or not aware cannot be done by studying the discriminative capacity alone—it can be good in the absence of awareness. In operational terms, within the blindsight mode, but in similar terms for all of the syndromes, we have to use something like the 'commentary key' in parallel with the ongoing discrimination, or (as in the animal experiments) obtain an independent classification of the events being discriminated. Finally, wherever the brain capacity for making the commentary exists, it is likely to reside outside the specialized visual processing areas, which are necessary but not sufficient for the task.

References

Barbur, J.L. (1995). A study of pupil response components in human vision. In Robbins, J.G., Djamgoz, M.B.A., and Taylor, A. (eds) *Basic and Clinical Perspectives in Vision Research*. New York: Plenum.

Barbur, J.L. and Forsyth, P.M. (1986). Can the pupil response be used as a measure of the visual input associated with the geniculo-striate pathway? *Clinical Vision Science* **1**, 107–111.

Barbur, J.L., Harlow, A.J., and Sahraie, A. (1992). Pupillary responses to stimulus structure, colour, and movement. *Ophthalmic and Physiological Optics* **235**, 137–141.

Barbur, J.L., Harlow, J.A., Weiskrantz, L. (1994a). Spatial and temporal response properties of residual vision in a case of hemianopia. *Philosophical Transactions of the Royal Society of London Series B Biological Sciences* **343**, 157–166.

Barbur, J.L., Harlow, J.A., Sahraie, A., Stoerig, P., and Weiskrantz, L. (1994b). Responses to chromatic stimuli in the absence of V1, pupillometric and psychophysical studies. In Vision science and its applications. *Optical Society of America Technical Digest*, **2**, 312–315.

Barlow, H.M. (1961). The coding of sensory messages. In Thorpe, W.H. and Zangwill, O.L. eds. *Current Problems in Animal Behaviour*, pp. 331–360. Cambridge: Cambridge University Press.

Bauer, R.M. (1984). Autonomic recognition of names and faces in prosopagnosia: a neuropsychological application of the guilty knowledge test. *Neuropsychologia* **22**, 457–469.

Berti, A. and Rizzolatti, G. (1992). Visual processing without awareness: Evidence from unilateral neglect. *Journal of Cognitive Neuroscience* **4**, 345–351.

Bisiach, E. (1992). Understanding consciousness: clues from unilateral neglect and related disorders. In Milner, A.D. and Rugg, M.D. (eds) *The Neuropsychology of Consciousness*, pp. 113–137. London: Academic Press.

Brain, R. (1954). Loss of visualization. *Proceedings of the Royal Society of Medicine* B **47**(4), 24–26.

Corkin, S. (1968). Acquisition of motor skill after bilateral medial temporal lobe excision. *Neuropsychologia* **6**, 255–265.

Cowey, A. and Stoerig, P. (1991). The neurobiology of blindsight. *Trends in Neuroscience* **29**, 65–80.

Cowey, A. and Stoerig, P. (1995). Blindsight in monkeys. *Nature London* **373**, 247–249.

Cowey, A. and Stoerig, P. (1997). Visual detection in monkeys with blindsight. *Neuropsychologia* **35**, 1929–1997.

Frederici, A.D. (1982). Syntactic and semantic processes in aphasic deficits: the availability of prepositions. *Brain and Language* **15**, 245–258.

Gabrieli, J.D.E., Milberg, W., Keane, M.M., and Corkin, S. (1990). Intact priming of patterns despite impaired memory. *Neuropsychologia* **28**, 417–428.

Humphrey, N. and Weiskrantz, L. (1967). Vision in monkeys after removal of the striate cortex. *Nature London* **215**, 595–597.

James, W. (1890). *Principles of Psychology*. London: Macmillan.

Knowlton, B.J., Ramus, S.J., and Squire, L.R. (1992). Intact artificial grammar learning in amnesias: Dissociation of abstract knowledge and memory for specific instances. *Psychological Science* **3**, 172–179.

Làdavas, E., Paladini, R., and Cubelli, R. (1993). Implicit associative priming in a patient with left visual neglect. *Neuropsychologia* **31**, 1307–1320.

Linebarger, M.C., Schwartz, M.F., and Saffran, E.M. (1983). Sensitivity to grammatical structure in so-called agrammatic aphasics. *Cognition* **13**, 361–392.

Marshall, J. and Halligan, P. (1988). Blindsight and insight in visuo-spatial neglect. *Nature London* **335**, 766–777.

Marzi, C.A., Tassinari, G., Aglioti, S., and Lutzemberger, L. (1986). Spatial summation across the vertical meridian in hemianopics: a test of blindsight. *Neuropsychologia* **30**, 783–795.

McAndrews, M.P., Glisky, E.L., and Schacter, D.L. (1987). When priming persists: Long-lasting implicit memory for a single episode in amnesic patients. *Neuropsychologia*, **25**, 297–506.

Milner, B., Corkin, S., and Teuber, H-L. (1968). Further analysis of the hippocampal amnesia syndrome: 14-year follow-up study of H.M. *Neuropsychologia* **6**, 215–235.

Milner, A.D. and Goodale, M.A. (1995). *The Visual Brain in Action I*. Oxford: Oxford University Press.

Mole, C. and Kelly, S.D. (2006). On the demonstration of blindsight in monkeys. *Mind and Language* **21**(4), 475–483.

Paillard, J., Michel, F., and Stelmach, G. (1983). Localization without content: a tactile analogue of 'blind sight'. *Archives of Neurology* **40**, 548–551.

Rossetti, Y., Rode, G., and Boisson, D. (1995). Implicit processing of somaesthetic information: a dissociation between where and how? *NeuroReport* **6**, 506–510.

Rossetti, Y., Rode, G., Perenin, M., and Boisson, D. (1996). No memory for implicit perception in blindsight and numbsense. Abstract of paper at conference, *Towards a science of consciousness (Tucson II)*, held at Tucson, Arizona.

Sahraie, A., Weiskrantz, L., Barbur, J.L., Simmons, A., Williams, S.C.R., and Brammer, M.L. (1997). Pattern of neuronal activity associated with conscious and unconscious processing of visual signals. *Proceedings of the National Academy of Sciences of the USA* **94**, 9406–9411.

Schacter D.L., Cooper, L.A., Tharan, M., and Rubens, A. (1991). Preserved priming of novel objects in patients with memory disorders. *Journal of Cognitive Neuroscience* **3**, 118–131.

Shallice, T. and Saffran, E. (1986). Lexical processing in the absence of explicit word identification: evidence from a letter-by-letter reader. *Cognitive Neuropsychology* **3**, 429–458.

Sperry, R.W. (1974). Lateral specialization in the surgically separated hemispheres. In Schmitt, F.O. and Worden, F.G. (eds), *The Neurosciences: Third Study Program.* Cambridge, MA: MIT Press.

Torjussen, T. (1976). Residual function in cortically blind hemifields. *Scandinavian Journal of Psychology* **17**, 320–322.

Torjussen, T. (1978). Visual processing in cortically blind hemifields. *Neuropsychologia* **16**, 15–21.

Tranel, D. and Damasio, A.R. (1985). Knowledge without awareness: an autonomic index of facial recognition by prosopagnosics. *Science* **228**, 1453–1455.

Tyler, L.K. (1988). Spoken language comprehension in a fluent aphasic patient. *Cognitive Neuropsychology* **5**, 375–400.

Tyler, L.K. (1992). The distinction between implicit and explicit language function: evidence from aphasia. In Milner, A.D. and Rugg, M.D. (eds) *The Neuropsychology of Consciousness*, pp. 159–179. London: Academic Press.

Warrington, E.K. and Weiskrantz, L. (1968). New method of testing long-term retention with special reference to amnesic patients. *Nature* **217**, 972–974.

Weiskrantz, L. (1986). *Blindsight. A case study and implications.* Oxford: Oxford University Press. Updated edition (1998).

Weiskrantz, L. (1990a). Outlooks for blindsight: explicit methodologies for implicit processes. The Ferrier Lecture. *Proceedings of the Royal Society of London Series B Biological Sciences* **239**, 247–278.

Weiskrantz, L. (1990b). Problems of learning and memory: one or multiple systems? *Philosophical Transactions of the Royal Society of London Series B Biological Sciences* **329**, 99–108.

Weiskrantz, L. (1987). Neuroanatomy of memory and amnesia: A case for multiple memory systems. *Human Neurobiology* **6**, 93–105.

Weiskrantz, L. (1997). *Consciousness Lost and Found. A Neuropsychological Exploration.* Oxford: Oxford University Press.

Weiskrantz, L. (1999). *Percepts, Brain Imaging, and the Certainty Principle: A Triangular Approach to the Scientific Basis of Consciousness.* (Herbert H. Reynolds Lecture in History and Philosophy of Science). Waco, TX: Baylor University Press.

Weiskrantz, L. and Warrington, E.K. (1970). Verbal learning and retention by amnesic patients using partial information. *Psychonomic Science* **20**, 210–211.

Weiskrantz, L., Warrington, E.K., Sanders, M.D., and Marshall, J. (1974). Visual capacity in the hemianopic field following a restricted occipital ablation. *Brain* **97**, 709–728.

Weiskrantz, L., Barbur, J.L., and Sahraie, A. (1995). Parameters affecting conscious versus unconscious visual discrimination without V1. *Proceedings of the National Academy of Sciences of the USA* **92**, 6122–6126.

Weiskrantz, L., Cowey, A., and LeMare. (1998). Learning from the pupil: a spatial visual channel in the absence of V1 in monkey and human. *Brain* **121**, 1065–1072.

Zeman, A. (2001). Consciousness. *Brain* **124**, 1263–1288.

Index

Note: *fn* = footnote; **bold** = illustrations; **CP** = Colour Plate

ability hypothesis 36–8, 38*fn*
'access' (A) and 'phenomenal' (P)
  consciousness 173–4
acquaintance and first-person point
  of view 16–18
action
  and consciousness 229–31
  joint 235
action awareness *see* awareness
affect *see* emotion
agency
  awareness of agency in self and
    others 233–4
  confusion 233
  and consciousness 231–8
    critical feature? 237–8
  importance of awareness of 234–8
agnosia 176, 178, 180, 186, 218
altruistic punishment, free riders 235–6
amnesia 145, 297, 324–5
amygdala
  back-projection systems 142
  body feedback 97–100
  direct connections to cortical areas 97
  evolution 154–5
  face-selective neurons 161–2
  fear conditioning **81**
  and hysteresis 155
  indirect influence via modulatory
    systems 97
  monitoring behaviour 100–1
  neural circuits **96**
  prefrontal cortex, and working memory
    circuits 95–6
  processing unlearned threats 82
  response prior to awareness 85
  reward-related processing 161–2
  two kinds of outputs 85
  unconscious processing 82–6
    functional implications in humans 84–6
    sensory pathways 83–4
  working memory circuits 95–6
anaesthesia 309
analogical reasoning 261–263, 265–9
animal consciousness 259–88
  alternative hypotheses / first-order
    representation 263–5, 269*fn*
  analogical reasoning / higher-order thought
    265–6, 269*fn*, 278–9

autonomy of the cognitive 281–2
beast machine, major constraint 283
belief—desire hypothesis 263–5
companion animals 260
consciousness as property of
  mental state 260
engaging cognitive—motivational interface
  279–80, **280**
examples 263–5
  alternative hypotheses / higher-order
    thought 266–8
first-order representation 65
instrumental incentive learning 282
manipulating affective experience 280–1
methods 261–3
  empirical approaches 261
mirror self-recognition 265–6
'Palermo' experiments 276–82
phenomenal consciousness 262–3, 279–80
  cognitive inference process 284
  evolutionary just-so story 282–5
  structure and evolution 285*fn*
  which animals are conscious? 286
S-R reinforcement mechanism
  280, 282, 284
anterior intraparietal (AIP) area
  'grasp' circuitry 178–9, 192–3
  left vs right 205
aphasia 328
arousal systems 73–4
attention
  allows use of emotional
    representations 90–1
  normal switching 175, 195–6
  selective neural enhancement 197
  and sensory events 74
  without awareness 198
attribution theory 101
autism 234
autonomic responses, functions
  of emotion 136
awareness
  of agency in self and others 233–4
  and conscious perception 195–201
  motor vs perceptual awareness 217
  neural basis 218–22
  in practice and theory 289–334
  and vision 215–24
  see also dorsal/ventral visual streams

behaviour, vs experience 12–13
behavioural responses, functions of
  emotion 136
binding, intentional 233, 233fn
binocular rivalry paradigm 228–9, 237, 246,
  248, 309
  animals 267fn
  functional neuroimaging 228–9
blindsight 1fn, 178, 198–9, 226, 249,
  252–3, 255, 267fn, 268, 308–9, 312,
  320, 331–2
body feedback from amygdala 97–100
brain
  mind—brain relation 1–48
  public understanding 294–5
brainstem, ascending activating system **305**
brainstem death 307

Cartesian ills 298–300
cerebellar cognitive affective syndrome 298
cerebellum 298
cerebral right/left hemispheres
  normal switching in attention 175
  visual form agnosia 176–80
  see also dorsal/ventral visual streams;
    split-brain patients
change blindness 246
chimaeric faces 173–5
cognitions, emotional colouration 72–5
coma 306, 307
commentary response procedure 275
commentary system, working
  memory 105
communication, facilitated 234, 234fn
conceivability argument 21
conceptions, descriptive and
  non-descriptive 40–1
conditioned stimuli, learning and storing
  information about threats 79–81
consciousness
  'access' (A) and 'phenomenal' (P) 173–4
  and action 229–31
  action-related vs perceptual aspects 217
  and agency 231–8
    critical feature? 237–8
  case for *a priori* physicalism 30–2
  causal role 151
  common ground 239
  contrastive analysis 308–9
  descriptive and non-descriptive
    conceptions 40–1
  disorders of conscious state 307
  dual phenomenal consciousness 175
  etymology 300–2
  evolution 312–15
  and free will 152
  interpreter theory 104–5
  introspective/reflexive 144
  and language 75–6, 102–6, 143–4

  limitation on promise of phenomenal
    concepts 42–3
  mind—brain relation 1–48
  network circuitry 110–11
  neural correlates of consciousness 225–8,
    245–58
  objective vs subjective measures 250–3
  outline 4–5
  positions in the philosophy of
    consciousness 2–4
  problem of report 228–9
  public understanding of the mind 294–5
  representationalism 32–4
  robots, sensate criteria 13–15
  science of consciousness 302–10
  senses of (wakefulness and
    content) 300–2
  for sharing 238–40
  social functions 225–44
  springs from the brain? 315–18
  structure of knowledge argument 22
  task performance as measure of
    consciousness 248–50
  theories of consciousness
    142–54, 310–13
    global workspace model 310–11
  three meanings 259–60
  unitary property 160
  what and why 1–2
  see also explanatory gaps; knowledge
    argument
contrastive analysis 308–9
convergence zones 74, 89–90
  prefrontal cortex 93–4
  sub-primate mammals 107
critical agenda 19–20
culture, and emotional colouration of
  cognition 102

danger, fear as a model system 70–2
Dennett, DC, on consciousness 19, 19–20fn
descriptive and non-descriptive
  conceptions 40–1
display rules 102–3
dorsal/ventral visual streams **170**, 177–201
  anterior intraparietal (AIP) area 178
  dorsal stream provides conscious visual
    percepts 177–95, 198
  lateral occipital (LO) area **179**, 193
  neural basis of action awareness 218–22
  obstacle avoidance 181–6, **182**
  selective neural enhancement 197
  subjective differences 172
  ventral-stream awareness
    and conscious perception 195–201
    implications for consciousness 215–18
  visual extinction **182**
dual phenomenal consciousness 175
dual routes to action 153–4, 159

dualism 21, 47*fn*
dualist intuitions
  feeling of a gap 59
  intuition of distinctness 57–9, 65–7
dysexecutive syndrome 296
dyslexia 328
dysphasia, Wernicke's area damage 296–7

electroencephalogram (EEG) 302–5
emotional arousal 74
emotional colouration of cognition 72–5
  central/cognitive theories 88
  and culture 102
  what is a feeling? 86–8
emotional colouration of consciousness 69–130
  brain mechanisms mediating threat processing in animals 78–82
  brain mechanisms of working memory 93–5
  defining emotions 69–70
  episodic buffer 92–3
  feelings and consciousness 69–76
  role of language in feelings and other forms of consciousness 102–6
  sensory pathways mediating unconscious processing in humans 83–4
emotional regulation 82
emotional representations
  attention allows use in thought and behaviour 90–1
  integration of sensory, memory, and emotional information 88–90
  transfer of information? 95
emotions 131–67
  comparisons with other approaches to consciousness 159–62
  consciousness in different types of processing initiated by emotional states 140–2
  functions of emotion 136–9
  reinforcement contingencies 133–5, **133**
  reward, and punishment,
    in brain evolution 139–40
    as states 132–6
epilepsy 297
epileptic amnesia 297
epileptic 'split-brain' patients 173
epistemological intuition 35–6
  and ability hypothesis 35–8
evolution of consciousness 312–115
  emotions, reward, and punishment 139–40
  feelings and consciousness 106–10
evolution of linguistic system 149
experience
  featureless experience 15–16
  from third-person point of view 15–18
  representational properties 16
  transparency 33*fn*
  vs behaviour 12–13

explanation of consciousness 1
explanatory gaps 2, 4, 19, 24
  dangers of confusion 60–1
  and dualist intuitions 55–68
  epistemology or semantics 64–5
  feeling of a gap 59
  intuition of distinctness 57–9, 65–7
  methodological limitations 61–2
  structure and substance 63–4
  subjects' reports 62
  terminology 56
  too many candidates 62–3

face-selective neurons 191*fn*
  amygdala 161–2
faces
  chimaeric faces 173–5
  Rubin face/vase figure, visual processing in human brain **192**
facilitated communication 234*fn*
Farrell, Brian 4*fn*
  on behaviour and experience, Martians and robots 10–18
  contrast argument 18
  Dennett, and the critical agenda 18–20
  type-A materialism 20, 21
fear 86–112
  language of 103
fear as a model system 70–2
  emotional control over instrumental actions 81
  James—Cannon debate 86–7
  see also emotion
feelings
  emotional colouration of cognition 72, 86–8
  minimal requirements (raw feels) 111–12
  see also emotional colouration of consciousness
fMRI *see* functional neuroimaging
free riders, altruistic punishment 235–6
free will 152
frontal cortex 82, 84, 93–5, 97, 99, 106–7, 109–12, 131, 137–8, 141–2, 148, 149, 151, 155, 157–61, 201, 227, 233, 255, 298, 305, 310
functional neuroimaging
  binocular rivalry 228–9
  and visual awareness 190–5, **194**, 221–2, **CP**
fusiform face area (FFA) 191

global workspace model of consciousness 310–11
'grasp' circuitry
  anterior intraparietal (AIP) area 178–9, 192–3
  error signals 204
grief 151–2

hallucinations 308
haptic exploration 218
higher-order thoughts 143–4, 145–6, 266
　syntactic thoughts (HOSTs) 147–8, 153–4
hippocampus, neuro-evolutionary aspects,
　primates and sub-primates 108–9
hysteria 291–3

imagination, failure of 324
implicit processing 83, 94, 159
instrumental incentive learning 282
instrumental reinforcers 134
intentional binding 233, 233fn
intentional/unintentional condition 236
interpreter theory of consciousness 104–5

Jackson, F
　epiphenomenalism 27, 29fn
　knowledge and epiphenomenalism 28–9, 39
　physicalism 29–32, 29fn, 31fn
　rejection of knowledge argument 4, 28–32
　water 23fn, 39
joint action 235

knowledge argument 4, 21–6, 21fn, 28–32
　argument against type-B materialism 43–5
　argument for physicalism 26–8
　causal argument for physicalism 26–7
　and epiphenomenalism 28–9
　intentionality 39fn
　Jackson on 39
　objection to second premise 23–4
　old fact, new guise response 38–40
　options 25fn
　propositional knowledge 7
　rejection 28, 32
　second premise and type-A materialism 24–5
　simplified knowledge argument 25–6
　strict and relaxed versions of physicalism, identity or supervenience 27–8
　structure 22
　two premises 22
Kripke, S, identity theorists 57

language
　confabulation 143
　and consciousness 75–6, 102–6, 143–4
　insufficient for consciousness 146
　split-brain patients 104
　syntax and parsing 103
　and working memory 105
language disorders 296–7
lateral occipital (LO) area 179, 193, CP
Levine, J, explanatory gap 24
locked-in syndrome 237, 307

McCollough illusion 178
McGinn, C, on acquaintance 16fn, 44fn

McLeod syndrome 296
masking
　backwards masking 309
　meta-contrast masking **254**
　visual masking 245–6, 247–8
material/mental duality 3
materialists
　type-A materialism 20, 21
　type-B materialism 43–4
　types A and B 3–4
medicine, and the mind 293–4
memory
　autonoetic 73, 94–5
　declarative 142
　episodic buffer 92–3
　episodic/ semantic 73, 89
　executive control 93
　input processing systems 92
　long-term 92
　short-term 131–2
　see also working memory model
mid-dorsolateral prefrontal cortex (mid-DLPFC) 255
mind—brain relation 1–48
　public understanding 294–5
　see also explanatory gaps
mirror neurons 205–6, 233fn
mirror self-recognition, animal consciousness 265–6
modulatory systems 73
motor control, without awareness 230
motor vs perceptual awareness 217
Müller—Lyer illusion 58

Nagel, T
　distinction between subjective and objective conceptions 5–6
　explanatory gap 2, 4, 19, 24
　knowing what it is like 175
　for a bat to be a bat 6–8
　physical theories and explanation of consciousness 8–9
neglect, spatial 181, 186, 197, 200, 227, 327
neural basis of action awareness 218–22
neural correlates of consciousness 225–8, 245–58
　frontoparietal network 246–7
　meta-contrast masking **254**
　objective vs subjective measures of conciousness 250–3
　performance matching 253–5, **254**
neural processing see processing;
neuro-evolutionary aspects of feelings and consciousness 106–10
neurology
　attitudes to mind and brain 291–5
　vs psychology, and psychiatry 295–8
numbsense 331

objective vs subjective conceptions of conciousness 5–10
objective vs subjective measures of conciousness 250–3
Ockham, William of, law of parsimony 46*fn*
optic ataxia 176

PANIC theory (Tye) 36*fn*
Papineau, D
  ability hypothesis 37*fn*
  causal argument for physicalism 26–7
  phenomenal concepts, and physicalism 42*fn*
parahippocampal place area (PPA) 191
parietal cortex 84, 94–5, 100, 108, 110–1, 132, 170, 179, 186, 192–3, 195, 203–4, 206, 220–1, 227–8, 246–7, 251, 255, 310, 327
Pavlovian fear conditioning circuitry 80
perception without awareness 1*fn*
performance confound, vs stimulus confound 245–8
performance matching 253–5
persistent vegetative state (PVS) 237, 306–10
phenomenal concepts, and physicalism 41–6
phenomenal consciousness 109, 175, 301
  'access' (A) and 'phenomenal' (P) consciousness 173–4
  animals 262–3, 279–80
  dual phenomenal consciousness 175
physical and conscious realms, gap *see* explanatory gaps
physical properties, material/mental duality? 3
physicalism 9–10, 21–2, 26–8
  a priori 30–2
  causal argument for physicalism 26–7
  conceptually deflationist 25
  Jackson 29–30
  methodological limitations 61–2
  options 45–6
  and phenomenal concepts 41–6
  properties in new guises 38–41
  and representationalism 32–5
  strict vs relaxed versions 3*fn*
primates 62, 64–5, 78, 106–8, 138, 154, 191–3, 196, 204–5, 233*fn*, 266–9, 271, 275, 286, 330–1
propositional knowledge 7
proprioceptive signals, status 233
prosopagnosia 297, 307, 324, 327
psychology, and psychiatry
  attitudes to mind and brain 291–5
  vs medicine 293
public understanding of mind—brain relation 294–5

qualia
  comparisons with other approaches to emotion and consciousness 159–62
  epiphenomenal 31
  evolution of linguistic system 149
  higher-order belief 149
  seat of 5
  unconscious processing 141

raw feels 109, 111–2
reinforcement contingencies **133**, 134
REM sleep 304, 306
representational properties of experience 16
  vs instantiated properties 35
representationalism
  emotional representation, integration of sensory, memory 88–90
  and epistemological intuition 35–6
  and phenomenology 34–5
  and physicalism 32–5
  sensory—memory—emotion 74
reward/punishment
  dual routes to action 154–9
  emotion in brain evolution 139–40
robots 13–14
Russell's principle 6*fn*

S-R reinforcement mechanism **280**, 282, 284
Sapir—Whorf hypothesis 102
second-order thoughts 143–4, 145
sensory processing 150
sleep 302–6
social functions of consciousness 225–44
  action 229–31
  agency 231–8
  awareness of agency in self and others 233–4
  experimental study of consciousness 225–9
  neural correlates of consciousness 225–8
  problem of report 228–9
  sharing 238–40
  what is consciousness 225
  importance of awareness of agency 234–8
split-brain patients
  actions performed by 'non-dominant' hemisphere 142
  and attribution theory 100–1
  chimaeric images 173
  epilepsy 173
  language 104
stimuli
  reinforcing/nonreinforcing 135–6
stimulus confound, vs performance confound 245–8
stimulus onset asynchrony (SOA) 253–5

stimulus—reinforcer association learning 136
stroke, visual extinction testing 181–90
subjective conception of consciousness 5–10
    Nagel's distinction (subjective/objective) conceptions 5–6
subjective conceptions, and physicalism 9–10
supervenient properties 27–8
syntactic processing 160

task automatization 309
task performance, as measure of consciousness 248–50
theory of consciousness 142–54, 311–3
theory of mind 23, 31, 310
threat processing in animals 78–82
threats (in humans)
    fear conditioning circuitry **80**
    learning and storing information 79–81
    processing unlearned threats 82
    unconscious processing 76–8
    amygdala 82–6
TMS 227
treadmill walking 231
trust games 235–6
Tye, Michael
    PANIC theory 36*fn*
    subjective conception of experience 7

unconscious processing
    fear-arousing events 76–86
    human amygdala 83
    qualia 141
    threat, in human amygdala 82–6
    threats in humans 77–8
unconscious thought 1

vegetative state 237, 306–10
vision
    action, and awareness 215–24
    ventral-stream awareness, implications for consciousness 215–18
    change blindness 246

visual extinction
    cortical dorsal/ventral visual streams **182**
    testing in stroke (subject V.E.) 181–90, **CP**
visual form agnosia 176, 177–80
visual masking 245–6, 247–8
visual processing in human brain 169–214
    brief afterthoughts 201–6
    cortical dorsal/ventral visual streams **170**
        subjective differences 172
    dorsal stream provides conscious visual percepts 177–95
    evidence from functional neuroimaging 190–5
    evidence from visual extinction 180–90
    evidence from visual form agnosia 176–80
    obstacle avoidance 181–6, **182**
    Rubin face/vase figure **192**
    ventral stream and conscious perception 195–201
    visual feedback during reaching 186–90
visuomotor control 169, 176*fn*

Wernicke's area, damage causing dysphasia 296–7
working memory model 75–6, 91–102
    brain mechanisms 93–5
    commentary system 105
    executive control 93
    and feeling 101–2
    as a functional framework of cognition, consciousness, and feeling 75
    input processing systems 92
    and language 105
    manipulation of items 159–62
    neural circuits **96**
    representation of emotional information 95–101
    see also memory

zombie 3